THE PETROLEUM ACIDS AND BASES

H. L. LOCHTE, PH.D
Professor of Chemistry
The University of Texas

AND

E. R. LITTMANN, PH.D
Enjay Company, Inc.

CONSTABLE AND COMPANY LTD.
10 Orange Street London, W. C. 2

First published . . . 1955

Printed in the United States of America

PREFACE

Although there are two important monographs on naphthenic acids available both of them were written before use had been made of modern methods of separation and before any individual acids had been identified. No attempt has been made to collect and organize what is known about the petroleum bases. Since the main problem involved in the study of both acids and bases — the problem of fractionation — is practically identical and since most of the known acids and all of the known bases were isolated and identified in the same laboratory, it seems logical for a worker connected with this laboratory to attempt a combined monograph on petroleum acids and bases.

Since the author is not connected with the commercial isolation and utilization of the only class of these compounds which has so far found extensive industrial use — the naphthenic acids — the generous coöperation of Dr. Littmann was sought and obtained to the extent of contributing the chapters on isolation and utilization of naphthenic acids.

An enormous amount of work has been done on the naphthenic acids since 1874 when Hell and Medinger published the first paper based on research on these acids. Any attempt to prepare a monograph dealing with these acids runs immediately into the difficulty of determining, after critical examination of data presented, which of the contributions should be omitted, which merely mentioned, and which stressed. Any attempt to include all published results would lead to many pages of discussion of physical properties of complex mixtures of acids, phenols, and perhaps even hydrocarbons. While at the time of publication these papers appeared to represent real contributions to our knowledge of petroleum acids, it is now obvious that the relatively crude methods of fractionation used could not separate all of the phenols and hydrocarbons from the acids and failed completely in the separation of aliphatic from naphthenic acids.

While the physical properties reported for such impure acids are not stressed in this book, an honest effort has been made to point out and credit all important advances in separation methods, chemical reactions of the acids and bases, and suggested theories. In the case of aliphatic acids and of phenols, so little has been published on the identification of individual compounds that most of such contributions have been included.

The petroleum bases have been the subject of so few research projects that practically all of the nonpatent publications were summarized. This seems appropriate since there has not even been a review paper on this topic except a section by Bailey in the *Science of Petroleum*.

There may have been serious errors of inclusion as well as of exclusion of results and data reported by various workers. Those interested in more detailed discussions of results published prior to 1920 should consult Budowsky's *Die Naphthensäuren* and Naphthali's *Chemische Technologie und Analyse der Naphthensäuren* and its 1934 supplement. Shorter sections on the acids may be found in Engler's *Die Chemie und Physik des Erdöls*, in Sachanen's *The Chemical Constituents of Petroleum* and in Brooks' *The Chemistry of Petroleum Derivatives*. Readers interested in separation methods will find much of interest in the excellent volume by van Nes and van Westen: *Aspects of the Constitution of Mineral Oils*.

The authors have, of course, consulted all of these as well as the original literature freely and are obligated to them in many ways particularly in preliminary selection of significant contributions to the voluminous literature.

It is hoped that this volume will be of interest not only to chemists working in these and related fields but also to all workers engaged in petroleum technology who may want to know more about the mixture with which they work. Finally it is hoped that workers in such fields of utilization of the naphthenic acids as paint and varnish manufacture and formulation and use of fungicides and insecticides may find parts of the book of value.

One of the authors (H. L. L.) gratefully acknowledges financial and other assistance given by the University of Texas Research Institute in the preparation of the manuscript for this monograph.

TABLE OF CONTENTS

PART II. THE PETROLEUM BASES

Part I

THE PETROLEUM ACIDS

CHAPTER ONE

INTRODUCTION

It is a remarkable fact that from the acidic and basic compounds in petroleum which form salts and derivatives readily, only a few dozen individual compounds have been isolated and identified, while many more hydrocarbons which, in general, can only be separated by physical methods have been identified. This peculiar situation is due to the following facts:

1. Fifty years of research has shown that isolation and identification of acids and bases is harder than might appear because, after removal of hydrocarbons and other neutral compounds, there still remains the difficult task of separating the complex mixture of isomers and homologues.
2. Most of the work leading to identification of compounds in petroleum was done during the last 25 years — a period during which there was a very keen interest in fuels and the hydrocarbons which make up the fuels from petroleum.
3. Several large groups of scientists have been liberally supported by the American Petroleum Institute in work on separation projects and have achieved brilliant results with newly developed methods of separation.

While industry has always been interested almost exclusively in a single type of acidic compound in petroleum — the naphthenic acids — there are actually several types of alkali-soluble compounds obtained in the petroleum industry:

1. Simple aliphatic acids
2. Naphthenic acids — carboxylic acids containing one or more alicyclic rings

3. Aromatic acids — presumably homologues of benzoic acid
4. Sulfonic and other sulfur-containing acids
5. Phenols

Of these, the sulfur-containing acids will not be considered in this monograph because the only such compounds, aside from mercaptans, about which we know anything, are produced during refining with sulfuric acid and so should be classified with the petrochemicals.

Phenols are treated only briefly since it is hoped that, with the rapid progress now being made not only in the industrial utilization of these compounds but also in separation and refining operations, as indicated by numerous patents, this field will be treated at a future date by some worker closely identified with this rapidly developing industry.

Seventy-five years ago — long before there was a rich and elaborately developed petroleum hydrocarbon industry — chemists were busy studying the naphthenic acids. They felt that isolation and identification of these compounds should be much simpler than the separations involved in the study of complex mixtures of closely related hydrocarbons. The separation of acids from hydrocarbons, phenols, and other compounds found in petroleum and its refinery products proved more difficult than had been anticipated since the acids, after conversion to their salts (soaps), formed solutions which were good solvents for phenols and even for hydrocarbons. The mixture also showed a strong tendency to emulsification so the separation of the acids was often a tedious operation. Even after the phenolic and neutral compounds had been more or less completely removed, the separation of the closely related carboxylic acids was very difficult because of variable and largely unpredictable degrees of association between acid molecules which caused large boiling point deviations. In view of this difficulty, chemists soon followed rough fractional distillation of the acids with more careful fractionation of the methyl esters which boil about 50°C below the corresponding acids and show nearly normal boiling-point behavior. The lowering in boiling point incidentally removed most of the remaining other types of compounds originally present since these were left as residue in the still pot, if the fraction of acids which had been esterified had a sufficiently narrow boiling range.

In spite of these refinements, few individual acids were obtained in a high state of purity, and none, except a few of the lower aliphatic acids and a few of the higher solid fatty acids, was identified definitely until relatively recently (1931). This was, in part, due to the use of inefficient distillation equipment; in part, to failure to combine different physical methods of separation; and in part, to the fact that the earlier workers, who studied the true naphthenic acids almost exclusively, were able to get only the commercial naphthenic acids which consist mainly of compounds with more than ten carbon atoms and so represent a much more complex mixture than would the first members of the series. It is extremely difficult even today to obtain adequately sized samples of straight-run acids including appreciable concentrations of the simple six- and seven-carbon naphthenic acids.

Although there are many uses for petroleum acids today, they almost all involve the application of the relatively crude mixture known technically as naphthenic acids. These consist very largely of numerous acids of the series $C_nH_{2n-1}COOH$ and $C_nH_{2n-3}COOH$. The aliphatic or fatty acids, $C_nH_{2n+1}COOH$, appear to be the only petroleum acids in the range of acids with less than six carbon atoms and make up a large fraction of the carboxylic acids with six to ten carbon atoms. This series of acids has been found also in Japanese, Polish, and Californian acids boiling much higher than ten-carbon acids and such high-boiling acids may occur more generally in other petroleums than is now assumed. The monocyclic acids start with the simplest naphthenic acid, cyclopentanecarboxylic acid, and continue to above twenty carbon atoms. In some cases at least in the case of acids with twelve or more carbon atoms they are mixed with bicyclic acids of the series $C_nH_{2n-3}COOH$.

Since present uses of naphthenic acids do not require highly fractionated and refined material and since it is probably true that the isolation of individual acids as pure compounds will rarely be possible except at great cost, there is only slight hope that any great industrial interest will develop in the exact nature of the individual acidic compounds in petroleum. This lack of industrial interest, coupled with the fact that even the best-known methods of isolation of individual compounds from these mixtures are so tedious and time consuming that this field of research is not very attractive to graduate students in the universities where this work should probably be carried out, have brought about the situation depicted in the first paragraph of this chapter.

CHAPTER TWO

EARLY INVESTIGATIONS

Seventy-five years ago, Hell and Medinger[4] reported what is now considered the first research work on the nature of petroleum acids. They suspected that these acids belonged largely to a single homologous series, but complained that conclusive results could not be obtained because the acids could not be converted to crystalline salts and derivatives which could be purified by recrystallization and analyzed. Analysis of what must have been mixtures of a number of different acids from Roumanian oil yielded the molecular formula $C_{11}H_{20}O_2$ or $C_{10}H_{19}COOH$. In spite of their difficulties in the purification of the acids, they arrived then in this first work at the molecular formula that is now known to be the correct one for monocyclic naphthenic acids. They also concluded that the acids must be cyclic since they showed none of the properties of unsaturated acids of the same molecular formula. For almost 60 years, there was no more definite information available on the structure of these acids, but later work contributed greatly to our knowledge of the nature of the ring involved and of other types of acids present in petroleum.

Also, in 1874, Eichler,[3] working with acids from Baku petroleum, isolated twelve different fractions of acids and showed that the analysis of one of them gave the formula $C_{11}H_{20}O_2$ and Markownikoff, in 1883,[5] recognized the acids definitely as carboxylic acids and suggested the name: naphthenic acids.

An enormous amount of work was done, often by outstanding organic chemists, on the nature of the acids isolated from various European and Russian crude oils and their distillation products. Some of this early work merely eliminated such compounds as lactones and unsaturated acids. During part of this early period, the cyclohexane ring was supposed to be present in all naphthenic acids and was assumed to have been obtained through hydrogenation of the corresponding aromatic acids. Since so many of the compounds in coal tar had been found to be aromatic in nature, this assumption was a very natural one. For years, this view was defended by Aschan[1] and most of the chemists working on petroleum products during this time took an active part in the debate. After development of Zelinsky's catalytic dehydrogenation technique,[11] a logical application of the method was in the elucidation of the nature of the ring in naphthenic acids. He found that little or no hydrogen was evolved when the hydrocarbons obtained by indirect reduction of the naphthenic acids were tested by his procedure.

Since Zelinsky's results seemed to eliminate hydroaromatic acids in significant concentration in naphthenic acids and since neither three- and four-membered rings nor seven- and eight-membered rings are common in natural compounds, investigators began to assume that the naphthenic acids consisted almost wholly of a series of compounds containing the cyclopentane ring and this view is still generally held, although we now think that a few crude oils also contain fair concentrations of hydroaromatic acids.

Since 1880, petroleum products — rarely crude oils — from many different fields in Eurasia, America, and Japan were studied and it was found that the concentration of naphthenic acids which can be isolated varies from negligible amounts from Pennsylvania fields, for instance, to more than 1%, which has been reported for a few petroleum products. Shipp[9] lists a southern California crude with approximately 3% of acids, but such extreme concentrations seem to be rare. In general, the European, Russian, and California fields show the highest concentrations of acids. In the great majority of cases, total acidities range from 0.1 to 1.0% or from 1,000 to 10,000 barrels per million barrels of crude oil processed. Since when the naphthenic acids are isolated at all, this is done only from the fractions boiling in the kerosene and distillate range, the actual amount isolated would be much less than the given estimate even if all refineries attempted to isolate naphthenic acids. In addition, since almost universally cracking processes are used which affect the yield of true naphthenic acids, further changes in the estimate are needed. However, in view of the total volume of petroleum refined, the potential tonnage of naphthenic acids is very considerable.

The ratio of the concentration of naphthene hydrocarbons to that of naphthenic acids present in various oils has been carefully studied. Von Braun, for instance, was convinced that there is a definite genetic relationship here, i.e., that oils high in naphthenes are also high in naphthenic acids because one is derived from the other.[2] It is generally assumed that high concentrations of naphthenic acids are obtained only from asphalt-base crudes, but apparently not all asphalt- or naphthene-base oils show high concentrations of naphthenic acids. In this connection, Shipp[9] presented data illustrating the fact that paraffin-base oils generally have the lowest concentrations of naphthenic acids while the highest yields are obtained from asphalt-base oils. His data show that the naphthenic acid content of crudes varies widely even from field to field in the same general area. Some of these variations are shown in Table 1.

Table 1. **Naphthenic Acid Content of Crude Oils from the Same Region**

Region	*Naphthenic Acid Content %*
TEXAS	
Winkler County	0.30
Howard County	0.07
Runnels County	0.03
Gulf Coast (Mixed)	0.60
BAKU	
Balakhany	1.05
Binagady	0.85
Bibi-Eibat	0.55
Ramani	0.40
Surakhany	0.20

In 1938, Schmitz[7] presented results obtained over a period of years in his laboratory on samples of crude oil from most of the important producing fields of the world. Each sample was subjected to a modified Engler distillation operating at atmospheric pressure up to 250°C (480°F) and at 1 to 4 mm pressure from there on, until only a heavy tar remained as a residue. His results again emphasize the great variation in the naphthenic acid content of crudes, but a more interesting set of data was obtained when the acidity of each distillation fraction was determined. He plotted the acidity of each fraction against the volume distilled and found that most of the oils (Baku, Poland, Kettleman Hills, Peru, Columbia, and Equador) show a sharp peak in the curve at a boiling point within 250 to 300°C (480 to 565°F) in the "solar oil" range. Both the gasoline range and the heavy lubricating oils have very low acidity.

Another group of oils gives abnormal curves with two or three maxima or with a peak in the normal region followed by a minimum and then a gradual rise to the end of the distillation. This group includes some samples from Germany, East Texas, Venezuela, California, Peru, and Equador. In general, the highest concentration of naphthenic acids is found in and just above the kerosene range, as stated previously, and this is the material from which naphthenic acids are generally isolated.

Some of the early workers used modifications of the method of Spitz and Honig [10] in which the crudely fractionated acids were converted to sodium salts, diluted with alcohol, and extracted with petroleum ether to remove high-boiling neutral impurities. The sodium salts were then treated with mineral acids and the liberated organic acids fractionated as such or esterified and carefully fractionated as the methyl esters. Fractionation consisted of distillation, in vacuum, from distillation columns equivalent to only a few plates. To obtain better fractionation, this operation was repeated one or more times. Petrov [6] described an acid fraction which seemed to be very similar to, if not identical with, the C_7 naphthenic acid of Markownikoff [5] and of Aschan.[1] Petrov was one of the few workers who reported both density and index of refraction for his acid, which were 0.9295 and 1.4305. The product of these is 1.3296 which is well within the range of values found for aliphatic acids and lower than that of naphthenic acid known (see Chapter 8). It is apparent that the C_7 acid of Petrov must have contained a high concentration of fatty acids or a mixture of hydrocarbons, aliphatic acids, and naphthenic acids and since the other C_7 acids were very similar to that of Petrov, they probably also represented mixtures rather than pure compounds.

The early workers isolated some of the lower fatty acids since they could be separated from the naphthenic acids by distillation. Markownikoff and Oglobin [5] found acetic acid and small amounts of other low molecular weight aliphatic acids in Caucasian oil, in 1881. In 1899, Shidkoff [8] reported a number of the lower fatty acids. Since most of the early work was done on Eurasian oils which appear to contain very low concentrations of aliphatic acids, it is not surprising that only a few positive isolations of these acids were reported. As far as the higher aliphatic acids are concerned, fractional distillation alone would not separate these from naphthenic acids which boil at the same temperature. Therefore, they could not be isolated by earlier workers unless they happened to be solid acids that crystallized out of fractions which occurred in several cases a little later.

Bibliography

1. Aschan, O., *Ber.* **25,** 3661 (1892) and **31,** 1803 (1898).
2. Braun, J. von, *Öl u. Kohle* **14,** 283 (1938).
3. Eichler, *Bull. soc. natur., Moscou* **46,** 274 (1874); taken from M. Naphthali's *Chemie, Technologie, und Analyse der Naphthensäuren,* Wissenschaftliche Verlagsgesellschaft, Stuttgart, 1927, page 10.
4. Hell, C., and E. Medinger, *Ber.* **7,** 1216 (1874).
5. Markownikoff, W., and W. Oglobin, *J. soc. russe de physicochimie* **13,** 34 (1883).
6. Petrov, I., *C. A.* **6,** 598 (1912).
7. Schmitz, P. M. E., *Bull. assoc. franç. techniciens pétrole* **46,** 93 (1938).
8. Shidkoff, N., *J. Soc. Chem. Ind.* **1899,** 360.
9. Shipp, V. L., *Oil Gas J.* (March 1936), page 56.
10. Spitz and Honig, from Naphthali's book (see reference 3), page 21.
11. Zelinsky, N., *Ber.* **57,** 42 (1924).

CHAPTER THREE

AVAILABILITY AND METHODS OF ISOLATION OF NAPHTHENIC ACIDS

Availability

The availability of naphthenic acid as an industrial raw material is dependent on the type of crude oil being processed at a refinery and the quantity of acid contained in the oil. Not all crude oils contain naphthenic acid in sufficient quantity to make its isolation an economic process. Although the acid may be extracted from an oil incidental to its refining, its recovery may still be uneconomical unless sufficient quantities are originally present. This situation develops when it is uneconomical basically to provide fresh chemicals for the process. As a consequence, the availability of naphthenic acid is limited by the properties of crude oil and by the ultimate processing. The crude oil will be a desirable source of naphthenic acid if it belongs to the class of naphthenic-base oils. The distribution of naphthenic acid sources is geographical as well as physical (or chemical). There are the following major and minor sources.

Major Sources:

Western United States (California)
Northern South America (Venezuela, Colombia)
Southern Europe (Roumania)

Minor Sources:

Southern United States (Louisiana, Texas Gulf Coast)
Western South America (Peru)
Mexico
Eastern and Southeastern Europe (Poland, USSR, Germany)

The bulk of the naphthenic acids is obtained from gas-oil distillates boiling in the range of 400 to 700°F (204 to 371°C). It is, therefore, interesting to compare relative theoretical acid yields, calculated from the inspections of the gas oils obtained from typical crude oils of the geographical areas listed in Table 2.

Table 2. **Typical Source Stocks for Naphthenic Acid**

		400° to 700°F (204° to 371°C) *Gas Oil*			
Location	*Crude Oil*	*Gravity (API)*	*Aniline Point (°F)*	*Neutralization Number*	*Calculated Yield Pounds Barrel*
West Venezuela	Bachaquero	26.3	129	2.10	2.60
	Cabimas	27.9	129	2.40	3.00
	Tia Juana Med.	33.8	154	0.62	0.77
	Tia Juana Hy.	27.2	133	4.60	5.80
	Tia Juana 102	32.6	145	1.30	1.60
	Lagunillas	28.5	136	1.50	1.90
	La Rosa	29.4	137	2.60	3.20
East Venezuela	Quiriquire	25.4	124	3.60	4.50
Colombia	De Mares	31.6	147	3.00	3.80
	Casabe	30.5	150	1.80	2.20
Mexico	Panuco	28.0	121	1.70	2.10
Roumania	Nonparaffinic	29.5	135	6.10	7.60
Peru	PCLT	30.9	151	0.66	0.82
South United States	Coastal	31.1	146	0.42	0.53
West United States	Gato Ridge	29.2	117	2.10	2.60
(California)	Santa Maria	24.2	104	2.20	2.70

In addition to acids obtained from gas oils, smaller but appreciable quantities are also derived from lighter distillates, such as kerosene and heavy naphtha. These acids, however, seldom play a major part in the overall picture of naphthenic acid production.

The major reason for using gas-oil distillates for the production of commercial naphthenic acids is not only their high acid content but

Table 3. **Distribution of Naphthenic Acid in Colombian Distillate**

Distillate °F (°C)	*Acid Number of Naphthenic Acid*	*Yield (Volume %)*
300 to 400 (149 to 204)	348	0.20
400 to 500 (204 to 260)	287	0.61
500 to 600 (260 to 315)	227	1.16
600 to 700 (315 to 371)	170	1.82

also the fact that these distillates can be readily treated by normal refinery processes. Typical of this situation are the data presented in Table 3 for Colombian crude-oil distillates.

Similar data have been reported by Blagodorov and Terteryan.[3] The variations in yield and availability are further illustrated in Table 4. A detailed study of the apparently typical distribution of acids within a gas-oil fraction has been made by Coles [5] who shows that peaks of availability exist within a distillate as well as between distillates.

Table 4. **Distribution of Naphthenic Acid in Distillates from Several Crude Oils**

Crude Oil	*Quiri-quire*	*Colombian*	*Panuco*	*Tia Juana-102*	*Casabe*	*La Rosa*	*Lagu-nillas*
Gas Oil (400° to 700°F) (204° to 571°C)							
Percent of Crude	39.20	28.00	19.10	26.80	29.70	24.80	22.40
Neutralization Number	3.65	3.00	1.75	1.30	1.70	2.30	1.39
Calculated Yield of Acid (%)	1.45	1.23	0.72	0.52	0.70	0.95	0.57
Narrow Lube (835° to 950°F (446° to 510°C)							
Percentage of Crude	9.40	7.40	7.50	9.00	5.40	7.30	5.90
Neutralization Number	2.63	0.60	0.74	0.30	0.80	0.50	0.52
Calculated Yield of Acid (%)	1.17	0.27	0.33	0.13	0.36	0.22	0.23

Isolation

The basic process for the isolation of naphthenic acid from petroleum involves extraction of the oil with an alkali and acidification of the aqueous extract with a mineral acid to liberate the combined acids. Commercial, that is, primarily gas-oil, acids will be considered in detail, while the isolation steps proposed for other petroleum products will be described only briefly.

Crude oil or heavy distillates are distilled from caustic soda, yielding a residue containing resins, asphalts, oils, and sodium naphthenate. Soda sludge, obtained by neutralizing the residual acidity in a sulfuric-acid-treated oil after the acid sludge has been removed, yields naphthenic acid, since the true naphthenic acids remain unsulfonated during the acid treating process. The relatively low concentrations of naphthenic acid in these sources and the tendency of the higher acids to form emulsions are the difficulties which generally make the crude oils and heavy distillates unattractive sources of acid. Nevertheless, many attempts have been made to recover the naphthenic acids from heavy petroleum products. Two general methods have been reported for the handling of still residues. In one, the residue is diluted with a light oil as a solvent before recovery of the sodium naphthenate, while in the other, the residue is acidified directly with a mineral acid. In all cases, the processes were devised to avoid emulsion formation. Accordingly, Powell [13] diluted still bottoms with an oil and discharged the hot mixture on a stream of flowing water to extract the water-soluble sodium naphthenate. Diluents, such as gas oil or cracking-oil tar, were used to reduce the viscosity of the bottoms. The soaps were extracted by water hot and without turbulence. The acids may be recovered by acidification of the aqueous extract. Becker and Sloane [2] used a more volatile solvent naphtha than did Powell for viscosity reduction of still bottoms. This had the advantage of reducing the tendency of the mixture to emulsify because of the lower density and viscosity of the diluent and also of yielding a raffinate from which the residual oil may be readily recovered after treatment. Andrews and Lauer [1] devised a different scheme in which the still bottoms were extracted with 80 to 95% alcohol. Alcohol is a solvent for the soaps but not for the oil and acts also to reduce or eliminate emulsification. After removal of the alcohol from the extract by distillation, the residual sodium naphthenate may be converted to acid by acidification with a mineral acid.

The direct acidification of still bottoms with a mineral acid is the simplest method of recovery for naphthenic acid, but the acids so prepared and those obtained from prepurification of the bottoms often contain oil which must be removed later. Although the steps involved in the acidification are obvious, several patents have been issued covering minor improvements in the operation. Cook,[6] for example, acidifies the bottoms after dilution with hot water. The, acids are then concentrated by vacuum distillation of the oil layer, collecting a volume of distillate equivalent to the volume of acids known to be present. A very similar process has also been described by Kaufman and Lauer.[10] Various combined operations, covering the liberation of naphthenic acids from still bottoms and their use, have also been described.[4, 16, 17, 18]

In general, the removal or recovery of naphthenic acid or naphthenates from still bottoms is incidental to the recovery of the oil which may be returned to process feed stocks or burned as fuel. In either case, the removal of the sodium naphthenate is desirable.

Little or no effort has been made to extract naphthenic acids on a commercial scale from whole crude oils, since the low percentage of acid in the oils and the large volume of fluids to be handled make this process uneconomical. Naphthenic acids are often removed from crude oils by treatment with caustic soda prior to and during their distillation. Acids so removed are, however, probably unrecoverable since they are mixed, as salts, with rather large quantities of residual fuel, asphalt, or pitch.

The recovery of naphthenic acid from heavy distillates is further limited by the tendency of the acids to form emulsions with the alkali solution and the oil. To avoid emulsification, it is necessary to use alcoholic or similar alkali solutions which, with the attendant solvent loss and recovery costs, make the process extremely expensive. If instead of the more desirable weak soda solutions, concentrated solutions are used, the sodium naphthenates formed may separate as a third phase containing high percentages of oil or may actually remain in solution in the stock. In either case, practical recovery of the soap becomes both difficult and expensive.

Undoubtedly, naphthenic acids derived from either crude oils or heavy distillates are obtained and used in small volume, but since their total commercial availability is negligible, the following discussion on isolation will be devoted primarily to gas-oil acids.

Commercial naphthenic acids, as mentioned earlier, are generally obtained from the middle distillates, gas oil or kerosene, by extraction with caustic-soda solutions. The process is essentially simple in that it involves only contacting the oil with the soda solution, separation of the phases, and regeneration of the acid from the spent soda. Since, however, the extraction of naphthenic acid is seldom the purpose for which a distillate is prepared, it is obvious that the extraction must be coordinated with normal refinery operations and, at the same time, be so designed as to yield an acid meeting with trade requirements. An additional factor to be taken into consideration is the properties of the distillate itself with regard to its behavior toward caustic soda.

The Distillate

In determining the type of extraction procedure to be employed and its place relative to the normal processing steps for a gas oil, these steps have to be examined in the light of the end use of the gas oil. Table 5 shows operating steps which may be required for certain end uses.

Table 5. **End Use of Gas Oils and Possible Sequence of Operations * Involved**

End Use	*Water Wash*	*Soda Wash*	*Acid Treat*	*Sweeten*	*Rerun*	*None*
Domestic and Diesel Fuel						
A	2,[illegible]	1				
B	2,4	3	1			
C	2,5	3	1	4		
D	2	3	1		4	
E						1
Thermal and Catalytic Cracking Feed Stock						
A	1					
B	2	1				
C						1
Lubricants and Specialties						
A						
B	2	1				
C		1			2	
D	2	3	1		4	
E	2	2	1	4		1
Heavy Fuel Diluent						
A	1					
B						1

(*) Numbers refer to order of treatment employed.

The processing of the gas oil and thus the recovery of naphthenic acid would be materially simplified were it possible to run a single crude oil or distillate continuously. This is seldom the case, since few refineries operate on single crude oils. It is, therefore, essential to design any process for the recovery of naphthenic acid so as to be adaptable to the most complex sequence of operations anticipated at any installation. This may well involve the treatment of oils not currently processed, as in the case of cracking coil-feed stocks where certain minimum requirements must be imposed on impurities to assure efficiency of subsequent operations. A further consideration in the processing of oils for naphthenic acid extraction is the anticipated yield of acid. Even a moderate acid treatment, for example, may halve the naphthenic acid yield if it is followed by an alkali extraction.

It may be more economical to sacrifice the acid lost at the initial stage of recovery if, by this, an over-all improvement in oil yield is obtained or if there is a saving in reprocessing the oil recovered from an extraction performed at some other point in the process.

Caustic Soda Solutions for Extraction

The single reagent in general use for the extraction of naphthenic acid is caustic soda. While other alkalis, such as caustic potash and ammonia may also be used, cost and availability exclude them from serious consideration for large-scale operations. After determining the most desirable place for the naphthenic acid extraction, the concentration of the soda and the operating temperature have to be decided on. As a rule, it is most economical to use soda solutions of high concentrations for washing. Small volumes of high-density solutions require a minimum of power for pumping, may be recycled to exhaustion, and need little make-up heat. However, strong caustic solutions as extractive media for naphthenic acid have the disadvantages of promoting emulsification, a salting-out effect on dissolved sodium naphthenate, and increasing the solubility of oil in the dissolved or salted-out sodium naphthenate. All these factors contribute not only to a lowering of the quality of the naphthenic acid, but also to high oil losses which, when debited against the acid, materially increase its cost. The same factors also determine, to some extent, the degree to which the soda is spent. Experience

Table 6. **Solubility of Distillates in 10° Bé Caustic Soda Solutions Neutralized with Naphthenic Acids from the Source Fractions**

Distillate	*Degree Spent* (%)	*Solubility of Distillate* (%)
Kerosene	54	0.48
	65	0.72
	85	0.88
	98	1.08
Light Gas Oil	44	1.08
	62	1.52
	93	1.92
Heavy Gas Oil	34	2.60
	68	5.44

has shown that a soda strength of 3° to 5° Bé (3.3 to 6.9%) may be used generally for kerosenes or gas oils and 15° Bé (10.5%) soda, on light naphthas. The effect of the degree of spending or exhaustion on the solubility of the distillate in the extracting solution is shown in Table 6.

There is an increase in the solubility of the distillate in the aqueous phase not only with increasing concentration of dissolved soap, but also with increasing molecular weight. Similar results are observed when the degree of spending is kept constant and the concentration increased from 3 to 10° Bé (2 to 8%). The inclusion of large quantities of oil with the naphthenic acid necessitates its later removal or a severe reduction in marketability of the product. For some. uses, low oil content is essential regardless of price.

The temperature at which naphthenic acid is extracted from an oil is not critical for the formation of sodium naphthenate. It is, however, an important consideration in the separation of the oil and water phases after neutralization of the acid and in the conservation of heat during processing the oil. In principle, the higher the temperature the greater the ease of separation of phases in an oil-water mixture since there is a relatively greater reduction in the density and viscosity of the oil phase than of the aqueous phase. Two practical limits, however, are imposed on the maximum temperature of the extraction and subsequent settling. Since most extractions and settlings are carried out at atmospheric pressure, the upper limit of temperature is the boiling point of the lowest-boiling constituent in the presence of water. The other limitation is the violence of convection currents which may be set up because of rapid cooling of the tank walls, rapid evaporation from the liquid surface within the tank or even by the movement of the droplets of oil or water through the liquid during separation of the phases. Pressure can be applied to the system to increase the operating temperatures, but this not only adds to the cost but also increases the mutual solubility of the phases and thus the operating difficulties. The temperatures found suitable are roughly as follows:

	°F	°C
Naphtha	100	37.8
Kerosene	90 to 100	32.2 to 43.3
Gas Oil	110 to 150	43.3 to 65.6
Lubricating Oil	120 to 180	48.9 to 82.2

Extraction Processes and Flow Plans

The literature contains many references to methods for the extraction of naphthenic acids from a variety of source stocks. As mentioned earlier, the discussion will be limited to the recovery of the acids from gas oils and kerosenes and typical flow plans will be presented for processes for the handling of these stocks. Those interested in variations of these processes regarding both source material and operational details may refer to the bibliography at the end of this chapter.

Naphthenic acid may be removed from a distillate by a batch or a continuous process. The choice of one or the other will depend on the refining operations rather than on the naphthenic acid requirements alone. If, for example, it is impractical or uneconomical to interrupt the flow of distillate, a continuous extraction system is required, whereas a batch process may be called for if an intermediate storage stage is required in the processing of the distillate or if there is only an intermittent demand for the oil after the acid has been removed. The specifications placed on the extracted oil may also influence the mode of extraction, since a batch process permits holding the oil and retreatment if necessary should an off-specification product be obtained. With proper manipulation, equal extraction efficiency can be obtained by either a batch or continuous process.

From an operating standpoint, conditions other than soda strength and temperature are not the only important factors. Of particular importance is a thorough contact between the caustic soda solution and the oil and the subsequent separation of the aqueous and oil phases. Other factors will become apparent from the discussion of the operations themselves.

Batch Process

Figure 1 illustrates a batch process for the extraction of naphthenic acid from a gas oil. In this process, the oil contained in an agitator is treated with a predetermined quantity of caustic soda solution. Contact may be obtained by spraying the soda on the surface of the oil and allowing it to settle out at the bottom or by introducing the soda as a stream from the bottom and securing contact by air agitation of the mixture. The first method is quite efficient and has the advantage of reducing emulsification to a minimum. It has the disadvantage of being slow because of the time required for

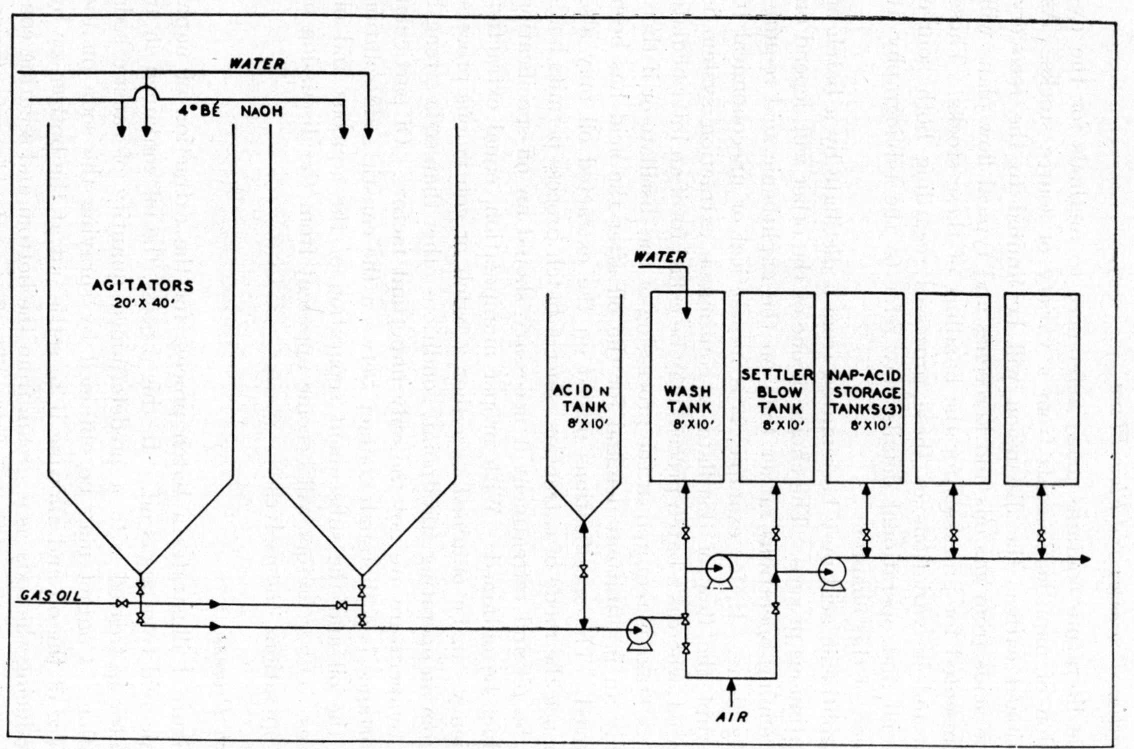

Figure 1. Flow Plan of Naphthenic Acid Production by a Batch Process

small droplets of the aqueous phase to settle out and because settling is often interrupted by convection currents. Where emulsion formation is not a problem, the second method of extraction is more desirable because of its speed, especially during the settling period. The strength of the soda solution is governed by the previously discussed considerations.

The temperature of the extraction process is adjusted to prevailing conditions. When sufficient agitator capacity is available, the oil may be charged from the stills at a few degrees above the desired treating temperature and then extracted with soda. Since extraction is usually effected in 1,000 to 5,000 barrel batches, the heat loss is not high. If the oil is processed from storage, it may be necessary to reheat it and this may be accomplished either by the use of a heat exchanger in the feed line or by steam coils placed in the agitator. The first method is preferable because of speed, economy, and the reduction in convection currents induced by localized heating.

On completion of the extraction, the two phases are allowed to separate and the aqueous phase segregated for further processing. The oil phase is washed with water to remove dissolved or entrained soap and salts. Then it is either stored or further processed, as required. The wash water is usually sent to the sewer as its naphthenic acid content is too low for profitable recovery. The completion of washing is the end of the actual extraction operation.

Continuous Processes

The number of modifications possible in a continuous process for the extraction of naphthenic acid from a distillate seems to be limited only by the ingenuity of the designer. As a rule, the processes are designed to yield a salable oil or one suitable for immediate further processing.

Figure 2 illustrates a continuous process for the extraction of heating-oil distillate in which the extraction is the final step. To assure the specifications of the finished oil, it is essential to make the extraction sufficiently complete to reduce its neutralization number to the proper value, to maintain the ash resulting from the inclusion of soap and salts below a predetermined value, and to meet haze and cloud specifications. The quality of the finished oil, therefore, imposes the following requirements on the extraction process: 1. The correct amount of soda has to be used at a predetermined concentration to achieve economy and rapid separation

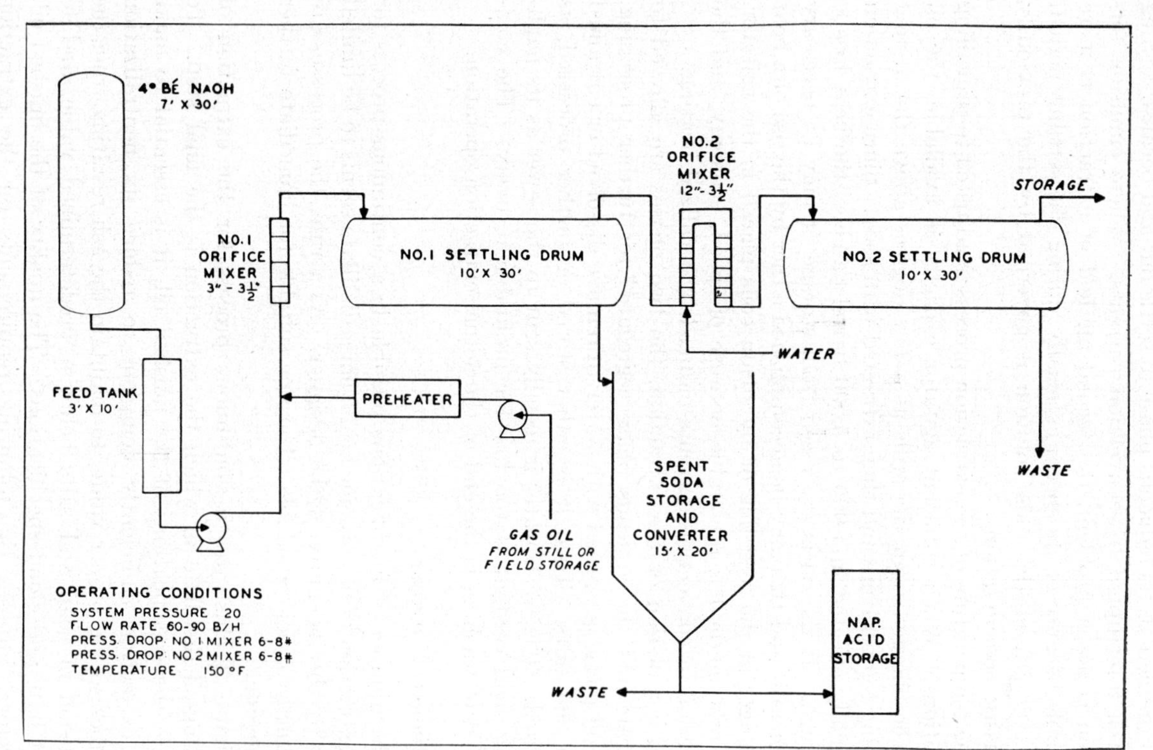

Figure 2. FLOW PLAN OF NAPHTHENIC ACID PRODUCTION BY A CONTINUOUS PROCESS

of the phases. This may involve incomplete extraction of the naphthenic acid and a determination of the soda strength permitting maximum settling rate. 2. Continuous equipment must be available to promote thorough contact between phases without undue emulsion formation. 3. Settling tanks or drums of adequate size and shape should be employed for efficient settling. 4. Continuous washing facilities must be provided to obtain almost complete removal of all water-soluble products retained in the oil.

The conditions of operation shown in Figure 2 apply to a specific type of gas oil and variations would probably be required for different oils. It will also be seen that the system shown is continuous only with respect to the treatment of the gas oil in storage. This situation arises because not all gas oils are considered suitable for naphthenic acid production and thus require segregation. Figure 3 represents a process similar to that described previously, except that the feed comes directly from the stills and needs no preheating.

In both methods described, the operation consists of the addition of soda at a single point in the process. Such an operation permits the production of only a single grade of naphthenic acid, the quality of which is governed primarily by the specifications imposed on the resultant oil. Where conditions and economics justify, the continuous extraction may be performed by a "split" process in which a portion of the acid is removed in one stage and the remainder extracted at a second. The net effect of this mode of operation is to produce a naphthenic acid of high acid number from the first half of the "split" extraction and a lower one from the second half. The function of the "split" operation is to obtain a selective extraction of the lower-molecular-weight and stronger acids by applying only a part of the total soda at the first step in the extraction. It is obvious that this type of operation illustrated in Figure 4 is relatively expensive because of the increased equipment cost and the difficulties encountered in maintaining proper segregation throughout the entire operation.

The very low naphthenic acid content of naphthas and most kerosenes generally does not permit the installation of extraction facilities solely for the recovery of the acid. It is, therefore, economical only to collect spent soda as it becomes available from the normal treating or refining operations. Figure 5 illustrates a process in which spent soda from the treating of light distillates is collected in a central system for later processing.

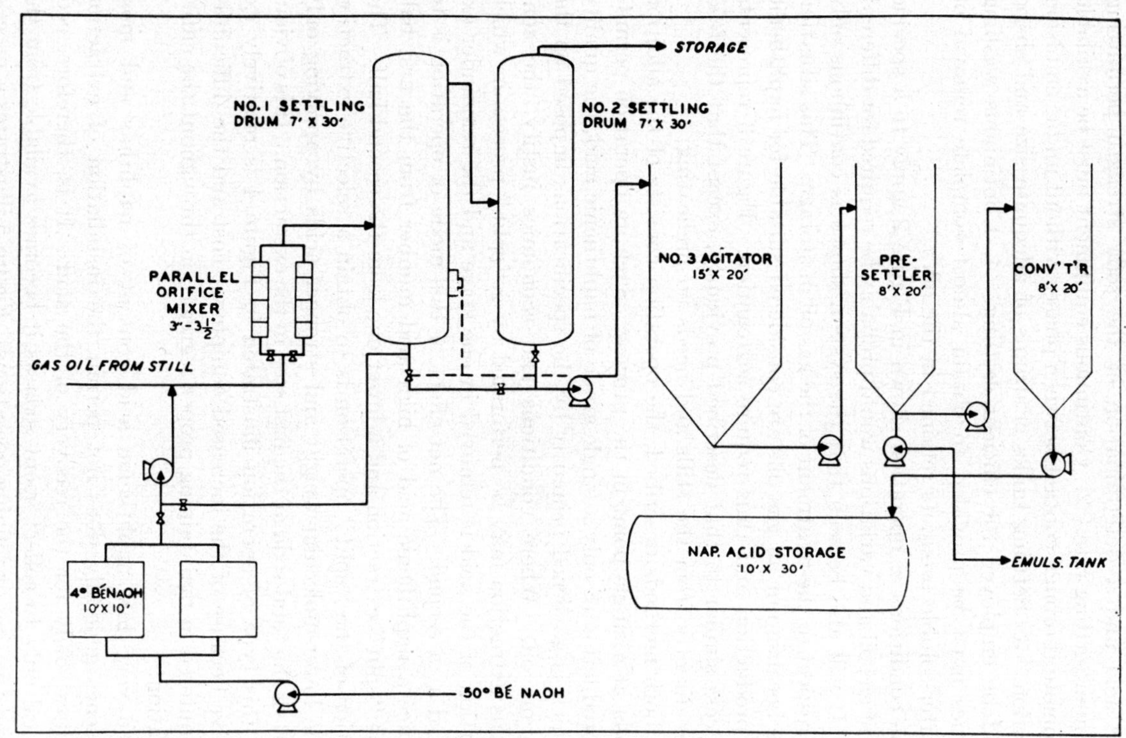

Figure 3. Naphthenic Acid Production by a Continuous Process Without Preheating

Regeneration of Naphthenic Acid from Spent Soda

The regeneration of naphthenic acid from spent soda or sodium naphthenate solution is a very simple process in itself and consists of acidifying the solution with sulfuric acid and collecting the supernatant layer of oily acid after separation of the two phases. Prior to acidification, it may be necessary to remove dissolved or entrained oil from the spent soda to reduce the oil content of the naphthenic acid. Where dilute sodas are used on gas oils or lighter distillates, the spent soda may be allowed to settle for some hours at 100° to 150°F (37.8° to 65.6°C) and the oil skimmed off. With heavier distillates, the settling process is not satisfactory for low-oil-content naphthenic acid and some further processing, such as extraction with a light solvent or centrifuging, must be employed.

To regenerate naphthenic acid from the previously prepared spent soda, sulfuric acid is applied to the aqueous solution. During this step, efficient agitation is maintained. In a batch process, agitation by air blowing is adequate. Where a continuous process is used, the sulfuric acid is introduced into a stream of soda and the mixture passed through a mixing device, such as an orifice mixer, to obtain an intimate contact. Typical operations are illustrated in Figures 1 to 5, which show coordinated processes.

The strength of the sulfuric acid used in the regeneration process is not critical since the naphthenic acids appear to be unaffected even by 95 % acid under normal operating conditions. It is important, however, to employ clean acid, since black or restored acid introduces "carbon" into the process which interferes with the clean separation of the phases and with any subsequent purification of the naphthenic acid. It is also important to add sufficient sulfuric acid during acidification to obtain a positive excess. This excess will be obtained when the pH of the aqueous phase is 4 or less. After separation of the aqueous solution from the naphthenic acid, the acid is washed with water to remove mineral acid and blown bright. Typical analyses of crude naphthenic acids obtained by the just described process are shown in Table 7.

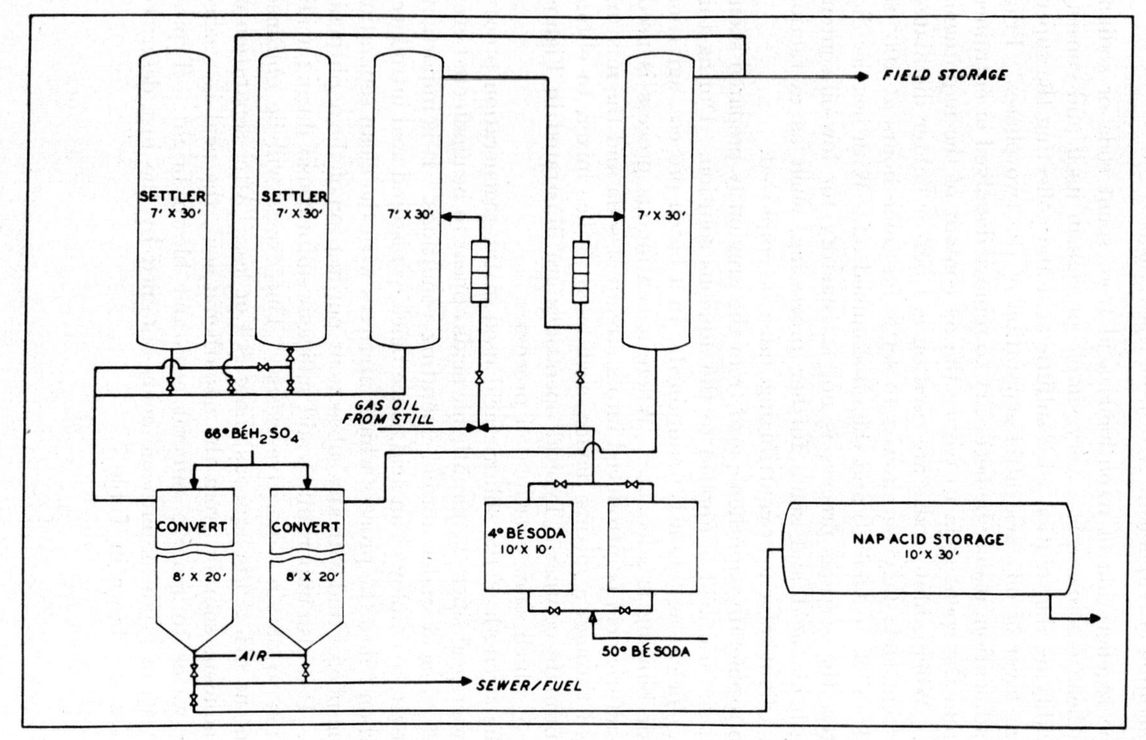

Figure 4. Split Extraction of Naphthenic Acid

Table 7. **Characteristics of Naphthenic Acids Obtained from Several Distillates**

Distillate	*Acid Number*	*Unsaponified Matter (%)*
Kerosene	275–300	2–5
Light Gas Oil	220–240	5–10
Medium Gas Oil	200–220	8–15
Heavy Gas Oil	180–200	10–25

Purification of Naphthenic Acids

For many practical uses, the crude naphthenic acids are too impure. The impurities associated with the acids are chiefly oil, iron salts, phenols, basic nitrogen, and sulfur-containing compounds. Numerous methods have been reported [11] for the purification of the acids though only two are in wide commercial use: acid treatment and distillation. Purification by acid treatment is probably the cheapest method regarding initial investment and operating cost though the degree of purification obtained is usually lower than that achieved by distillation. In this procedure, the acid is contacted with 1 to 5% strong sulfuric acid (92 to 95%) which removes iron salts, phenols, nitrogen, and most sulfur compounds, but has no effect on the oil content. Following the treatment with acid, the solution is usually mixed with sufficient activated clay to absorb all mineral acid and filtered. To facilitate handling, the naphthenic acid may first be dissolved in a solvent, such as mineral spirits, but this is not essential. In the absence of solvent, treatment and filtration are performed hot, at 200° to 250°F (93.3° to 121.1°C). The operation is carried out at room temperature, when a solvent is used. The bleaching obtained by the acid-clay treatment is, e.g., as follows:

	Color-Tag Robinson
Crude Acid	$7\frac{3}{4}$
Bleached Acid	9–10

The purification of naphthenic acid by distillation, while somewhat more costly than by acid treatment, is a more flexible process and permits, to some extent, the removal of oil as well as of other impurities. The distillation may be of the flash type in which there is little or no fractionation or of a more complicated type in which fractionating columns are employed. The flash distillation method

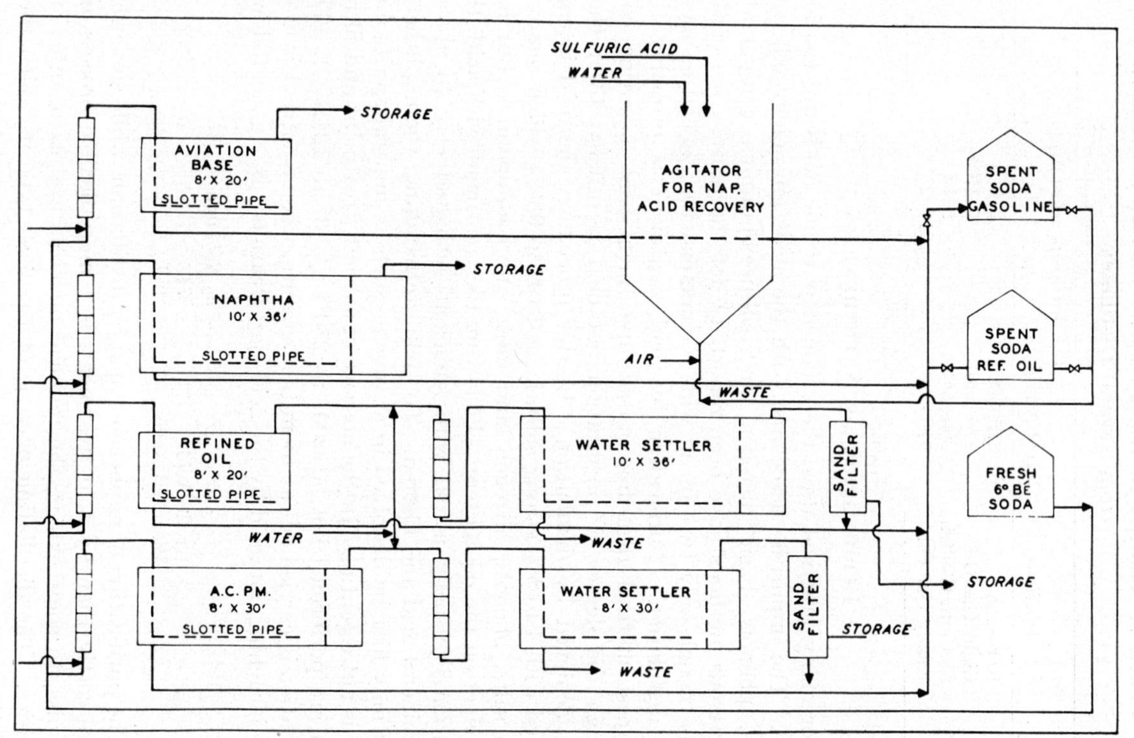

Figure 5. Collecting Spent Soda from Light-Distillate Treatment

is reasonably satisfactory for color improvement and if the first runnings are removed from the total distillate, improvement in acid number or reduction in oil content is obtained. From a properly isolated feed stock, yields of an about 80 % product are obtained. Typical results of this method of purification are listed in Table 8.

Table 8. **Improvements in Naphthenic Acid Obtained by Flash Distillation**

Original			*Distilled*		
Acid Number	*Oil %*	*Color Tag-Robinson*	*Acid Number*	*Oil %*	*Color Tag-Robinson*
256	8.0	3/4	263	7.4	12–13
205	13.5	3/4	235	7.9	11–12

The use of fractionating columns permits a closer control of the oil content and color of the distillate. The results obtained by the use of an efficient fractionating column are listed in Table 9.

Table 9. **Improvement in Naphthenic Acid Obtained by Fractional Distillation**

	Acid Number	*Oil %*	*Yield %*	*Color Tag-Robinson*
Original	223	11	—	$\frac{3}{4}$
Distillate	234	5	87.62	$17\frac{1}{2}$
Bottoms	158	—	8.56	—
Loss (Oil, Etc.)	—	—	3.82	—

Both methods of distillation must be carried out under reduced pressures to avoid decomposition of the acid. As a rule, it is inadvisable to permit temperatures to go above 550°F (287.8°C) in any part of the equipment. Industrially, pressures of about 10 mm of mercury are employed. Both flash distillation and column fractionation may be operated either on a continuous or batch basis, but the first is preferable both from an economical and product-quality standpoint when a sufficient volume of raw material is available. A recent study on the purification of naphthenic acids by batch distillation [12] shows the maximum purification which may be expected reasonably when operating at constant reflux ratio. For additional general discussion see references 7, 8, 9, 11, and 14.

NAPHTHENIC–ACID PATENTS

U. S. Patent No.	*Year*	*Topic*
1,425,882 — Maitland	1922	Process of treating hydrocarbon oils
1,681,657 — Bransky	1928	Oil-soluble naphthenic compound
1,694,280 — Powell	1928	Process for reclaiming distillation residues
1,694,462 — Alleman	1928	Mineral-oil derivatives and processes of making them
1,694,463 — Alleman	1928	Mineral-oil derivatives and process of making them
1,720,821 — Coleman	1929	Extraction and purification of naphthenic acid
1,784,262 — Wheeler and Prutzman	1930	Alkaline treatment of petroleum vapors
1,785,242 — Becker and Sloane	1930	Process of reclaiming residues
1,802,336 — Cook	1931	Naphthenic acid
1,804,451 — Andrews and Lauer	1931	Process of treating petroleum
Re. 19,179 — Wheeler and Prutzman	1931	Alkaline treatment of petroleum vapors
1,886,647 — Coleman	1932	Process of making naphthenic compounds
1,916,805 — Meidert and Schatz	1933	Production of siccatives on the base of naphthenic acids
1,931,855 — Alleman	1933	Process of purifying fatty acid derivatives of mineral oil
1,931,880 — Angstad	1933	Process of treating mineral oil still bottoms to recover asphalt and fatty acids
1,938,513 — Brunck, Kreuzer and Boeck	1933	Process for working up naphthenic acid soap
1,938,515 — Brunck, Kreuzer and Boeck	1933	Process for working up naphthenic acid soap
1,984,432 — Robinson	1934	Method of neutralizing a petroleum oil
1,986,775 — Kaufman and Lauer	1935	Process of treating hydrocarbon oil containing naphthenic acids
1,998,765 — Logan	1935	Process for neutralizing organic acidity in petroleum lubricating oils
2,000,244 — Merrill and Blount	1935	Process for separation and recovery of naphthenic acids and phenols
2,003,640 — Wunsch	1935	Recovery of naphthenic acids
2,007,146 — Rogers	1935	Removal of naphthenic acids from hydrocarbon oils
2,014,936 — Hendrey	1935	Dehydrating oil

NAPHTHENIC-ACID PATENTS (*Continued*)

U. S. Patent No.	*Year*	*Topic*
2,026,073 — Swerissen	1935	Naphthenic acid
2,035,696 — Ewing	1936	Purification of naphthenic acids
2,035,741 — Ewing	1936	Purification of naphthenic acids
2,035,742 — Ewing	1936	Purification of naphthenic acids
2,035,747 — Ewing	1936	Purification of naphthenic acids
2,039,106 — Nelson and Zapf	1936	Method for purifying naphthenic acids
2,056,913 — Terrell, Hughes and Carter	1936	Process of removing soaps from asphaltic still bottoms and purifying them
2,072,053 — Hendrey	1937	Purification of naphthenic acids
2,081,475 — Carr	1937	Process for purifying naphthenic acids
2,093,001 — Blount	1937	Refining naphthenic acids
2,108,448 — Rutherford	1938	Purification of naphthenic acids
2,131,938 — Donker	1938	Purification of naphthenic acids
2,133,765 — Ewing	1938	Purification of naphthenic acids
2,170,506 — Reiber	1939	Process for the purification of naphthenic acids
2,186,249 — Lazar and Galstaum	1940	Process of recovering organic acids
2,191,449 — Blount	1940	Purification of organic acids
2,200,711 — Berry	1940	Process for the refining of naphthenic acids
2,214,438 — Rich and Cannon	1940	Process for purifying naphthenic acids
2,220,012 — Bruun	1940	Method of obtaining naphthenic acids
2,220,013 — Bruun	1940	Method of obtaining naphthenic acids
2,245,548 — Ralston and Harwood	1941	Process of purifying naphthenic acid through conversion to nitriles
2,295,065 — Vesterdal	1942	Separation of soaps from oil
2,296,039 — Knowles and McCoy	1942	Refining naphthenic acids
2,301,285 — Kellog and Gaylor	1942	Production of naphthenic acids
2,301,528 — Ewing	1942	Refining naphthenic acid
2,331,244 — Strickland	1943	Refining mineral oils
2,337,467 — Hewlett	1943	Refining mineral oils
2,339,889 — Strickland	1944	Refining mineral oils
2,346,734 — Dempsey	1944	Refining mineral oils
2,357,252 — Berger, Nygaard and Angel	1944	Process for recovery of phenols

NAPHTHENIC–ACID PATENTS (*Continued*)

U. S. Patent No.	*Year*	*Topic*
2,375,596 — Strickland	1945	Refining mineral oils
2,381,729 — McCorquodale	1945	Purification of naphthenic acids
2,391,729 — McCorquodale	1945	Refining naphthenic acids
2,420,244 — Henderson and Ayers	1947	Recovery of naphthenic acids
2,422,794 — McCorquodale, Magill and Hagy	1947	Extraction of saponifiable acids
2,424,158 — Fuqua and Lovell	1947	Process of refining petroleum oil containing naphthenic acids
2,477,190 — Linford	1949	Purification of naphthenic acid
2,610,209 — Honeycutt	1952	Refining of naphthenic acid

Bibliography

1. Andrews, T. M., and C. E. Lauer, U. S. Patent 1804451 (1931).
2. Becker, A. E., and R. G. Sloane, U. S. Patent 1785242 (1930).
3. Blagodorov, I. F., and A. B. Tertaryan, *C. A.* **28,** 7491 (1934).
4. Brunck, R., A. Kreutzer, and W. Boeck, U. S. Patent 1938515 (1933).
5. Coles, F. K., *J. Inst. Petroleum* **33,** 325 (1947).
6. Cook, L. W., U. S. Patent 1802336 (1931).
7. Gruse, W. A., and D. R. Stevens, *The Chemical Technology of Petroleum,* McGraw-Hill, New York, 1942.
8. Jezl, J. L., *Pet. Processing* **8,** 89 (1953).
9. Kalichevsky, V. A., and B. A. Stagner, *Chemical Refining of Petroleum,* Reinhold, New York, 1942.
10. Kaufman, H. L., and C. E. Lauer, U. S. Patent 1502956 (1924).
11. Littmann, E. R., and J. Klotz, *Chem. Rev.* **30,** 97 (1942).
12. Littmann, E. R., and J. Klotz, *Ind. Eng. Chem.* **41,** 1462 (1949).
13. Powell, R. E., U. S. Patent 1694280 (1928).
14. Sachanen, A. N., *The Chemical Constituents of Petroleum,* Reinhold, New York, 1945.
15. Shipp, V. L., *Oil and Gas J.* **34,** No. 44, 56 (1936).
16. Swerissen, H. T., U. S. Patent 2026073 (1935).
17. Terrell, H. T., E. M. Hughes, and P. L. Carter, U. S. Patent 2056913 (1936).
18. Wunsch, J. A., U. S. Patent 2003640 (1935).

CHAPTER FOUR

DISTRIBUTION AND CONCENTRATION OF ACIDS IN PETROLEUM

Throughout the literature of petroleum acids, one finds references to concentration of acids or the acidity of crude oil or fractions obtained in refining. The total acidity of crude oil varies from negligible amounts in some highly paraffinic crudes to 2 % or slightly higher for a few exceptional fields. Acidities as high as 2.3% for a Roumanian oil and 3% for a California crude were reported by Naphthali [1] and Shipp.[4]

It is often difficult to determine the basis on which concentrations are reported, i.e., whether the concentration is in terms of weight per cent or volume per cent and whether weight per cent is calculated on the basis of weight of acids isolated or of acidity titrated and the use of an arbitrary average molecular weight. Sometimes it is difficult to determine whether the concentration refers to the original crude oil or to the distillation fraction studied. Obviously an error in interpretation of results reported may be quite misleading. The weight of *purified* naphthenic acids isolated from kerosene may amount to only 0.003 to 0.010% of the weight of the original *crude,* but the concentration of crude acids isolated from the kerosene may amount to several per cent of the *kerosene* fraction.

Probably the best method of reporting acidity of any type of petroleum material studied, at least when the acidity is determined by titration, is in terms of milligrams of KOH required to neutralize 1 g of the sample. This is the familiar acid number generally used by workers in the naphthenic acid field, which may be calculated readily to any other basis desired. The acid numbers of crude oils are quite low (0.1 to 7.0).

Instead of collecting the numerous existing reports on the acidity of various oils, Tables 10 to 12 of data from Naphthali, Shipp, and Schmitz are presented because they give a correct picture of the range of acidities reported and of the wide range of acid concentrations observed in crude oils from neighboring fields.

Since some of the concentrations were obtained by weighing the acids isolated while others are based on titration data, all results listed have been calculated to a weight per cent basis so as to permit ready comparison. The weight per cent data of Table 10 were calculated from the acid numbers and an assumed average molecular weight of 280 for the acids listed.

Table 10. **Acid Content of Various Eurasian Oils (Naphthali)**

Source	*Acidity of Crude Oil Weight %*
Galicia, Boryslav, Paraffinic	0.07
Galicia, Potok, Asphaltic	0.420–1.19
Roumania, Paraffinic	0.049–0.49
Roumania, Asphaltic	1.190–2.38
Russia, Balachany	0.700
Russia, Bibi-Eibat	0.500
Russia, Surachany, Light	0.420
Russia, Surachany, Red	0.280
Russia, Surachany, Heavy	0.217
Russia, Binagady	0.952
Russia, Swiatoy Island	0.868

The weight per cent data of Table 11 were calculated by the method used for Table 10. This table, given by Shipp, and that of Naph-

Table 11. **Acid Content of Various American Crude Oils (Shipp)**

Source	*Acidity of Crude Oil Weight %*
Texas, Winkler County	0.30
Texas, Howard County	0.07
Texas, Runnels County	0.03
Louisiana, Hosston	0.70
Michigan, Saginaw	0.03
Arkansas, Smackover	0.34
Mid-Continent	0.03
Pennsylvania	0.03
Gulf Coast	0.60
California, Various	0.10–3.00

thali [1] which he credits to F. A. Hessel of New York, lists an acidity of 0.03% for the Smackover crude, while that calculated from the acid number is 0.345 as given in Table 11.

The concentrations given in Table 12 were calculated from the acid-number data of Schmitz. An average molecular weight of 250 was used since this is approximately the molecular weight of acids at a boiling point of 250°C (480°F) near which the maximum acidity of most oils was observed by Schmitz.[3] The acid-number determinations were all made in the same laboratory.

Table 12. **Acid Content of Various Crudes (Schmitz)**

Source	*Acidity of Crude Oil Weight %*
Russia, Balachany	0.90
Germany, Pechelbronn	0.18
Germany, Nienhagen	0.35
Germany, Volkenrode	0.04
Poland, Potok	0.11
Poland, Kosmacz	0.90
Poland, Urycz	0.26
Irak	0.03
California, Kettleman Hills	3.01
East Texas	0.05
Peru, Lobitos	0.37
Peru, Low Cold Test	0.54
Peru, High Cold Test	0.06
Peru, Lobitos, Low Cold Test	0.28
Columbia	0.59
Venezuela, Lagonillas	1.20
Venezuela, Tocuyo	0.22
Equador	0.29
Equador, High Cold Test	0.09

The data presented in these tables show that no paraffinic crude yields a high concentration of acids, nor do all asphaltic crudes have a high acidity. The data also indicate that acid concentrations vary widely from field to field, often in the same state. It would be interesting to see whether or not wells in the same field producing from different horizons show different acidities.

In one of the most important and complete surveys of acid content and of distribution of acids within each crude oil, Edmond Schmitz [3] obtained samples of crude oils from nearly all important fields in production between 1931 and 1938. He determined the total acidity of each crude oil as received and then subjected each sample to a modified Engler distillation at atmospheric pressure to 250°C (480°F) and at 1 to 4 mm pressure from there on until only a viscous tar remained as residue. He then titrated each fraction and plotted the data to show the relation between the acid number and the cut

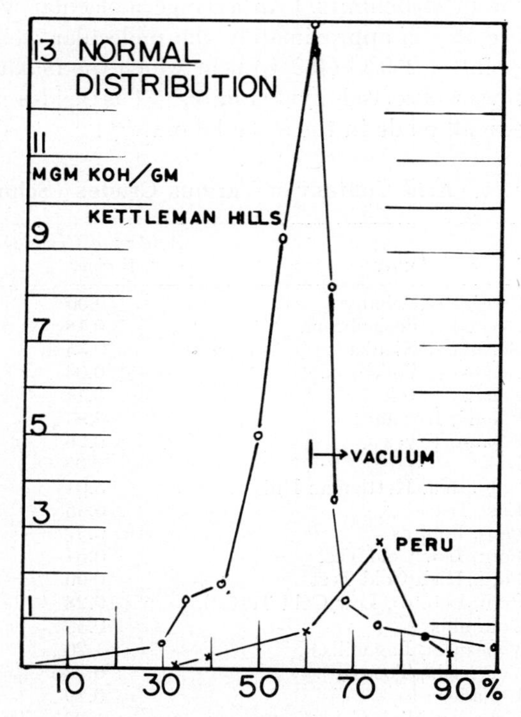

Figure 6. Normal Types of Boiling Point — Acidity Curves

number. Unfortunately he did not collect constant-volume or constant-temperature-range fractions so that it is difficult to determine the acidity remaining in the residue. He did however indicate the point at which he changed to vacuum distillation (250°C) in each case and this makes it possible to estimate the amount distilled below and above this point in most cases. His survey has the advantage that all results were obtained in the same laboratory, by the same procedure, and calculated to the same basis.

As would be expected from the data reported by others, the total acidity varies widely from crude to crude. More interesting are the distribution curves which may be divided into two main classes.

The common or normal curve given by most oils of high acidity has a rapid rise in acid content from cut to cut to a sharp maximum in the kerosene or distillate range, usually near a boiling point of 250°C (480°F), then drops off equally rapidly to a small fraction of the

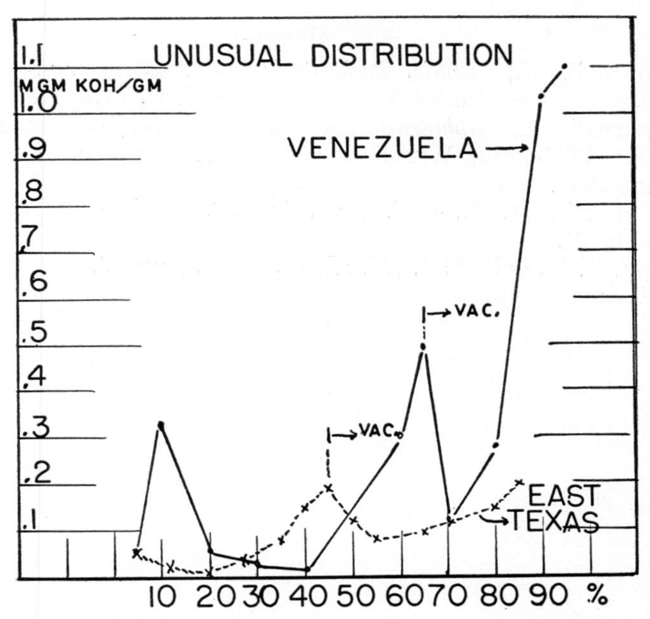

Figure 7. Abnormal Types of Boiling Point — Acidity Curves

maximum value. The abnormal and less common curve obtained apparently only with oils containing relatively low concentrations of acids also shows a maximum in the kerosene or distillate range but then has one or more additional maxima at higher boiling ranges or goes through a single minimum and then gradually rises to the end of the distillation. Figures 6 and 7 show typical curves of both types reported by Schmitz.

The results of Schmitz prove, what had been known at least qualitatively for years, that the acidity of the gasoline-range fractions is very low, that the kerosene and distillate fractions have the highest concentrations of acids, and that the heavy lubricating-oil cuts usually show a relatively low acid content. In the case of the abnormal types of oils, which generally do not have a high total acid content, the acidity of the high-boiling fractions may be appreciable.

Schmitz pointed out to the Russians that they should be able to obtain the best yields of acids from the "solar oil" fractions. In American practice, acids are isolated from this range since it gives a good yield of acids of the correct boiling point.

Bibliography

1. Naphthali, M., *Chemie, Technologie und Analyse der Naphthensäuren*, Wissenschaftliche Verlagsgesellschaft, Stuttgart, 1927.
2. Naphthali, M., *Naphthensäuren und Naphthensulfosäuren*, Supplement 1927–1933, Wissenschaftliche Verlagsgesellschaft, Stuttgart, 1934.
3. Schmitz, P. M. E., *Bull. assoc. franç. techniciens du pétrole* **46,** 93 (1938).
4. Shipp, V. L., *Oil Gas J.* (March 19, 1936), page 56.

CHAPTER FIVE

PURIFICATION OF PETROLEUM ACIDS

In a few special projects in which the acids present as such in crude oil were to be studied, the laboratory supply of acids was obtained by extraction of crude oil, but in a great majority of the cases, the acids studied were isolated from distillation products at a refinery. The acids are usually liberated from the alkaline solution obtained in washing the kerosene or distillate with alkali. Unless the refinery is operating this part of its processing on material obtained from several different crudes, it can also supply information on the field from which the crude oil came. Since a study of acids produced during cracking processes should probably succeed and not precede the simpler study of acids from straight-run operations, investigators generally avoid working with samples obtained from cracking-process materials or from a mixture of cracking-process and straight-run fractions. This is very difficult to do, since American refineries generally employ some type of cracking process and mix straight-run and cracking-process products prior to any alkali-washing stage. Probably even straight-run kerosenes or gas oils contain a certain amount of material formed through slight cracking at superheated areas of the stills, but these compounds should be present in such low concentrations that they will not interfere with the study of the regular acids and certainly will not be isolated.

If an attempt is made to neutralize a dilute solution of ordinary naphthenic acids with 0.1 equivalent of a dilute sodium hydroxide solution, much less than 0.1 equivalent of sodium salts will be found in the water layer. If a second 0.1 equivalent of base is then added and stirred, the amount of sodium salts found in the water layer will now be much nearer to 0.1 equivalent. The third 0.1 equivalent of base will carry at least 0.1 equivalent and usually more of sodium salts into the lower layer, while the next will perhaps carry 0.2 equivalent and finally the fifth or sixth aliquot will produce a homogenous solution. Evidently, at this stage, the sodium salts formed are just able to solubilize the remaining acids.

Where equipment is available, liquid propane or butane may be used for the extraction by a nonpolar solvent, as was done by Harkness and Bruun.[5]

Steam distillation has been used as another method of removing inert substances, while avoiding troublesome emulsion formation that often makes extraction by solvents very tedious. This method fails in the case of the higher acids, because the inert impurities which boil at the same temperature as the acids do not undergo steam distillation to any appreciable extent and thus cannot be removed from the sodium salts in this manner.

For many years, a standard method has been the Spitz and Honig method [13] * or some modification of it. This tries to avoid emulsion troubles by using a mixture of water and some alcohol containing an excess of alkali as one phase and petroleum ether as the other. In this method, the acids are treated with dilute alkali to convert all acids to salts, an equal volume of ethanol is added, and the mixture washed repeatedly with petroleum ether to remove hydrocarbons and most of the phenols. Unless the petroleum-ether solution is washed with water to remove dissolved sodium salts, there may be a considerable loss of salts in working with the higher acids. Isopropyl alcohol has been used instead of ethanol and other nonpolar solvents in place of petroleum ether.

The alcoholic solution of the salts is then treated with an excess of dilute sulfuric acid and the resulting acid layer separated as such or after solution in petroleum ether. The petroleum ether layer is then washed repeatedly with water or with a solution of a salt, such as sodium sulfate, to remove entrained sulfuric acid. The petroleum-ether layer is then dried, filtered, and distilled to remove the petroleum ether. For higher-boiling acids, drying is not needed since a petroleum ether boiling above 80°C can now be used and the

* This method is mentioned in most discussions on purification of naphthenic acids, but the author has not been able to find an original paper which is obviously the one referred to as the Spitz and Honig method. A paper by Spitzer and M. Honig, *Monatshefte* **39**, 1 (1918), may be the one referred to, but it deals with separation of ligninsulfonic acids and seems to be only distantly related to naphthenic acids.

water will distill over with the ether. If the aliphatic acids are also to be studied or if the lowest of the naphthenic acids are present, the petroleum-ether solution should be dried before distillation to remove the solvent.

The acids obtained from the refinery contain 3 to 30% hydrocarbons, other neutral compounds, and phenols. The removal of these is the first problem confronting the worker in any study of petroleum acids.

Part of the hydrocarbons and other inert substances can be removed by extracting dilute aqueous or alcoholic solutions of the sodium salts formed on adding an excess of alklai to the crude acids. The extraction is done by means of nonpolar solvents, like petroleum ether, benzene, or ethyl ether. Simple extraction of this type is only partially successful because:

1. Some of the acids will be removed along with the hydrocarbons and phenols, if the pH of the solution is too low.
2. Some of the phenols will remain in the acid mixture if the pH is too high.
3. Under all conditions, both inert matter and sodium salts will be distributed between the two phases involved, unless careful cross-extractions by both solvents are carried out or some system of countercurrent extraction is employed.

The difficulties mentioned under 3 are due to the solubilizing and emulsifying effect of the sodium salts (soaps) which has been stressed frequently by workers in this and related fields.[1, 6, 7, 8, 9, 10, 11] In the case of very high molecular weight acids, the sodium salts are somewhat soluble in oil and are not easily extracted with an aqueous solution as pointed out by Schmitz.[12] Golumbic [3, 4] has made a careful study of the separation of phenols and his results should be applicable to the separation of phenols and the higher acids.

Von Braun [2] was convinced that it is impossible to remove all of the hydrocarbons and other impurities from the liquid acids and advocated conversion of the acids to a solid derivative, or preferably degrading the acids to the next lower amine by the hydrazoic acid method and converting the amines to solid derivatives. In either case, thorough washing of the solid derivatives should now remove all impurities boiling above the boiling point of the solvent.

If the refinery operations include treatment with concentrated sulfuric acid and thus formation of various sulfo acids, vacuum distillation is said to remove these by decomposition. Tanaka and

Nagai [14] added a small amount of copper oxide during distillation to promote the decomposition of sulfur compounds.

Most of the difficulties met in the preliminary purification of petroleum acids arise from the tendency of the sodium-salt solutions to foam or emulsify. The use of water-alcohol mixtures as one phase in the Spitz and Honig method and the application of steam distillation are attempts to overcome these difficulties. Most of these troubles in attempted purification of unfractionated acids seem to be due to the presence of dark and tarry high-boiling substances in the crude acids.

A very considerable saving of time has been achieved in the author's laboratory by the use of the following series of steps which do not attempt to remove impurities except tarry residues in any one operation.

1. Rough distillation of the acids diluted with petroleum ether up to 150° to 175°C at atmospheric pressure, distillation at water-pump vacuum to about 250°C (480°F), followed by vacuum distillation at 2 to 5 mm to a heavy, tarry residue. Most of the troublesome substances remain as residue.
2. Approximately five plate distillation of the acids at atmospheric pressure to 250°C (480°F) and at 2 to 5 mm pressure to a tar residue.
3. Recombination of the acids to yield fractions boiling over 10° ranges.
4. Esterification of the individual fractions by methanol and hydrogen chloride.
5. Fifty-plate fractionation of the individual ester fractions. Since the esters boil about 50° below the boiling points of the original acids, neutral impurities and most of the phenols remain as residues. Any remaining impurities are removed during later extraction operations.
6. Recombination of ester fractions from various fractionations according to their boiling point so that finally fractions boiling over a 2° to 5° range are obtained.
7. Saponification of selected ester fractions to determine the equivalent weight and physical properties of the acids.
8. Efficient countercurrent fractional neutralization or liberation of the acids obtained on saponification of the ester fractions. This removes the remaining hydrocarbons and phenolic compounds and yields fractions which have gone through both purification and

separation operations. For higher-boiling fractions, step 8 encounters the same difficulties as were mentioned in connection with the Spitz and Honig

Although this appears to be a tedious series of steps it saves time since it combines separation with purfication while earlier schemes often took almost as much time to merely remove impurities from the acids — incompletely at that. For acids with less than ten carbon atoms all steps can be carried out successfully and individual acids, in most cases, can be converted to derivatives or salts and finally isolated and identified.

The last step of the series becomes very tedious when applied to acids with ten or more carbon atoms so that fractional neutralization may have to be confined to 3 to 5 stages — sufficient to remove phenols and neutral impurities and to effect a partial separation of aliphatic from naphthenic acid. The use of organic bases, like triethanolamine, appears to lead to less emulsification and emulsions that form break more rapidly, but additional work remains to be done to determine how efficient such systems are in separation of types of acids.

Fractionation by chromatographic columns, plain or ion exchange, and by thermal diffusion of the acids or their methyl esters has not been studied enough to determine whether these will generally separate acids with approximately the same boiling point into types or into individual compounds. They have the additional disadvantage of requiring the use of large equipment if samples of considerable size are to be studied so as to permit identification of individual acids.

Bibliography

1. Bhagwati, M. V., M. Varma, and K. G. Hermalkar, *J. Indian Chem. Soc.* **19,** 363 (1942).
2. Von Braun, J., *Ann.* **490,** 104 (1931).
3. Golumbic, C., *Anal. Chem.* **20,** 951 (1948).
4. Golumbic, C., *Anal. Chem.* **23,** 1210 (1951).
5. Harkness, R. W., and J. Bruun, *Ind. Eng. Chem* **32,** 499 (1940).
6. Klevens, H. B., *Chem. Rev.* **47,** 1 (1950).
7. Kolthoff, I. M., and W. Bosch, *J. Phys. Chem.* **36,** 1685 (1932).
8. Osol, A., and M. Kirkpatrick, *J. Am. Chem. Soc.* **55,** 4430 (1933).
9. Palit, S. R., *J. Indian Chem. Soc.* **19,** 272 (1942).
10. Pilat, S., *Petroleum Z.* **34,** 6 (1938).
11. Ross, J. D. M., T. J. Morrison, and C. Johnstone, *J. Chem.Soc.* **1938,** 264.

12. Schmitz, P. M. E., *Bull. assoc. franç. techniciens pétrole* **46,** 93–148; *C. A.* **33,** 9297 (1939).
13. Spitz and Honig, from M. Naphthali's *Chemie Technologie und Analyse der Naphthensäuren,* Wissenschaftliche Verlagsgesellschaft Stuttgart, 1927, page 21.
14. Tanaka, Y., and S. Nagai, *J. Am. Chem. Soc.* **45,** 754 (1923).

CHAPTER SIX

ANALYTICAL METHODS

In work with petroleum acids, both quantitative and qualitative analyses are often required. The odor of the acids, and especially of the ethyl esters, is frequently used as a rough qualitative test for naphthenic acids. The fruity odor of the ethyl esters is said to be absent in the case of fatty acids boiling in the naphthenic acid range. In either case the odor is often masked by that of other compounds present.

Several allegedly specific tests have been proposed for naphthenic but it is doubtful whether any one of them can be depended on to distinguish between aliphatic and naphthenic acids and particularly between rosin and naphthenic acids.

One of the oldest and best-known qualitative tests is the Charitschkow copper-salt test.[4] In this, a 10% aqueous solution of copper sulfate is added to a neutral sodium or potassium naphthenate solution to yield the copper salt of the acid. When petroleum ether is now added, it dissolves the copper naphthenate to yield a green solution. Tutunnikoff,[18] Schindler,[15] and Davidsohn,[6] all found that the copper salts of unsaturated acids of the same molecular-weight range are also soluble in petroleum ether to yield a green solution. To eliminate troubles caused by unsaturated acids, Tutunnikoff treated the acids, purified by the Spitz and Honig method, with a 1.5% solution of potassium permanganate. This produced lower-molecular-weight acids whose copper salts were insoluble in petroleum ether. Naphthali [17] reported that the oxidation is fairly fast, but even then, the total time required is excessive. Therefore, he recommended the addition of sodium thiosulfate at the end of the oxidation to remove the excess of oxidizing agent and to hasten precipitation of the manganese dioxide. He considered the test fairly satisfactory even in the presence of 10% unsaturated acids.

Spalming [16] found that the test is positive at 3%, but negative at 1% concentration of naphthenic acids in oil. Miss Luft [11] reported a solution of naphthenic acids in gasoline to show a negative test at concentrations less than 0.1% of acids. Often the free acids also appear to yield a positive test, but Pyhala [13] observed that this was due to sodium salts dissolved in the free acids.

Davidsohn [5] attempted to distinguish between aliphatic and naphthenic acids through the solubility of the magnesium naphthenates in hot water in which the magnesium salts of the fatty acids were said to be insoluble. Later workers claim that even many of the long-chain saturated aliphatic acids form soluble magnesium salts. It is not surprising that qualitative tests of this type are not satisfactory, since it would be unusual indeed if all naphthenic acids showed one type of solubility while all non-naphthenic carboxylic acids showed another. There seems to be a possibility, however, of using differences in solubility of salts in the systematic separation of aliphatic from naphthenic acids instead of depending entirely on other lengthier methods.

One of the early methods most commonly used for the semi-quantitative determination of naphthenic acid is that of Spitz and Honig.[17] It is not specific for naphthenic acids nor is it specific even for carboxylic acids, since the strong sulfonic acids would be included, if present, and since at least a portion of the phenols would also be included.

In this method, as described by Naphthali, a 5-g to 10-g sample is refluxed for 30 to 40 minutes with 25 to 30 ml of normal alcoholic potassium hydroxide, then cooled, and diluted with an equal volume of water. The mixture is now placed in a separatory funnel and extracted with 40- to 50-ml portions of petroleum ether until no additional extract is obtained. In the quantitative method, the petroleum-ether extracts are treated repeatedly with dilute alkali to remove all acids dissolved or suspended in the petroleum-ether layer. The sodium salts obtained in the alkali extracts are then combined with the main aqueous-alcohol layer and the resulting solution diluted with water, if not clear, placed in a separatory funnel, and acidified to methyl orange with dilute sulfuric acid.

The milky emulsion that forms if acids are present is extracted with petroleum ether. This organic layer is washed repeatedly with a concentrated solution of sodium sulfate and finally dried with anhydrous sodium sulfate. The petroleum ether is evaporated and the residue dried at 105°C for 5 minutes. Weighing this dry acid completes the determination.

As pointed out before, this method determines not only the naphthenic and other carboxylic acids but also the sulfonic acids, if any are present, and part of the phenolic compounds, since the solution is distinctly basic so that only a fraction of the phenols would be extracted with petroleum ether. Among numerous modifications proposed and used in various laboratories, that of Klotz and Littmann [9] is probably one of the best. In this scheme, the sample is titrated to the phenolphthalein end point with 0.5 normal sodium hydroxide solution and the naphthenic acids are determined in this solution in much the same way as in the Spitz and Honig method. A number of precautions are taken to avoid losses at various stages of the extraction and liberation steps and the acids finally obtained are dried to constant weight instead of only for 5 minutes. Since the solution is only slightly basic at the end of the titration to the phenolphthalein end point, practically all phenols should be removed by the petroleum-ether extraction. After weighing the dry acids, they are again titrated to determine the acidity due to carboxylic acids. The original titration of total acidity and final determination of carboxylic acids permits calculation of the phenol content. The difference between the sum of the weights of phenols and acids and the weight of the original sample used yields the weight of inert material present. Obviously, the weight of phenols must be calculated on the basis of an assumed average molecular weight of the phenols present and the total acidity is known only as accurately as is the titration of acids and phenols to the phenolphthalein end point. To check on this point, it should be possible to titrate the total acidity of a separate sample in nonaqueous solution by one of the modern methods.[7, 8, 14] Since any method of this type includes all carboxylic acids, such fatty acids as stearic, palmitic, and arachidic would evidently be reported as naphthenic acids, if present, as they are in some samples. Klotz and Littmann published a table showing that acids from seven different sources contained 74 to 94.7% of naphthenic acids.

Sometimes low concentrations of unsaturated acids are found in naphthenic acids as received in the laboratory and the iodine number is determined to measure the amount of unsaturation. Most pure naphthenic acids have an iodine number of zero, but commercial acids contain a small amount of unsaturated compounds. The iodine number is determined by methods used in the fat and oil industry.

In most cases, in industry, the results obtained in titrating a sample of acid by standard alkali to the phenolphthalein or α-naphtholbenzein end point are calculated to yield the so-called acid number of the acid. This is defined as the number of milligrams of potassium hydroxide required to neutralize 1 g of the acid. This value is readily converted to equivalent weight by the equation:

$$\text{Equivalent weight} = \frac{56{,}000}{\text{acid number}}$$

A high acid number indicates a low molecular or equivalent weight and at the square root of 56,000 (approximately 237), the two values are the same. A sample of crude oil would have a very low acid number, since the acidity is low.

The titration is usually carried out in alcoholic or isopropyl alcohol-benzene solution to avoid two-phase titrations which are always tedious and usually less accurate than titration of a homogeneous solution. Electrometric titration, using a glass electrode, is often used as the shape of the titration curve indicates whether or not appreciable amounts of phenols are present.

Since no dicarboxylic acids have been isolated from naphthenic acids so far, we may assume that the equivalent is also the molecular weight of these compounds and this is one of the most important determinations run in naphthenic acid analyses. Whether future work will show that some dicarboxylic acids occur in the high-boiling fractions of naphthenic acids remains to be seen. The anhydride of dimethyl maleic acid has been isolated from two different samples of petroleum acids in the author's laboratory.

Instead of obtaining the equivalent weight by titration, it is sometimes more convenient to get it through the analysis of silver, barium, or sodium salts of the acids. Since the silver and barium salts are insoluble, these are most often employed in analysis. Ignition of the silver salt to free silver, by micro methods if desired, is a particularly rapid and convenient procedure.

When a determination of moisture in naphthenic acids is required, the simplest procedure consists of adding an excess of benzene, xylene, or petroleum ether (boiling to at least 100°C) and distilling the solvent and moisture, using the customary apparatus and precautions in regard to droplets of water adhering to the condenser or emulsified in the distilled solvent.[1c] For the high-boiling commercial acids, the loss in weight on drying to constant weight, as in the procedure of Klotz and Littmann, can, of course, be determined, but will include part of the solvent used and thus will not give a true measure of water content.

Sometimes ash determination is called for which may be a difficult task if ash resulting from suspended inorganic matter is to be distinguished from that due to dissolved salts of organic acids. If the ash is to be determined by merely ashing the coke left on distillation of the sample, the procedure is essentially that of determining ash in a coal sample.

The sulfur content of petroleum acids is usually determined qualitatively or quantitatively and presents no special problems aside from the fact that the sample analyzed must be large enough to permit accurate determination of the small amount of sulfur present. The peroxide bomb, the oxygen bomb, or even the Carius method may be employed, although the last method is not well suited for use with relatively large samples.[1a] The lamp method is often used when the sulfur content is low.[1b]

In research work, the determination of carbon and hydrogen is required, but usually not until the substance analyzed is essentially a pure compound. As with many petroleum compounds, these analyses must be run with considerable care since, if they are carried out too rapidly or at too low a temperature in the combustion tube, pyrolysis seems to yield gaseous products — methane and acetylene have been suggested — which are not completely burned in an ordinary furnace.[10, 12] If a silica tube is used at bright-red heat and if the sample is volatilized carefully, correct results are obtained.

For most purposes, the determination of hydrogen alone would be sufficient, because the equivalent weight will normally give the same information as the analysis for carbon, while a determination of hydrogen is the best way of deciding whether a given acid is monocyclic, bicyclic, or even aromatic. The lamp method, which has been developed to determine sulfur in liquids, is also used for hydrogen determination in oils and can be used apparently with liquid naphthenic acids.[20] For satisfactory results, the weight of the burned sample should be greater than 1 g which may be a serious objection to this method in the analysis of very highly purified and, therefore, very expensive substances.

Another analytical method, that will be discussed in greater detail in connection with methods used in the elucidation of structure of acids, is the chlorine-number determination of von Braun.[2,3] He found that when the ethylamide (RCONHEt) of an acid is heated with an excess of phosphorus pentoxide, the three different types of ethylamides encountered in naphthenic acid chemistry react as follows:

1. $RCH_2CONHEt \rightarrow RC(Cl)_2C(Cl)=NEt$
2. $RR'CHCONHEt \rightarrow RR'CClC(Cl)=NEt$
3. $RR'R''CCONHEt \rightarrow RR'R''CC(Cl)=NEt$

The tertiary acids (type 3) are rare in petroleum and can easily be separated during the usual purification procedure, because they are not esterified readily by methanol and hydrogen chloride so that in the ordinary research operations, only acids of types 2 and 1 need be considered.

Type 1 acids are converted to compounds with two chlorines on the α carbon atom, while type 2 acids can have only one chlorine on the α carbon and thus this method can be used to distinguish between acids in which the carboxyl group is separated from the ring by at least one CH_2 group (type 1) and those in which the carboxyl is attached directly to the ring (type 2) or in which there is a branching at the α carbon atom.

While the types of compounds described before can sometimes be isolated and analyzed as such, more satisfactory results are obtained if they are refluxed with water to convert them to the α-chlorinated ethylamides, $RCCl_2CONHEt$ and $RR'CClCONHEt$, which are more stable and are analyzed to yield directly the number of chlorine atoms mentioned earlier. In the first case, von Braun said the chlorine number was 200, while it was 100 in the second. He attempted to calculate the ratio of type 1 to type 2 acids from the value obtained in his determination, but such ratios are of doubtful value since the reproducibility and accuracy of results are not high enough to yield reliable ratios. The real value, and it is great, of the method lies in its ability to decide whether the α carbon atom carries one or two hydrogens. If there are two hydrogens, the carboxyl group obviously cannot be directly connected to the ring.

Von Braun gave few details of procedure, but on the basis of his general directions, workers in the author's laboratory carried out this determination according to the following steps:

1. The acid is converted to the acid chloride by either the thionyl chloride or the phosphorus pentachloride procedure.

2. The acid chloride is converted to the ethylamide by reaction with a slight excess of ethylamine.

3. A solution in anhydrous benzene of 1 mole of the ethylamide is treated with 4 moles of phosphorus pentachloride and refluxed for 30 minutes. The resulting mass is then distilled under water-pump vacuum to yield a mixture which is fractionated carefully. Sometimes, a "heart cut" was analyzed for chlorine, but usually the results were not satisfactory and the chloroimide was refluxed for 1 hour with water to convert it to the chloroamide which was purified and analyzed by any of the standard methods for chlorine. The number of chlorine atoms found per mole of the chloroamide was multiplied by 100 to obtain the chlorine number.

The results are not entirely satisfactory for several reasons. In the first place, more of any admixed monochloro compound may be lost in the procedure than of the dichloroamide and thus yield results that indicate a greater concentration of acids with a chlorine number of 200 than was present in the sample, if it was a mixture. In the second place, even von Braun found that a repetition of the phosphorus pentachloride treatment raised the chlorine number so that apparently either the conversion is not complete or part of the halogen on the α carbon is eliminated during hydrolysis. However, the method has been used by a number of different investigators and has been found very valuable in the determination of the structure of pure compounds.

Bibliography

1. Am. Soc. Testing Materials, *Testing Methods and Standards of Committee D-2 on Petroleum Products*, Philadelphia, a. Sulfur by Bomb Method, D129–49; b. Sulfur by Lamp Method, D1018–49T; c. Moisture, D95–46; d. Acid Number, D974–48.
2. von Braun, J., F. Jostes, and W. Munch, *Ann.* **453,** 116 (1927).
3. von Braun, J., *Ibid.* **490,** 100 (1931).

4. Charitschkoff, K. W., *C. A.* **3,** 1699 (1909).
5. Davidsohn, J., *Seifenfabrikant* **34,** 323 (1914); from Naphthali (see reference 3 in Chapter 2), page 102.
6. Davidsohn, J., *Seifensieder Ztg.* **36,** 1591 (1909); *C. A.* **5,** 798 (1911).
7. Ferguson, H. P., *Anal. Chem.* **22,** 289 (1950).
8. Fritz, J. S., *Ibid.* **22,** 578 (1950).
9. Klotz, J., and E. Littmann, *Ind. Eng. Chem. Anal. Ed.* **12,** 76 (1940).
10. Kramer, G., and W. Böttcher, *Ber.* **20,** 599 (1887).
11. Luft, A., *Petroleum* **28,** No. 24, 16 (1932).
12. Markownikoff, W., *Ann.* **307,** 367 (1899).
13. Pyhala, E., *Chem. Ztg.* **36,** 869 (1912).
14. Rescorla, A. R., F. L. Carnahan, and M. R. Fenske, *Ind. Eng. Chem. Anal. Ed.* **9,** 574 (1937).
15. Schindler, H., *Petroleum Z.* **33,** No. 15, 1 (1937).
16. Spalming, G., *Chem. Ztg.* **1903,** 196.
17. Spitz and Honig in M. Naphthali, *Chemie, Technologie und Analyse der Naphthensäuren,* Wissenschaftliche Verlagsgesellschaft, Stuffgart, 1927, page 21.
18. Tutunnikoff, B., *Seifensieder Ztg.* **50,** 591 (1923).
19. Tutunnikoff, B., *Ibid.* **50,** 603 (1923).
20. Wilder, C. R., M. A. Thesis, The University of Texas, 1952.

CHAPTER SEVEN

METHODS OF SEPARATING PETROLEUM ACIDS

As in the study of petroleum hydrocarbons and bases, the task of separating the acids from nonacids, phenols, and finally from other carboxylic acids is a formidable one which requires much time and patience, but it is the most important step in the isolation and identification of petroleum acids with more than six carbon atoms.

The preliminary refining or purification of petroleum acids, which is a separation process in itself, has been discussed in Chapter 5. It normally involves a rough separation of acidic compounds from neutral and phenolic compounds and is based almost entirely on the fact that the acids react with sodium hydroxide to produce stable sodium salts while the phenols and neutral compounds do not; thus the phenols and neutral compounds can be removed, more or less completely, by extraction with a nonpolar solvent or by distillation or steam distillation.

The finer separation of the acids from neutral compounds, phenols, and other types of acids, and even from homologues and isomers, may be based on one or more of the following differences of these compounds:

1. Vapor pressure
2. Acidic strength
3. Distribution between immiscible solvents
4. Solubility of salts or derivatives in a suitable solvent
5. Reaction rate with a suitable reagent
6. Adsorbing tendency on a solid adsorbent

When the problem is as complex as the separation of naphthenic acids from aliphatic and aromatic acids and finally the separation of the naphthenic or aliphatic acids from each other, it is obvious that no one method of separation can be expected to separate mixtures into individual compounds unless they have some freak property. For any single method, two or more of the compounds in the mixture studied will be sure to have so near the same vapor pressure or acid strength or adsorbing tendency that separation simply cannot be attained, no matter how efficient the apparatus.

This has, of course, been recognized at least vaguely since the earliest studies on the acids when distillation fractions were converted to amides and/or other derivatives, and separation by difference in solubility (recrystallization) was attempted. Usually it was impossible to even obtain solid derivatives and much less to be able to separate these. In a few cases, a combination of methods has been found sufficient to identify a few acids, especially when the lower acids, which represent a rather simple mixture in the first case, were studied.[11, 21, 36]

Fractional Distillation

The simplest and most easily applied method of separation has been fractional distillation or separation by difference in vapor pressure or boiling point. It is also roughly a method of separation by molecular weight as far as closely related compounds are concerned, since, in general, isomeric compounds of the same class have boiling points which are not far apart.

The earliest workers in the field of naphthenic acids used a modified distilling flask equivalent to, at most, four or five plates. To get better separation, they simply repeated the fractionation until the properties seemed to indicate that fair separation had been accomplished. When the boiling point reached about 250°C, they usually changed to water-pump vacuum and continued until decomposition was again noticeable.

In the hands of Aschan,[2, 3] Markownikoff,[33] and others, this separation appeared to yield good results, but analyses indicated that the fractions were still mixtures, as became clear when a fraction was converted to the ester and again fractionated. Since the methyl esters boil approximately 50° below the boiling point of the parent acid and since in mixtures, they boil closer to the boiling point of the esters concerned, this procedure has become standard practice, i.e., the acids are roughly separated by fractionation as such, then the fractions are converted to the methyl esters and these esters very carefully fractionated.

There is one situation, however, when the acids as such should be carefully fractionated into narrow boiling cuts which are then individually esterified and again separately fractionated as esters. This procedure removes a very high percentage of phenols and hydrocarbons which may not have been removed by preliminary purification, because the phenols and neutrals boil at the boiling point of the original close-cut fraction while the esters boil about 50° lower, so that all of the impurities remain in the still pot when the esters are distilled. If a second method of separation, like separation by acid strength, is employed this precaution is not required, since the phenols and neutral compounds are easily separated.

This combination of fractionation of acids and esters was employed in practically all past work and, until recently, was the only really efficient method available for isolating individual compounds. In addition to the earlier studies, those of von Braun,[7] Nenitzescu,[36] Muller and Pilat,[35] Tanaka,[50] Kuwata,[27] and the Texas group employed this method either as the only method or as the method used as long as the mixture was a simple one and did not require further systematic separation.

Except through the combination of acid and ester fractionation, it is, of course, impossible to separate acids from hydrocarbons, phenols, or any other class of compounds by the use of fractional distillation alone, since many of these nonacid compounds will have the same vapor pressure as some of the acids.

Von Braun [7] used fractionation of acids and esters, and finally of the amines formed, by degradation of the acids to amines with n-1 carbon atoms. The amines appear to boil over a wider range than the esters of the original acids, but while he hoped to be able to isolate individual compounds in this way, he found it better to identify the amines by converting them to a solid derivative — the oxalate. Even this combination yielded no individual amine.

Nenitzescu [36] fractionated as acid and as ester and then converted the esters to solid amides. He was able to isolate and identify a few aliphatic and a few naphthenic acids in this way — the acids identified accounted for was only about 8% of the total acids and most fractions could not be purified.

Chichibabin [11] attempted to use the same route as Nenitzescu (who worked later), but decided that the amides were not suited for this work and used the differences in solubility of cadmium salts, as we shall see later.

Holzman and Pilat [40] used the combination of distillation of acids and their esters, followed by fractional precipitation of the magnesium salts. They added a methyl-alcoholic solution of magnesium acetate to a cold, nearly saturated, solution of a fraction of the acids. The magnesium salts of myristic, palmitic, stearic, and arachidic acids precipitated out and were filtered and liberated by hot dilute hydrochloric acid. The acids crystallized, were purified by recrystallization, and identified. In this case, the fatty acids gave the less soluble salts. They did not determine the per cent yield of solid acids, but apparently each of the four fractions obtained from each of the distillation cuts yielded solid acids.

In general, the use of fractional distillation of the acids and their esters, combined with recrystallization of solid derivatives, has been successful in but a few cases and then the mixtures involved were relatively simple, because only the lower acids were present.

In the separation of the higher acids, fractionation has been carried out at 0.1 to 12 mm pressure, thus making possible to distill acids which would be decomposed at water-pump vacuum. However, most types of fractionating columns are less efficient at higher vacuum and the throughput is so much lower that the actual fractionation is not as practical as at 12 to 30 mm pressure, which still permits distillation of acids boiling above 300°C at atmospheric pressure.

Distillation in the high vacuum of a molecular still has been used a few times, notably by Harkness and Bruun, when only a moderate amount of separation was required or expected.[22]

Steam distillation, at atmospheric pressure, under vacuum, or with superheated steam, has not been successful except as a means of separating hydrocarbons and phenols from the sodium salts of carboxylic acids. Ney *et. al.*[37] tried to separate a difficult mixture by steam distilling a mixture of sodium salts in water solution to which dilute mineral acid was added slowly, but the results were not as good as with other methods.

Azeotrope formation as a means of separating isomers has not been tried, probably because it was felt that the chance of obtaining an azeotrope with one acid and not with its isomer was too remote and if both formed azeotropes, they would probably not be much more easily separated than the original pair, with the added problem of removing the azeotrope former completely.

Amplified distillation developed by Bratton and Bailey [8] was used by them with very good results in the separation of small amounts of the alkylpyridines from petroleum bases. In this method, a carefully washed hydrocarbon mixture, boiling over the whole range of the bases to be separated, was added to the fraction of bases and then carefully fractionated. The main advantage, aside from differential azeotrope formation, appears to lie in the fact that if several compounds boiling a few degrees apart are present, each of them will distill along with part of the hydrocarbon boiling at the same temperature; then, for a short interval, there will be only hydrocarbon distilling, followed by the second compound again mixed with hydrocarbon. The base, in the case of the work of Bratton and Bailey, was then converted to the picrate and filtered and was found to be essentially pure.

The method has been tried on mixtures of acids, but in the case of the lower acids, the simple hydrocarbon mixture was separated into plateaus and with the higher acids, no special separation was obtained. Weitkamp,[55] in his brilliant work on the acids from wool fat, used this method for the separation of the methyl esters and the author has used it on mixtures of methyl esters of acids produced by high-pressure condensate wells.[28, 31]

Finally, there is the recent method of extractive distillation in which a different compound is again added, this time one that boils higher than any of the components of the mixture. Normally, this involves the use of a continuous still and the third compound is added near the top of the column in the hope that it will influence the vapor pressure of one component more than it does that of the other or others. As far as the author is aware, the method has not been tried in laboratory fractionation of petroleum acids, perhaps because the control of a continuous laboratory column is more difficult than that of a batch still. Possibly, the addition of a high-boiling weak base, like carbazole, would produce good results here with acid mixtures boiling below the base.

Whenever, as in the case of the lower acids, fractional distillation of the acids and then of their methyl esters, followed by formation and purification of a solid derivative, or of a metal salt yields a pure compound that can be identified, this amount of fractionation is normally all that is used. The procedure has the serious defect in practice that only one or, at most, two acids, out of a fraction containing perhaps five or six compounds, are isolated and identified. Other fractions yield no solid derivatives, or the derivatives formed cannot be recrystallized to constant melting point. The filtrates, in all cases, contain acids or derivatives of acids which cannot be identified and only a small fraction of the total acidity of a batch will be identified.

To overcome this weakness, and especially to isolate even a single compound out of a complex distillation fraction, an additional fractionation method, depending not on separation through difference in vapor pressure, but on some other property which does not vary directly with the molecular weight, is used. Only in this way may we expect to be able to isolate not only one but most of the individual compounds that make up a mixture of the lower acids and to get at least one of the acids of a complex mixture in a fraction of the higher acids.

To illustrate what is meant by this statement, we list the following general example: A preliminary fractionation gave us a mixture containing *abcde*. Fractionation method A yields *ab* and *cde*. Probably neither of them can be identified at this stage. Method B on another portion of the original mixture yields *abc* and *de;* again probably no individual can be purified. If method C now yields *ae* and *bcd* on the original fraction, we still are unable to purify an individual except through good luck. Three different methods of separation each by itself did not enable us to isolate a single one of the five compounds in the mixture.

Now, let us assume that we start with another batch of the original sample, separate by method A, and get, of course, *ab* and *cde*. If method B is now applied to *ab*, it yields simply *ab*, but when applied to *cde*, it yields *c* and *de* and we are able to purify *c*. Method C now applied to the fractions from method B yields *a* and *b* from *ab;* both can be purified. From *de*, we get *d* and *e*, both of them again nearly pure. Thus we have isolated not one, or none, but all of the five compounds of the mixture.

We know of no combination of methods which would be able to separate a closely related mixture of five acids into five individuals, but the mixture obtained through a combination of two or three independent efficient fractionation methods should be much simpler than that obtained by only one method and there is a good chance that some additional method of limited applicability, such as derivative formation and recrystallization, will now make it possible to isolate individual acids in satisfactory yield where before none or such a small yield was obtained that the acid could not be identified.

Separation by Extraction

One method of separation that does not depend primarily on vapor pressure or molecular weight is fractional neutralization. This method can be used before or after fractional distillation and consists, in its simplest form, of adding, with thorough stirring or shaking, aliquots of dilute alkali to a dilute solution of the acid mixture in petroleum ether or another solvent that is immiscible with water. If the aliquot of base added is 0.1 equivalent of sodium hydroxide, we should expect to find 0.1 equivalent of sodium salts in the bottom layer. Actually, with higher acids (after the emulsion breaks), we find perhaps 5% of the total acidity as sodium salts. The other 5% is held in the top layer. In spite of this behavior, by repeated operations, all of the acids have been converted to sodium-salt solutions which were collected as cuts. Such a series of fractions will show a fair degree of separation if the mixture consists of aliphatic and naphthenic acids boiling over the same narrow range.

A much higher degree of separation is obtained if, at the end of each step of neutralization, the top layer is extracted with water and the bottom layer by petroleum ether, the new extracts being combined with the proper layer of the extraction step involved. Although the results are much better, this stepwise neutralization still has the serious defect that if a weaker acid once gets into a stronger fraction by equilibrium with a high concentration of weak acid in the mixture, it will never be removed and the stronger acid will be contaminated while the amount of the weaker acid isolated will be reduced by this amount.

To overcome this weakness, the principle of countercurrent contact was used by Jantzen [23] in his brilliant work on the separation of coal-tar bases. The operations and principles involved are given by Morton [34] and by a number of chemical engineers, particularly by Fenske and others in the mid-thirties.[1, 10, 53, 54]

Countercurrent fractional neutralization or liberation — quite similar in principle and often combined in a single series of fractionations — can be carried out in separatory funnels, carboys, flasks, or any other containers that can be stirred or shaken and can be emptied conveniently. Many analogous apparatus have been used for this type of separation.[23, 48, 30, 38, 43, 52]

For this separation, which, of course, may also be purely liquid-liquid extraction not involving acid strength, various arrangements can be used, in which the individual vessels are connected in such a way that all stages of the series can be operated by one adjustment, or even automatically.[20, 44, 56] Craig and coworkers have worked out not only the apparatus, but also the theory concerned, and their equipment is widely used in biochemical and pharmaceutical work.[9, 12, 13, 14, 15] Golumbic has pointed out its applications in the separation of phenols.[18, 19]

Finally, countercurrent extractions or neutralizations are carried out in packed columns of various types, ranging from large industrial equipment used in refining lubricating oil to various laboratory type columns.[32, 45, 46] The most efficient column that was used extensively by both petroleum research groups at the University of Texas was developed by Jantzen [23] and Tiedcke [51] and used extensively at Hamburg on coal-tar bases. The Tiedcke column was a continuous type in which the bases, in petroleum ether solution, were added at the bottom and rose to the top, where any bases not neutralized by the dilute acid added at the top flowed into a receiver to yield one fraction for every time the whole batch was passed through the column. If the 10-m tubing which served as extraction column had not been packed, the efficiency would have been perhaps one stage. Actually, the packing consisted of a rapidly spinning rod or closed tubing which was rotated at such a rate that the descending or ascending phase moved in the form of nearly flat visible spirals. Contact between phases was practically all film contact and a very high efficiency resulted.

Jantzen's continuous column was difficult to regulate and workers at the University of Texas found it more convenient to operate batch columns in which the upper end of the column proper terminated in a distilling flask that could be heated to remove the solvent as it accumulated. Since the whole batch of acids was placed in the upper flask at the start, countercurrent contact in the spinner column was provided through addition of a convenient ratio (usually 0.5 equivalent) of dilute hydrochloric acid through a side connection near the bottom of the column. The organic acid liberated by the mineral acid was dissolved in the ascending spirals of petroleum ether and came in contact with the descending film of sodium salt solution, liberating any weaker acid in this water layer and forming the sodium salt of the stronger acid. Fractions were collected at the bottom. The petroleum ether was removed by distillation at the top and recycled as fast as it reached the top flask.

Various forms of this spinner column were used in separation of the pyridine bases from petroleum and in separation of difficult mixtures obtained in the acid studies.[4, 5, 6, 29, 37, 47] In some cases involving simple acid mixtures, a modified stepwise neutralization was used, but this was not efficient enough for use with pairs of naphthenic acids or other very difficult mixtures for which the spinner columns were used. It was the combination of fractional neutralization of this type with efficient fractionation that made the isolation and identification of a number of acids possible in the work of the Texas group.

Unfortunately, it is difficult, for mechanical reasons, to operate a column with a diameter in excess of about 15 mm or longer than 4 feet and the maximum throughput of the small columns used was only about 400 cc an hour of combined phases. With the dilute solutions which had to be used, this meant that only a few grams of acid mixture could be treated in an hour. This would not have been objectionable if it had been possible to obtain dependable constant rates of fluid flow into the column. While some pumps and other schemes were found to give a very constant flow for hours, they could not be depended on, and the apparatus required frequent attention or checking. Pumps which were said to be reliable were so expensive that they could not be bought. An additional serious complication in the use of spinner columns for the separation of acids is the tendency to form emulsions whenever higher acids are treated. This trouble was not encountered with the bases which gave very satisfactory results at all times.

In the study of bicyclic acid mixtures in the C_{13} to C_{15} range, an attempt was made to overcome this objection by working with alcoholic sodium hydroxide solutions containing just enough water to obtain two phases and by carrying out the separation in a rotating rack of twenty-four tall-form liter separatory funnels.[25] In this range, however, even the alcoholic solutions emulsified to such an extent, in spite of gentle rotation, that a twenty-four-stage fractional neutralization required several days, but the operations were simple and not subject to errors in flow rate and other difficulties of the spinner columns. The method gave clean-cut separations on shale-oil bases.

In work with petroleum acids, the Texas group used the simple stepwise neutralization with cross extraction for simple mixtures [21, 41] and the batch spinner column for the separation of the higher aliphatic and naphthenic acids. It was interesting to observe that the naphthenic acids were always found in the first fractions (strongest acids), while the fatty acids came through last.

Although in this discussion, separation by acid strength was assumed, it was actually based partly on solubility, because a mixture of normal hexanoic and heptanoic acids with practically the same K_a values is readily separated, the hexanoic acid being obtained first, as if it were the stronger acid. The separation of naphthenic from aliphatic acids depends probably on the same property.

Pure liquid-liquid extraction, not depending on acidity of the acids, has not been used successfully — possibly because it has not been studied seriously since it was always felt that separation based on reaction between acids and bases should be simpler. In view of the emulsion difficulties mentioned, true liquid-liquid extraction should be tried on mixtures of the acids or their methyl esters. Recent experiments in the author's laboratory showed that some of the organic bases, particularly triethanolamine, in water solution can be used in fractional neutralization of acids in petroleum-ether or benzene solution without very stable emulsion formation, but even these operations should be carried out in batch equipment in which settling time can be adapted to stability of emulsions.

Separation by difference in solubility of salts or derivatives of acids is, of course, involved in all methods of recrystallization of derivatives, like amides. It has also been used for the lower acids by fractionally precipitating the silver salts and filtering them as they formed.[21, 48] It was also used by Chichibabin [11] who found that primary and secondary acids, i.e., those with the carboxyl group at the end of a chain and those with the group on a ring, could be separated by adding small portions of cadmium chloride solution to an aqueous solution of the sodium salts. He obtained fairly satisfactory separations, especially when he repeated this operation. He found a similar use of copper salts not satisfactory, although additional work should probably be done along this line, using more dilute solutions.

An apparently very powerful method of separation depends on the difference in esterification rates of acids or on the difference in saponification rates of esters. Since equilibria are involved, countercurrent schemes suggest themselves, but attempts to use a heated spinner column for fractional esterification indicate that the rate is too slow for use in columns. This method was, however, found successful when applied to a mixture of ordinary and hindered acids, since in this case, the hindered acids were not esterified while the others were and the unesterified acids were simply removed by extraction with cold alkali. 1,2,2-Trimethylcyclopentanecarboxylic acid (camphonanic acid) [21] was isolated and identified in this way, as were both of the *cis-trans* forms of the very highly hindered 2,2,6-trimethylcyclohexanecarboxylic acid.[37] It was hoped that it would be possible to separate primary from secondary acids by fractional esterification with isopropyl alcohol, since the esterification of a secondary acid by a secondary alcohol should be slow. There was some separation, but other methods were found more satisfactory.[37]

Finally, aside from the use of thermal diffusion-column separation which has apparently not been tried, there is the whole field of chromatographic separation, which has been a very powerful tool in biochemical work, where a few milligrams of mixtures are sometimes studied. Methods have been developed for separation of the fatty acids from the lowest to the C_{18} type.[26, 42] Petroleum laboratories have long used a somewhat similar procedure in decolorizing oils and are now using a percolation technique in separating aromatic from nonaromatic hydrocarbons. A simple scheme has been used in removing polar compounds and aromatic hydrocarbons from condensate well hydrocarbon by passing the mixture through a pipe filled with silica gel.[49]

By a suitable choice of solvents or solvent mixtures and of adsorbent, it should be possible to adsorb the acids from a distillation or extraction cut and then either continue to add the same solvent until the acids pass through the column, one by one, or probably better use an eluting solvent to move the different acids down the column. This method has not been used in the author's laboratory in separating the acids in the C_{10-12} range, but it has been tried with mixtures of bicyclic and monocyclic C_{13-14} acid mixtures. Jones apparently obtained extensive separation, but has so far not been able to isolate a solid bicyclic acid by this method.[25]

Since chromatographic methods often lead to better separations with derivatives than with the original compounds, good results may possibly be obtained with the methyl esters, which would not be as tightly held as the highly polar acids. The main difficulty with the use of chromatographic columns is the fact that laboratory columns of the usual dimensions are able to process only a gram or two of mixture, which would not be sufficient to permit identification of any acid isolated. Larger columns are in use, however, and the method is so powerful that it should be seriously considered for use after fractional distillation or after both fractional distillation and extraction.

With space and equipment and technical forces capable of operating the types of separating equipment discussed in this chapter, a combination of fractional distillation with one or both of the other methods mentioned now makes it possible to attain fairly complete separation of acids through the ten-carbon range. What progress will be made when properly and skillfully applied to acids with thirteen or more carbon atoms, including aliphatic, monocyclic and bicyclic, naphthenic and possibly aromatic acids, remains to be seen, but the outlook is good.

Bibliography

1. Appel, F. J., and J. C. Elgin, *Ind. Eng. Chem.* **29,** 451 (1937).
2. Aschan, O., *Ber.* **24,** 2710 (1891).
3. Aschan, O., *Ibid.* **25,** 3661 (1892).
4. Axe, W. N., and J. R. Bailey, *J. Am. Chem. Soc.* **60,** 3028 (1938).
5. Axe, W. N., and J. R. Bailey, *Ibid.* **61,** 2609 (1939).
6. Axe, W. N., and A. C. Bratton, *Ibid.* **59,** 1424 (1937).
7. von Braun, J., *Ann.* **490,** 100 (1931).
8. Bratton, A. C., and J. R. Bailey, *J. Am. Chem. Soc.* **59,** 175 (1937).
9. Bush, M. T., and P. M. Denson, *Anal. Chem.* **20,** 121 (1948).
10. Cannon, M. R., and M. R. Fenske, *Ind. Eng. Chem.* **28,** 1035 (1936).
11. Chichibabin, A. E., et al., *Chim. et ind.* **17,** 306 (Special Number) (1932).
12. Craig, L. C., *J. Biol. Chem.* **188,** 304 (1951).
13. Craig, L. C., and O. Post, *Anal. Chem.* **21,** 500 (1949).
14. Craig, L. C., *Ibid.* **22,** 1346 (1950).
15. Craig, L. C., *Ibid.* **23,** 41 (1951).
16. Feitelson, J., *J. Am. Oil Chemists Soc.* **27,** 4 (1950).

17. Fischer, W., and O. Jubermann, *Chem. Eng. Tech.* **23,** 299 (1951).
18. Golumbic, C., *J. Am. Chem. Soc.* **71,** 2627 (1949).
19. Golumbic, C., *Anal. Chem.* **23,** 1210 (1951).
20. Grubhofer, N., *Chem. Eng. Tech.* **22,** 209 (1950).
21. Hancock, K., and H. L. Lochte, *J. Am. Chem. Soc.* **61,** 2448 (1939).
22. Harkness, R. W., and J. Bruun, *Ind. Eng. Chem.* **32,** 499 (1940).
23. Jantzen, E., *Dechema Monograph, Vol. 5, Das fraktionierte Destillieren und das fraktionierte Verteilen als Methoden zur Trennung von Stoffgemischen,* Verlag Chemie, Berlin, 1932.
24. Johnson, J. D. A., *J. Chem. Soc.* **1950,** 1743.
25. Jones, W. A., Sun Oil Co., Fellow, University of Texas, Unpublished Data on Bicyclic Acids.
26. Kirchner, J. G., A. N. Prater, and A. J. Haagen-Smit, *Ind. Eng. Chem. Anal. Ed.* **18,** 31 (1946).
27. Kuwata, T., *C. A.* **23,** 1390 (1929).
28. Lochte, H. L., and E. N. Wheeler, *Condensate Well Corrosion,* N. G. A. A. Condensate Well Corr. Comm., Nat. Gasoline Assn. of America, Tulsa, 1953.
29. Lochte, H. L., W. W. Crouch, and D. Thomas, *J. Am. Chem. Soc.* **64,** 2753 (1942).
30. Lochte, H. L., and W. G. Meinschein, *Petroleum Eng.* March, 1950, page 725.
31. Lochte, H. L., and H. W. H. Meyer, *Anal. Chem.* **22,** 1064 (1950).
32. Mair, B. J., and S. T. Schicktanz, *Ind. Eng. Chem.* **28,** 1446 (1936).
33. Markownikoff, W., *Ann.* **307,** 367 (1899).
34. Morton, A. D., *Laboratory Technique in Organic Chemistry,* McGraw-Hill, New York, 1938, page 22.
35. Muller, J., and S. Pilat, *Brennstoff-Chemie* **17,** 461 (1936).
36. Nenitzescu, C. D., D. A. Isacescu, and T. A. Volrap, *Ber.* **71,** 2062 (1938).
37. Ney, W. O., W. W. Crouch, C. E. Rannefeld, and H. L. Lochte, *J. Am. Chem. Soc.* **65,** 770 (1943).
38. O'Keefe, A. E., M. A. Dolliver, and E. T. Stiller, *Ibid.* **71,** 2453 (1949).
39. Perrin, T. S., and J. R. Bailey, *Ibid.* **55,** 4136 (1933).
40. Pilat, S., and E. Holzman, *Brennstoff-Chemie* **14,** 263 (1933).
41. Quebedeaux, W. A., G. Wash, W. O. Ney, W. W. Crouch, and H. L. Lochte, *J. Am. Chem. Soc.* **65,** 767 (1943).
42. Ramsey, L. L., and W. I. Patterson, *J. Assn. Off. Agri. Chemists* **31,** 441 (1948).
43. Raymond, S., *Anal. Chem.* **21,** 1292 (1949).

44. Rometsch, R., *Angew. Chem.* **62,** 24 (1950).
45. Scheibel, E. G., *Chem. Eng. Progress* **44,** 681 (1948).
46. Scheibel, E. G., *Ind. Eng. Chem.* **42,** 1048 (1950).
47. Schenck, L. M., and J. R. Bailey, *J. Am. Chem. Soc.* **63,** 1364 (1941).
48. Schutze, H. G., W. Shive, and H. L. Lochte, *Ind. Eng. Chem.* **12,** 262 (1940).
49. Shock, D. A., and N. Hackermann, *Ind. Eng. Chem.* **40,** 2169 (1948).
50. Tanaka, Y., *J. Am. Chem. Soc.* **45,** 754 (1923).
51. Tiedcke, K., Thesis, Hamburg, 1928.
52. Tschesche, R., and H. B. Koenig, *Z. angew. Chem.* **61,** 441 (1949).
53. Varteressian, K. A., and M. R. Fenske, *Ind. Eng. Chem.* **28,** 1353 (1936).
54. Varteressian, K. A., *Ibid.* **29,** 270 (1937).
55. Weitkamp, A. W., *J. Am. Chem. Soc.* **67,** 453 (1945).
56. Weygand, F., *Chem. Eng. Tech.* **22,** 213 (1950).

CHAPTER EIGHT

CHARACTERIZATION OF PETROLEUM ACIDS

Formerly, when the only method of separation of acids used was that of fractional distillation, fractions could be recombined according to boiling point, which was the only property used in the separation. Sometimes, in addition to boiling point, the determination of equivalent or molecular weight by titration of acidity or by determination of silver in the purified salt was employed as an indication of progress in the fractionation and for subsequent recombination of similar cuts. To avoid accumulation of innumerable fractions, many of which may be nearly identical, some basis of recombination of fractions is essential in the use of all methods of separation of acids.

Combination based on boiling point became worthless when methods of separation independent of boiling point or vapor pressure began to be employed. Methods depending on physical or chemical properties had to be developed, since distillation fractions boiling over a narrow range were now divided into dozens of different cuts still boiling over the same, or slightly expanded, boiling range but differing markedly in such properties as density and index of refraction.

Index of Refraction

The most easily and accurately determined physical property of these acid fractions is the index of refraction. It is sensitive to the C:H ratio, i.e., it is high for unsaturated and alicyclic compounds and low for acyclic saturated acids. Since the unsaturated acids appear to be absent or present only in traces, the index of refraction is a convenient indication of the presence or absence of aliphatic acids in naphthenic acid mixtures, if the hydrocarbons had been completely removed in previous operations on the mixture.

For any one series of homologous and isomeric acids, after the lowest members of the series, the index of refraction varies only slightly. This is true especially for the higher acids of similar type varying only in the length of a long side chain or in the size of alkyl substituents. Roger Adams and coworkers[2, 3, 4, 20] reported data for a number of series of synthetic acids. Table 13 shows some of their results.

Table 13. **Index of Refraction of Synthetic Long-Chain Naphthenic Acids**

Series	*Alkyl Range*	*Index of Refraction*	*Reference*
Cyclohexylalkylacetic Acids	n-C_5 to n-C_{12}	1.4640–1.4650	2
3-Cyclohexyl-2-Alkylpropionic Acids	C_2 to n-C_8	1.4621–1.4640	2
4-Cyclopentyl-2-Alkylbutyric Acids	C_2 to n-C_8	1.4590–1.4629	4
4-Cyclohexyl-2-Alkylbutyric Acids	C_2 to n-C_8	1.4613–1.4640	3

The first few members of various series, such as cyclopentane- and cyclohexanecarboxylic acids, have an index of refraction of 1.4520 to 1.4530, but succeeding members of the series show a rapid rise to a value of 1.46 or slightly higher and change only slightly with further increase in molecular weight. As will be pointed out later, the density falls rapidly with increasing molecular weight in this range.

Petroleum acids with twelve or more carbon atoms may contain considerable concentrations of bicyclic or polycyclic acids,[7, 8, 26, 29, 44, 47, 50] which have a higher index of refraction than monocyclic acids. Unfortunately, this mixture may also contain aliphatic as well as naphthenic acids and the resulting average index of refraction may lie anywhere between the low values observed for fatty acids and the much higher values of the bicyclic acids. We must conclude, then, that the index of refraction of acids fractionated by distillation alone is of little value for the characterization of these acids. However, fractions obtained by any method of separation usually show a considerable spread in index of refraction, and when two or more different methods of separation have been employed, it is often possible to arrive at tentative conclusions on the nature of acids studied.

Density

Density is much more sensitive to structural differences in acids than is the index of refraction, but, like molecular volume, the range in values for any homologous series is so great that densities of different series may overlap. The density of a mixture of acids may range from values slightly above 1.06 to about 0.94 with a few published values even outside of this range. It is obvious, then, that density alone cannot be used in determining what type of acid is being studied.

In the case of petroleum acids, the change in density with increase in boiling point or molecular weight appears to vary from field to field and it may be possible to determine the source of a commercial acid by studying the curve obtained when boiling point *vs* density curves are plotted. Von Braun [8, 10] found, for instance, that acid fractions from Roumanian kerosene show a steady rise in density from 0.9379 to 0.9924 with rise in boiling point, while the density of fractions of gas oil from the same source drops from 0.9941 to 0.9876. In the case of acids from California crudes, he found a steady rise from 0.9738 to 1.0538, while Polish acids rose from 0.977 to 0.999, then dropped to 0.938, and rose again to 0.9507 with increase in boiling point. Similar results have been reported by others [25, 48, 56, 64] and have also been obtained in the author's laboratory.[54, 55] This difference in density may be due to considerable concentrations of aliphatic acids in some samples and not in others or to the presence, in some fractions, of acids with long side chains which would lower the density. It should be possible, then, to determine the origin of a given sample of naphthenic acids by fractionating a suitably purified mixture, plotting the curve for density against boiling point, and comparing the resulting curve with those of acids of known source.

Optical Activity

For many years, isolated reports of optical activity of various naphthenic acids have appeared in the literature. In some of the earlier reports, the slight activity observed may well have been due to optically active hydrocarbons present as impurities, since the activity was reported as similar to that of accompanying hydrocarbons. Conversely, of course, in some cases, the activity of the hydrocarbons might have been due to the presence of optically active acids. The optical activity of petroleum compounds, whether they be hydrocarbons or acids, is one of the best proofs of the animal or vegetable origin of petroleum, since optical activity would not be expected of compounds of inorganic origin. Albrecht [5] reported optical activity in fractions of Texas acids. He found them about as active as the hydrocarbons from which they were isolated, i.e., he found that the activity of the crude acids isolated from a Texas machine oil was about the same as that of the acids obtained on purifying the sample of crude acid by the Spitz and Honig method. Budowsky [14] concluded from this that the acids were optically inactive. It seems that if any conclusion can be drawn from Albrecht's observations it is that, in this case, the acids and the hydrocarbons boiling in the same range have about the same optical activity. In view of the rarity of optical activity in petroleum, it would be a strange coincidence if acids and hydrocarbons of much higher molecular weight, which would boil at the same temperature, were to show the same optical activity.

Bushong and Humphrey [15] found a small but definite rotation in the C_6 to C_8 range of Baku acids and their methyl esters. They observed that this activity rose steadily from -0.70 to $+0.30$ during fractionation of the acids in this range, indicating that there must have been at least two optically active compounds of opposite rotation present in their material.

Chichibabin and coworkers,[16] also working with Baku acids, observed that several of the methyl-ester fractions boiling between 137° and 185°C (276° and 365°F) were slightly dextrorotatory. They also said that the fraction boiling at 167° to 170°C (332° to 338°F) was levorotatory, but did not include this in their tabulation of data. They found that the acids obtained on saponifying the active esters were also optically active, which eliminated the possibility that the activity was due to hydrocarbon impurities since the esters boil about 50°C below the acids from which they are derived.

As most of the naphthenic acids contain at least one asymmetric carbon atom, optical activity might be expected to be a common property of these acids if they are of direct animal or vegetable origin but not derived from fats or oils. The fact that active acids are rare is, however, not surprising when we consider the length of time during which they must have been in contact with substances that may cause rearrangements and racemization at the temperatures observed in oil wells.[12]

Since optical activity is rather rare, it is obvious that when it does occur, it is extremely valuable for characterization, if the acid is to be isolated or is to be completely removed. Unfortunately, this property is so rare that it can almost be neglected as a means of characterization of acids.

Surface-Film Areas

Much of the early work on surface films and the area occupied by molecules in monomolecular films was done with long-chain fatty acids,[1, 28, 31, 43, 53] but only one report deals with naphthenic acids. Stenhagen [58, 59] studied films of a number of known acids, including some of the acids synthesized by Adams and coworkers. The list included, however, only a few acids of the types probably existing in petroleum. Kovaleve and Terosov [38] determined the apparent cross-section and length of naphthenic acids in fractions and concluded that the area occupied in this case was determined by the hydrocarbon portion of the molecule rather than by the carboxyl group. Langmuir had concluded that, in the case of long-chain fatty acids and their esters with long-chain alcohols, the cross-section is determined by the carboxyl group and is practically constant for a series of such acids.[43] Kovaleve and Terosov found that the length of the molecules in monomolecular films of naphthenic acids increases with the molecular weight to about 9 Å, while the area increases steadily to about 42 $Å^2$. Details on the purification of their acids, the source, and the molecular weight are not included in the abstract available, but the acids were almost certainly complex mixtures.

Additional data are presented in a number of recent papers dealing with the orientation of molecules of acids in films on water or aqueous solutions.[70] These data show that branching of the hydrocarbon chain of fatty acids affects the surface area occupied to a degree depending on the size and position of the branch. They indicate that films of acids of eighteen or more carbon atoms are often quite unstable, but that films of acids in the C_{10} to C_{16} range tend to be stable enough so that fairly accurate surface-area measurements can be made with the Langmuir balance. This property should be of value in this range in distinguishing aliphatic, monocyclic, and bicyclic acids, but work with known acids of each type appears to be needed before use can be made of this method for characterizing the acids.

Combustion Analysis for Hydrogen

In the case of highly fractionated material on which several different and independent separation procedures have been used, combustion analysis for hydrogen and carbon is worthwhile. If a simple and accurate method by which hydrogen by itself, or along with another element, could be determined in perhaps 1 hour or less could be developed, it would undoubtedly be the most common method for determining the nature of an acid isolated or concentrated to near purity. Since the hydrogen content is, of course, dependent on unsaturation (absent in most acids) or ring formation, determining the hydrogen content is probably the best method of deciding whether an acid is monocyclic or bicyclic. As pointed out before, one must be reasonably certain that the sample does not contain several different types of acids, as this would yield a worthless average value here.

Several investigators in the petroleum industry have used modifications of the lamp method for determining sulfur and obtained very satisfactory results in hydrogen analysis. Grosse [27] has pointed out the value of the C:H ratio in type analysis for hydrocarbons. His data and those of Hindin and Grosse [30] indicate a reproducibility of better than 0.05% in hydrogen analysis by this method. No published report on the use of this method in analyses of acids has appeared, but if an equal accuracy can be obtained with petroleum acids which are nearly all liquids — at least prior to purification — one should be able to determine whether an acid is aliphatic, monocyclic, or bicyclic, and this method may prove to be very valuable in this particular field.

Paper Chromatography

Chromatographic methods are valuable not only as methods of separating two or more compounds or types of compounds (Chapter 7), but also in deciding whether a given substance is a pure compound or a mixture. A pure compound would yield only a single sharp band or a small series of cuts containing all of the sample tested. The reverse of this is, of course, not always true because an inseparable mixture will appear to be chromatographically pure.

The rapidly developing and very valuable paper-chromatography technique permits separation and sometimes, in favorable cases, identification of very small samples. Several procedures have been developed which make it possible to identify solid or nonvolatile acids, at least tentatively, by their R_F values or by comparison of the acid with a known acid placed on the same strip of paper at the same time. The R_F value is obtained by placing a tiny drop of the compound or mixture to be studied on a horizontal starting line near one end of a filter paper strip, bringing that end of the paper in contact with the eluting liquid, and permitting the liquid to move up or down the strip by capillarity until its front is near the other end of the paper. The sample will usually move along with the eluting liquid but not as fast. The rate of movement of the sample depends on the structure and nature of the compound and if two or more compounds moving at different rates are present, they will be found at different places on the paper. At the end of the run, the paper (either after drying or while still wet) is treated with some sort of indicator. Colored or decolorized spots will then show the position of each compound. The ratio of the distance traveled by the compound to that traveled by the eluting liquid is the R_F value of that compound. This value is reproducible for any given compound when the same eluting liquid is used on the same type of paper at the same temperature. If one does not wish to work under identical conditions from run to run and has a sample of the suspected acid, a drop of it is placed at another place on the starting line at the beginning of the experiment. If the acids are identical, they will move the same distance; if they are different but move at exactly the same speed, they will again move the same distance and this is often true or nearly true of isomers and adjacent homologues with six or more carbon atoms. In many cases, however, when compounds differ considerably in structure, they will move at a different rate and will show up as different spots after development.[13, 17, 42] Since most of the naphthenic acids are volatile, they have to be converted to nonvolatile derivatives either before application to the filter paper or prior to drying at the end of the test.

Chromatographic methods have been used extensively for fatty acids from various sources,[23, 35, 46, 52] but no report on their use with petroleum acids has been published so far. Further study with various types of naphthenic acids of known structure may yield valuable results. Preliminary results obtained in the author's laboratory show that the method is often valuable in deciding whether an acid is pure and, if not pure, whether it consists of only two or more than two different compounds.

Distribution Analysis

While chromatographic methods are most conveniently applied to samples of only a few milligrams — and that is sometimes one of their greatest advantages — another very valuable method can be applied, with proper equipment, to samples ranging from a few milligrams to any convenient large size. This method — distribution analysis — can be used in the separation of acids or bases (Chapters 7 and 23) or it will be useful in deciding whether a highly purified acid or base is actually pure or still consists of a mixture of compounds, if the compounds are not distributed between the two liquid phases in exactly the same ratio. The apparatus for and the theoretical considerations of this method were developed largely by Craig and coworkers at the Rockefeller Institute.[19, 24, 73] Various modifications of equipment, as well as a resurrection of the original apparatus of Jantzen,[32, 45] have been described in the last few years.

Distribution analysis is based on the fact that two acids of slightly different strength or solubility will be distributed in a different ratio between two immiscible solvents or solutions. As the solvent layers are contacted and one of them moved to the next stage, it will carry with it more of one acid than of the other. This occurs at each mixing and separating or settling stage so that, in effect, one acid moves faster than the other just as in the case of the chromatographic column. At the end of a suitable number of stages, the two acids may be completely separated, partially separated, or may prove to have moved at exactly the same rate.

In characterization of acids, distribution analysis is valuable if, at the end of the operation, it is found that the acidity is not symmetrically distributed with respect to the fraction showing the largest amount of acid, because this indicates that at least two compounds are present. If distribution is found to be symmetrical with respect to the cut of maximum concentration, the acid is either a pure compound or it is a mixture that cannot be separated by the two solvents used. Since only a fraction of a gram of acid is needed and as the sample can be recovered if desired, the test does not use up valuable material. Tentative identification or classification can be achieved by comparing the behavior of the sample with that of a known acid or type tested under the same conditions. While distribution has been used extensively in our laboratory in separation of acids, its systematic use in characterization does not seem to have been attempted.

Thermal Diffusion

One of the latest and most powerful methods — thermal diffusion — has apparently not been tried either in separation or in characterization of acids. Since the method separates certain mixtures very effectively, it may prove valuable for the characterization of acids and for separation. In the case of hydrocarbons, apparently, it cannot be used for distinguishing between aromatic and naphthenic compounds, but it does seem to have possible value in differentiating monocyclic from bicyclic hydrocarbons and also does definitely distinguish between aliphatic and naphthenic compounds of the same boiling range. Whether it can be used for separating or distinguishing between monocyclic and bicyclic naphthenic acids remains to be seen. The theory of the effect and a description of the apparatus used for certain mixtures of hydrocarbons have been presented in a number of recent publications.[33, 37, 39, 49, 51]

The Use of a Combination of Properties

Schemes similar to the familiar-type analysis extensively used in petroleum-hydrocarbon chemistry can probably be worked out for the petroleum acids. This will require collection and study of a number of different types of pure naphthenic acids which, in turn, will probably require the synthesis of several new acids and resynthesis and careful study of some of the known acids. One difficulty encountered here is the fact that definite structures have been reported only for a handful of the simpler petroleum naphthenic acids, so that it is hard to select the proper types of acids — especially for the group of bicyclic acids of which no member is known.

Chichibabin and coworkers[16] used both the index of refraction and density, but independently of each other, in estimating the percentage composition of mixtures found in their fractions. Assuming that only two types of acids are present and assigning a definite density or index of refraction to each type of acid, it is possible to calculate the composition of the mixture from either the density or the index of refraction. These workers used both properties and so calculated the composition by each method. Unfortunately, these calculations cannot be considered accurate, for, in the first place, there is no good reason (except in the case of the lower acids) for assuming that only two types of acids are present in any fraction and, in the second place, there is no such a value as *an* index of refraction or *a* density for any one type of acid. This is particularly true of density and, although the index of refraction does not show as large a range for any one type as does the density, the calculations of Chichibabin can only be considered as rough approximations.

Extensive work has been done with such properties as refractivity intercept, refractive dispersion, and solubility relationships similar to the aniline point of hydrocarbons and some practical scheme for acids may appear soon. Unless based on the properties of all of the more common types of acids found in petroleum, really valuable schemes can hardly be developed. Unfortunately, we know very little of the different types of naphthenic acids in petroleum; therefore, we cannot intelligently select types to be included in a scheme.

When fractional liberation and neutralization of acids began to be used at the University of Texas on highly fractionated material, it became absolutely essential to develop some fast and fairly reliable method of characterization to determine which of the many fractions obtained could be recombined without undoing separation work already accomplished. Obviously, the boiling point could not be used, because the fractions treated boiled over a range of only a few degrees. Depending on the scheme of separation by distribution used, both index of refraction and density generally rose or fell regularly with the fraction number. There were enough exceptions to this behavior to make the use of either property by itself less valuable than a combination developed at the time.

Schutze [55] experimented with plots of $\frac{n^2 - 1}{n^2 + 2}$ against density, $\frac{n - 1}{\text{mol. wt.}}$ against density, and finally with the simpler and purely empirical product of index of refraction and density, $n \times d$. In nearly all cases, recombination on the basis of either density or index of refraction and boiling point would have been of real value, but it was found that the product $n \times d$ (nd) tended to accentuate differences in properties in such a way that the different classes of petroleum acids present in the mixture of lower acids could be differentiated clearly.

Calculations of $n \times d$ values of a large number of aliphatic and saturated cyclic acids and of phenols, which might be present in a mixture of acids with ten or less carbon atoms, showed that there was a gradual rise in n × d value with an increase in molecular weight, but the ranges for the classes mentioned were so far separated that there was no overlapping. Aliphatic acids were found to have an n × d range of 1.25 to 1.35, the naphthenic acids a range of 1.39 to 1.48, while phenols showed n × d values well above 1.5. Among the lower acids, no other types are known to be present. Unsaturated acids, if they had been present, would have shown the same range as the naphthenic acids.

Density, index of refraction, and n × d values of acids with more than ten carbon atoms, as far as these properties have been listed for such acids, appear to depend much more on the structure than these values of the lower acids. The n × d values can apparently no longer be used to distinguish between highly branched fatty acids and certain types of naphthenic acids.[2, 3, 4]

In the case of the naphthenic acids carrying a number of methyl

Table 14. **Properties of Some Synthetic, Long-Chain Naphthenic Acids**

Alkyl	n_D^{25}	d_4^{25}	$n \times d$	*Reference*
CYCLOHEXYLALKYLACETIC ACIDS				
n-C_5	1.4640	0.9544	1.397	2
n-C_8	1.4642	0.9298	1.360	
n-C_{11}	1.4650	0.9166	1.343	
3-CYCLOHEXYL-2-ALKYLPROPIONIC ACIDS				
Ethyl	1.4623	0.9812	1.434	2
n-C_4	1.4620	0.9564	1.398	
n-C_6	1.4627	0.9448	1.383	
n-C_8	1.4640	0.9331	1.365	
4-CYCLOPENTYL-2-ALKYLBUTYRIC ACIDS				
Ethyl	1.4590	0.9602	1.400	4
n-C_4	1.4608	0.9435	1.378	
n-C_6	1.4616	0.9303	1.360	
n-C_8	1.4629	0.9210	1.348	
4-CYCLOHEXYL-2-ALKYLBUTYRIC ACIDS				
Ethyl	1.4613	0.9619	1.405	3
n-C_4	1.4624	0.9410	1.375	
n-C_6	1.4628	0.9283	1.358	
n-C_8	1.4640	0.9193	1.345	
5-CYCLOHEXYL-2-ALKYLPENTANOIC ACIDS				
Ethyl	1.4622	0.9509	1.390	3
n-C_4	1.4630	0.9317	1.364	
n-C_6	1.4638	0.9221	1.350	
6-CYCLOHEXYL-2-ALKYLHEXANOIC ACIDS				
Ethyl	1.4622	0.9447	1.382	
n–C_4	1.4631	0.9300	1.362	
n–C_6	1.4638	0.9191	1.344	

or other small radicals on the ring, the n × d values appear to continue high and in the usual naphthenic acid range, but no such acids with more than ten carbon atoms have been reported with both index of refraction and density; therefore, we have to depend either on the index of refraction or on the density alone in guessing at the n × d value. An additional complication is the fact that most of such acids are solids. The n × d values of commercial naphthenic acids in this molecular-weight range are usually found within the monocyclic naphthenic acid range and in the case of acids with ten to twelve carbon atoms, this is probably not due to the presence of bicyclic acids in sufficient concentration to raise the average value materially.

For monocyclic acids in which a simple cyclohexane or cyclopentane ring is merely one of the substituents of a fatty acid, the data of Adams and coworkers [2, 3, 4] show that the index of refraction is high and fairly constant, while the density and n × d values drop sharply with increase in molecular weight as shown in Table 14.

Table 15. **Properties of Some Synthetic Bicyclic Naphthenic Acids**

Acid	n_D^t	*t°C*	d_4^t	*t*	*n × d*	*Reference*
3-Cyclopentylcyclopentanecarboxylic Acid	—	—	1.0398	19	—	10
3-Cyclohexylcyclopentanecarboxylic Acid	1.4925	20	1.0343	20	1.54	10
3-(3-Ethylcyclopentyl)-Cyclopentane-carboxylic Acid	—	—	0.9980	—	—	11
2,3-Dicyclopentylpropionic Acid	—	—	1.0160	—	—	9
3-(3-Methylcyclohexyl)-Cyclopentane-carboxylic Acid	—	—	1.0135	—	—	9
(3-Cyclopentyl)-3-Cyclopentylpropionic Acid	—	—	1.0173	Isomers		9
	—	—	1.0105			
3-Decalylpropionic Acid	—	—	1.0368	—	—	9
2,6,6,10-Tetramethyldecalincarboxylic Acid	1.4738	24	—	—	—	18
4-Cyclohexyl)-Cyclohexylstearic Acid	1.4888	—	—	—	—	57
CYCLOHEXYL-$(CH_2)_x$-CH(COOH)-$(CH_2)_y$-CYCLOHEXYL						
Value of *x*, *y*						
0, 2	1.4747	—	1.0160	—	1.50	20
2, 2	1.4722	—	0.9931	—	1.46	
3, 3	1.4710	—	0.9811	—	1.44	
2, 4	1.4710	—	0.9810	—	1.44	

From what little is known about the physical properties of liquid bicyclic acids, it appears that the effect of the two rings on the index of refraction is sufficient to yield n × d values well above 1.4 for those with two separate cyclopentane or cyclohexane rings and much higher — usually above 1.48 — for those in which the rings are fused, connected to a common carbon, or separated by only one or two carbons. Table 15 illustrates this regularity.

The n × d values proved invaluable in the study of the lower naphthenic acids with less than eleven carbons per molecule, because this mixture contains none of the acids that would interfere by having the same range of values or by producing average values that would be misleading. Unsaturated acids were absent; bicyclic acids and aromatic acids were either absent or present only in traces and so would not give misleading values. In view of these facts, the n × d value, along with the boiling point, was used unhesitatingly in deciding which fractions could be recombined and in making a rough estimate of the concentration of fatty acids in a naphthenic acid sample.

An interesting use of the n × d values has been that of examining the data published by earlier workers to determine whether their fractions represented essentially pure naphthenic acids or consisted of a mixture of these with aliphatic acids or with hydrocarbons.

Von Braun rarely reported both index of refraction and density for the same acids, apparently because he felt that only the density showed enough variation with the structure to be worth determining. In the case of the two Roumanian kerosene acids which he studied most carefully [8] he reported:

(a) Density 0.9718 and index of refraction 1.4607, from which the calculated n × d value is 1.418. For the acid obtained from this by degradation to the next lower one, he reported density 0.9819 and index of refraction 1.4519, which gives an n × d of 1.425.

(b) For the acetic acid synthesized from his petroleum ketone, he reported $d_4^{20} = 0.9783$; $n_D^{15} = 1.4609$, yielding an n × d value of 1.430. All of these were within the naphthenic acid range that would be expected of highly purified acids.

Bushong and Humphreys [15] were among the first Americans to study naphthenic acids. They reported fractions obtained from Baku petroleum with the properties shown in Table 16.

Table 16. **Properties of Some Baku Acids**

Cut	B. P. °C	d_{15}^{23}	$n_D^{20.8}$	n × d Calculated
00	206–215	0.9246	1.4366	1.330
0	215–220	0.9369	1.4362	1.340
1	220–224	0.9467	1.4365	1.360
4	230–232	0.9580	1.4440	1.380
6	234–236	0.9630	1.4462	1.390
9	240–242	0.9629	1.4488	1.395
12	246–248	0.9625	1.4510	1.395
15	252–254	0.9622	1.4530	1.400
18	258–260	0.9624	1.4523	1.400

They purified their acids by distillation as acids and as methyl esters, then subjected the acids obtained from the fractionated esters to twelve distillations through a Wurtz dephlegmator, which should have been equivalent to a total of at least twenty-five plates. It is not clear if all of the acids were first converted to the sodium salts and purified by blowing steam through them to remove hydrocarbons and phenols, but this procedure is reported for the first ten cuts. In spite of this apparent care in purification, it is obvious that the first six fractions must have contained high concentrations of aliphatic acids, or hydrocarbons, or both. Probably in the case of the first ten fractions, the main impurity consisted of fatty acids.

There are a number of other cases reported in which density and index of refraction show clearly that the acids were obviously mixtures of naphthenic acids and fatty acids or hydrocarbons. Since many of the earlier workers may not have removed phenols completely, these may have had a tendency to raise values which otherwise would have been well below the naphthenic acid range.

Even within the range of 1.39 to 1.48 there are large variations in both the index of refraction and in the n × d value, as is shown by the data of Table 17.

All of the data included were obtained by workers who apparently purified their acids very carefully. For acids of the same molecular weight, densities may vary widely and even the index of refraction shows a range of 0.01 in some cases. The n × d value of any one series of fractions tends to change only slowly, but for fractions from different fields, both the index of refraction and the density may be high or low and the product shows an even greater range.

Table 17. **Properties of Various Petroleum Acids**

Cut No.	*Number of Carbon Atoms*	*Density*	*Index of Refraction*	*n × d*	*Crude from Field*	*Reference*
3	10	0.9642	1.4520	1.400	Grozny	56
9	11	0.9679	1.4590	1.410	Grozny	56
—	11	0.9876	1.4706	1.451	Roumania	25
—	11	0.9709	1.4624	1.420	Baku	36
17–19	12	0.9720	1.4655	1.425	Grozny	56
—	12	0.9712	1.4697	1.426	Japan	61
29–30	13	0.9756	1.4706	1.435	Grozny	56
—	13	0.9916	1.4784	1.465	Japan	60
35–36	14	0.9775	1.4745	1.440	Grozny	56
—	14	0.9930	1.4807	1.472	Japan	60
42	15	0.9764	1.4778	1.445	Grozny	56

The results of Smirnov and Buks [56] show that their Grozny acids reached a maximum in density at C_{14}, while the index of refraction continued to rise slowly. The first fractions may have contained small amounts of hydrocarbons, but more probably, they represented a mixture of fatty and naphthenic acids. They had purified their acids carefully and then fractionated them extensively by distillation, as usual. The results listed were selected from a large number of fractions studied.

Komppa's Roumanian acids [36] were carefully purified and fractionated in the usual manner and then were further fractionated by distillation of the acid chlorides. The final acids were converted to the hydrocarbons over the esters and alcohols.

Frangopol's results [25] were obtained in early work reported in a thesis which was not available to the author, but his data have been extensively quoted in previous monographs.

Tanaka [60–67] probably studied more different acids than any other worker and fractionated his material very carefully after purification by Spitz and Honig type methods.

Acids from different fields in Japan seem to be quite different in type as are some from California. This wide variation in properties is shown in Table 18.

Acids with n × d values ranging from 1.35 to 1.47 were here obtained by the same group of workers using the same procedure, so that the differences in properties can hardly be due to differences in purification and fractionation operations. They repeatedly re-

Table 18. **Properties of Different Japanese Acids**

Number of Carbons	*Density*	*Index of Refraction*	*n × d*	*From*	*Reference*
8	0.9417	1.4318	1.36	Nishiyama	41
—	0.9587	1.4707	1.41	Kubiki	62
—	0.9902	1.4887	1.47	Niitsu	63
—	0.9747	1.4796	1.44	Katsurane	64
12	0.9681	1.4678	1.42	Nishiyama	65
13	0.9916	1.4784	1.47	Kurokawa	60
13	0.9704	1.4708	1.43	Nishiyama	65
14	0.9732	1.4740	1.43	Nishiyama	65
14	0.9272	1.4636	1.35	Hokkaida	66
15	0.9744	1.4765	1.44	Nishiyama	65

ported appearance of solid acids in the distillation residues and finally isolated and identified palmitic, stearic, myristic, and arachidic acids from Ishikari acids.[67] In view of these results, Japanese acids with low n × d values probably contain considerable concentrations of aliphatic acids. From California acids, fatty acids through C_9 were isolated in the author's laboratory,[54] and there was no indication that higher aliphatic acids were not present in these mixtures.

Since none of the workers reporting before 1938 would probably have been able to separate fatty acids from naphthenic acids, low n × d values for acids with eleven or fewer carbon atoms can probably be explained on the same basis as in the case of the Japanese acids. In the C_8 Nishiyama acid reported by Kuwata,[41] there may have been as much as 50% of aliphatic acids.

Characterization by Type Analysis of Hydrocarbons Derived from Naphthenic Acids

Since a very large amount of work has been done on the type analysis of hydrocarbon mixtures in petroleum[21, 22, 40, 68, 69] and these methods are in daily use in the oil industry, it is not surprising that some attempts have been made to apply these methods to the hydrocarbons derived from the naphthenic acids and some valuable scheme may be forthcoming.

The over-all change involved in the conversion of the acids to hydrocarbons may be expressed by:

$$RCOOH \rightarrow RCH_3$$

Originally, the method of Wreden[71] was used for conversion, in which a sample of acid is heated in a closed tube with hydriodic acid and amorphous phosphorus to achieve the whole change in one step. Aschan [6] applied this method to petroleum acids in an attempt to prove that these acids could be hydrogenated to substituted cyclohexanes. His results were inconclusive and the confusion resulting from his work and that of others was not cleared up until 1897 when Kishner [34] was able to show that benzene is rearranged in part to methylcyclopentane on heating with phosphorus and hydriodic acid according to Wreden's procedure.

Since 1897, the conversion has always been carried out by the following series of steps:

$$RCOOH \rightarrow RCOOR' \rightarrow RCH_2OH \rightarrow RCH_2X \rightarrow RCH_3$$

This series was employed by Zelinsky in 1924,[72] by Komppa in 1929,[36] by Muller and Pilat in 1936,[47] and by others. Kuwata, in 1928,[41] reported a greatly improved yield by the use of the phenyl ester instead of the usual methyl or ethyl homologues. Modern type-analysis methods had not yet been developed and most of these workers simply compared their hydrocarbons with those isolated directly from petroleum fractions. Goheen,[26] in 1940, used modern characterization methods on the hydrocarbons obtained from C_{19} to C_{23} Texas Gulf Coast acids by this series of reactions.

The hydrogenation can be carried out somewhat more easily, and with at least as good a yield, over a copper chromite catalyst, according to the method of Adkins and Folkers, who report yields of 80 to 99% with, however, a few failures.

No report has appeared on the direct reduction of the acids by means of lithium-aluminum hydride, which is now generally employed for this type of reduction. The esterification step would be avoided, but so far, the cost of the reducing agent is a drawback when large batches of acids are to be reduced.

One of the difficulties involved in the use of these hydrocarbons for characterization of petroleum acids from which they were derived arises from the fact that a fairly large sample of acid must be used to obtain enough hydrocarbon to permit type analysis and such a large sample of carefully purified acid is rarely available after other important tests have been run. Another weakness is the fact that we do not know what different types of hydrocarbons are obtained on reduction of petroleum acids. Grosse [27] pointed out that some types of hydrocarbons do not fit into this scheme of analysis and would introduce a considerable error, if present.

It must be admitted that no generally applicable system of characterizing naphthenic acids, particularly those with more than twelve carbon atoms found in commercial acids, is available to us. For acids of a molecular weight of 200 or less, it can be predicted with some confidence what types of acids are present in fractions obtained by a combination of two or more different systematic methods of separation. However, for the higher naphthenic acids, which are the ones found in commercial products, no generally reliable method is known — a situation that hampers the use of modern methods of fractionation very seriously.

Bibliography

1. Adam, N. K., *Proc. Royal Soc.* **A142,** 401 (1933).
2. Adams, R., W. M. Stanley, and H. A. Stearns, *J. Am. Chem. Soc.* **50,** 1478 (1928).
3. Adams, R., W. M. Stanley, S. G. Ford, and W. R. Peterson, *Ibid.* **49,** 2939 (1927).
4. Adams, R., and G. R. Yohe, *Ibid.* **50,** 1508 (1928).
5. Albrecht, R., *Chem. Rev. Fett-u. Harzind.* **18,** 152 (1911). Through *C. A.* **5,** 3522 (1911).
6. Aschan, O., *Ber.* **25,** 3664 (1892).
7. Balada, A., and J. Wegiel, *C. A.* **31,** 2399 (1937).
8. Von Braun, J., *Ann.* **490,** 100 (1931).
9. Von Braun, J., *Oel u. Kohle* **13,** 799 (1937).
10. Von Braun, J., *Ber.* **70B,** 1750 (1937).
11. Von Braun, *Ibid.* **74,** 1109 (1941).
12. Brooks, B. T., *Science* **111,** 648 (1950).
13. Brown, F., *Biochem. J.* **47,** 598 (1950).
14. Budowsky, I., *Die Naphthesäuren,* Julius Springer, Berlin, 1922.
15. Bushong, F. W., and I. W. Humphrey, *8th Int. Cong. Applied Chem.* **VI,** 57 (1912).
16. Chichibabin, A. E. *et al., Chim. et ind.* (*Special Number*) **17,** 306 (1932).
17. Claborn, H. V., and W. I. Patterson, *J. Assoc. Off. Agri. Chemists* **31,** 134 (1948).
18. Collins-Asselineau, Mme., M. Lederer, and Mme. Polonsky, *Bull. soc. chim. de France* **1950,** 715.
19. Craig, L. C., and O. Post, *Anal. Chem.* **21,** 500 (1949).
20. Davis, L. A., and R. Adams, *J. Am. Chem. Soc.* **50,** 2297 (1928).
21. Deanesly, R. M., and L. T. Carleton, *J. Phys. Chem.* **46,** 859 (1942).

22. DeKok, W. J. C., H. J. Waterman, and H. A. von Westen, *J. Soc. Chem. Ind.* **55,** 225T (1936).
23. Dutton, H. J., *J. Phys. Chem.* **48,** 180 (1944).
24. Feitelson, J., *J. Am. Oil Chemists' Soc.* **27,** 4 (1950).
25. Frangopol, Thesis, Munich, 1910; from Naphthali. (See reference 48.)
26. Goheen, G. E., *Ind. Eng. Chem.* **32,** 503 (1940).
27. Grosse, A. V., *Refiner* April 1939, page 149.
28. Harkins, W. D., *Science* **102,** 294 (1945).
29. Harkness, R. W., and J. Bruun, *Ind. Eng. Chem.* **32,** 499 (1940).
30. Hindin, S. G., and A. V. Grosse, *Anal. Chem.* **19,** 42 (1947).
31. Hughes, A. H., *J. Chem. Soc.* **1933,** 340.
32. Jantzen, E., *Dechema Monograph, Vol. V, Das fraktionierte Destillieren u. das fraktionierte Verteilen als Methoden zur Trennung von Stoffgemischen,* Verlag Chemie, Berlin, 1932.
33. Jones, A. L., *Petroleum Processing* **6,** 132 (1951).
34. Kishner, N., *J. prakt. Chem.* (**2**) **56,** 364 (1897).
35. Kirchner, J. G., A. N. Prater, and A. J. Haagen-Smit, *Anal. Chem.* **18,** 31 (1946).
36. Komppa, G., *Ber.* **62,** 1562 (1929).
37. Korsching, H., and K. Wirtz, *Ibid.* **73B,** 249 (1940).
38. Kovaleve, L. L., and S. P. Terosov, *C. A.* **35,** 1385 (1941).
39. Kramers, H., and J. J. Broeder, *Anal. Chim. Acta* **2,** 687 (1948).
40. Kurtz, S. S., and C. E. Headington, *Anal. Chem.* **9,** 21 (1937).
41. Kuwata, T., *C. A.* **23,** 1390 (1929).
42. Landau, A. J., R. Fuerst, and J. Awapara, *Anal. Chem.* **23,** 162 (1951).
43. Langmuir, I., *J. Chem. Phys.* **1,** 756 (1933).
44. Lapkin, I. I., *C. A.* **34,** 611 (1940).
45. Lochte, H. L., and H. W. H. Meyer, *Anal. Chem.* **22,** 1064 (1950).
46. Manunta, C., *Helv. Chim. Acta.* **22,** 1156 (1939).
47. Muller, J., and S. Pilat, *Brennstoff-Chem.* **17,** 461 (1936).
48. Naphthali, M., *Chemie, Technologie u. Analyse der Naphthensäuren,* Wissenschaftliche Verlagsgesellshaft, Stuttgart, 1927.
49. Von Nes, K., and H. A. van Westen, *Aspects of the Constitution of Mineral Oils,* Elsevier, New York, 1951, page 156.
50. Neyman-Pilat, E., and S. Pilat, *Ind. Eng. Chem.* **33,** 1390 (1941).
51. O'Donnell, G., *Anal. Chem.* **23,** 897 (1951).
52. Peterson, M. H., and M. J. Johnson, *J. Biol. Chem.* **174,** 775 (1948).
53. Polgar, N., and R. Robinson, *J. Chem. Soc.* **1945,** 389.
54. Quebedeaux, W., G. Wash, W. O. Ney, W. W. Crouch, and H. L. Lochte, *J. Am. Chem. Soc.* **65,** 767 (1943).

55. Schutze, H. G., W. Shive, and H. L. Lochte, *Ind. Eng. Chem. Anal. Ed.* **12,** 262 (1940).
56. Smirnov, P., and Z. Buks, *Aserbaidzhan Oil Ind.* **1932,** No. 11, page 60; *C. A.* **27,** 1492 (1933).
57. Smith, N. L., and F. L. Schmehl, *J. Org. Chem.* **13,** 859 (1948).
58. Stenhagen, E., *Trans. Faraday Soc.* **36,** 597 (1940).
59. Stenhagen, E., and S. Stallberg, *J. Biol. Chem.* **139,** 345 (1941).
60. Tanaka, Y., and S. Nagai, *J. Am. Chem. Soc.* **45,** 754 (1923).
61. Tanaka, Y., *Ibid.* **47,** 2369 (1925).
62. Tanaka, Y., and S. Nagai, *C. A.* **18,** 2332 (1924).
63. Tanaka, Y., S. Nagai, and S. Ishida, *Ibid.* **19,** 486 (1925).
64. Tanaka, Y., and S. Nagai, *Ibid.* **19,** 1135 (1925).
65. Tanaka, Y., and S. Nagai, *Ibid.* **20,** 583 (1926).
66. Tanaka, Y., and T. Kuwata, *Ibid.* **21,** 1004 (1927).
67. Tanaka, Y., and T. Kuwata, *Ibid.* **23,** 4051 (1929).
68. Vlugter, J. C., H. I. Waterman, and H. A. van Westen, *J. Inst. Petroleum Technology* **21,** 661 (1935).
69. Ward, A. L., and S. S. Kurtz, *Ind. Eng. Chem., Anal. Ed.* **10,** 559 (1938).
70. Weitzel, G., A. M. Fretzdorf, and W. Savelsberg, *Hoppe-Seyler, Z. f. physiologische Chem.* **285,** 230 (1950).
71. Wreden, F., *Ann.,* **187,** 153 (1877).
72. Zelinsky, N., *Ber.* **57,** 43 (1924).
73. Zilch, K. T., and H. J. Dutton, *Anal. Chem.* **23,** 775 (1951).

CHAPTER NINE

THE STRUCTURE OF PETROLEUM ACIDS

As pointed out in previous chapters, the first and most difficult problem involved in the identification of acids is the separation of the individual acid, first from nonacidic impurities, then from other classes of acids, and finally from homologues and isomers of the acid concerned. Assuming that the separation and purification operations have been successful and crystalline salts and derivatives are obtainable, it may be possible to remove the remaining impurities by recrystallization of the solids. The chemist is then confronted with the problem of determining the structure and possibly the geometrical configuration of the pure acid.

In the past, workers have attempted to identify any acid isolated and purified by comparing its properties with those of acids of known structure and the same molecular formula. Usually there are so many possible isomers that this direct comparison has little chance of success and it is necessary to eliminate as many as possible of the numerous structural formulae that fit the molecular formula of the acid. Most of the discussion in this chapter deals with methods of eliminating possible structures so as to reduce the number of compounds that may finally have to be synthesized.

So little is known about the fatty acids in petroleum that not much can be said about methods of determining their structure. The lower acids up to C_7 are, of course, known and present no particular problems.[14, 30, 45, 52, 54, 64] A few higher solid acids, like palmitic, stearic, and myristic, have been identified by taking advantage in their isolation of their solid state.[49, 56] No attempt has been made to identify any of the higher liquid aliphatic acids that are probably present in some samples of petroleum acids. At present, any work in this field would probably start with the methods developed by Hilditch and coworkers in the study of acids from vegetable oils and fats [31] and by Weitkamp in his brilliant study of acids from wool fat and from human hair.[59,60] The complete separation of aliphatic from naphthenic acids would probably be required and would be one of the most difficult problems. If, as seems likely, the normal fatty acids continue to predominate in the higher acids as they do through the C^7 acids, the final identification may not be as difficult as feared.

Meanwhile, the determination of the structure of the alicyclic acids has been developed as a separate field of research, mainly through the work of von Braun and his coworkers, who tried numerous methods in the course of nearly a decade of research.

Preliminary Experiments

Tests for unsaturation, using bromine or dilute permanganate solution, are usually made just to eliminate the remote possibility that the acid is an unsaturated one which would, of course, fit the same molecular formula as the cyclic naphthenic acid. No unsaturated acids have been found in straight-run petroleum acids, although commercial acids usually have a low iodine number. This may be due to phenols and nonacidic impurities, as well as to unsaturated acids.

By the time a pure or nearly pure acid has been isolated by the ordinary procedure, the worker will have observed incidentally whether the acid is sterically hindered or not, since highly hindered acids are not esterified in the usual course of purifying acids. There are, of course, various degrees of hindrance, but most tertiary acids, and some secondary ones, cannot be esterified in a few hours of refluxing with methanol and hydrogen chloride. In previous work, such acids, if present, have probably been discarded along with hydrocarbons and other unesterifiable matter present in commercial acids. In the study of California acids, it was found that the mixture which was not esterified in the usual manner still contained organic acids. There were only a few such acids and as they were found in widely scattered distillation fractions, they were easily isolated on refractionation of the unesterified mixture. The first such acid identified was 1,2,2-trimethylcyclopentanecarboxylic acid.[30] This discovery started a search for other hindered acids, but only the *cis* and *trans* forms of 2,2,6-trimethylcyclohexanecarboxylic acid were found.[46, 55]

The last acid was first assumed to be another tertiary acid, i.e., to have the carboxyl tied to a carbon attached to three other carbons. After most of the plausible $C_{10}H_{18}O_2$ acids had been synthesized and found to differ from the petroleum acid, tests for tertiary acids were run. The Whitmore and Crooks [61] test seemed to indicate that the acid was tertiary in nature, but von Braun's chlorine-number test [12, 14, 22] indicated that there was one hydrogen on the α-carbon atom and the acid should be a secondary one. Von Braun had shown in another case [17, 20] that certain branched-chain secondary acids are fairly highly hindered. A search of the literature indicated that the properties of 2,2,6-trimethylcyclohexanecarboxylic acid agreed closely with those of the petroleum acid. Synthesis of this acid showed that they were identical. In the course of synthesis, the *cis* form of the acid had been obtained first and this was converted to the more stable *trans* acid. Rannefeld [46] later was able to isolate the *cis* acid from petroleum, so that in this case both the *cis* and *trans* form of an acid have been found in petroleum. Cason [24] has since shown that among the branched-chain fatty acids, one with branching on the β-carbon atom may be more highly hindered than the one with branching on the α-carbon atom. It should be realized, then, that the hindered acids found may be either secondary or tertiary.

Von Braun [12, 14, 22] developed and used the chlorine number to determine whether the carboxyl group is attached to a carbon that carries none, one, or two hydrogen atoms. In this test, the acid is converted to the acid chloride which is then reacted with ethyl amine. The amide formed is treated with phosphorus pentachloride, converting it to the imidechloride:

$$RC(Cl)_2\underset{\displaystyle Cl}{\underset{|}{C}}{=}NC_2H_5$$

which can be analyzed or, better, hydrolyzed to:

$$RC(Cl)_2CONHC_2H_5$$

Analysis of the last compound shows two chlorines which have replaced the two α-hydrogen atoms and the original acid is said to have a chlorine number of 200. If there had been a branching on the α-carbon atom, or if it had been one of the ring carbons, there would have been only one chlorine and the acid would have had a chlorine number of 100. If the acid had been tertiary, the chlorine number would, of course, have been zero, but this type of acid would have been separated at the esterification stage. Von Braun employed this method in the study of a number of acids not derived from petroleum [12] and the method has been also used by him and others as an aid in establishing the structure of naphthenic acids.[9, 12, 14, 22, 44, 46, 59] Since fatty acids may have been present in von Braun's fractions, their normal chlorine number of 200 may have contributed to values of 150 and over obtained by him. Chichibabin [25] degraded Baku acids by another method and came to the conclusion that they usually had the carboxyl attached directly to the ring, while von Braun concluded that Roumanian acids usually have at least one methylene group between the ring and the carboxyl.

Assuming that a naphthenic acid has been shown to have a chlorine number of 200, it would be desirable to determine whether any additional methylene groups separate the ring and carboxyl. Here again von Braun made use of a method which he tested previously. In this test, he prepared the α-bromo acid by the Hell-Volhard-Zelinsky method, eliminated hydrogen bromide, and heated the resulting a,β-unsaturated acid with sulfuric acid or, better, with benzenesulfonic acid, to rearrange it to the β,γ-unsaturated acid and to lactonize it at the same time to:

$$\underbrace{R{-}CH{-}CH_2CH_2CO}_{O}$$

On oxidation, this lactone yielded succinic acid, which indicated that there must have been at least three methylene groups between the ring and carboxyl group of the original acid. He obtained small amounts of succinic acid from several fractions studied and, therefore, concluded that these fractions contained acids with at least a four-carbon side chain, including the carboxyl.[14] He seems to have been the only worker who has used this method. If it is found reliable with pure acids, it should prove valuable in the special cases in which such a long side chain exists.

Derivative Formation

Much of the work of the past has been confined to attempts to determine the structure of acids, usually imperfectly fractionated, by comparing their properties with those of known synthetic naphthenic acids. As a rule, some derivative, like the amide or anilide, was used. Unfortunately, the derivative obtained with the petroleum acid was impure, because the acid fraction was a mixture and physical properties of the derivatives did not fit those of any pure synthetic acid or, if one derivative agreed, another did not. This was partly due to the fact that some of the derivatives, the amides in particular, tend to form mixed crystals when a mixture of acids is converted to the derivative.[25, 45]

Aschan [4, 5, 7, 8] and Markownikoff [37, 38] were among a number of workers who failed to identify definitely any individual acid, in spite of an enormous amount of tedious work.

Finally, in 1938, Nenitzescu and coworkers published results obtained in work with the lower naphthenic acids and the aliphatic acids associated with them.[45] They simply added concentrated ammonium hydroxide to cuts of carefully fractionated methyl esters. At the end of several weeks, a number of the stoppered flasks were found to contain crystals. In a few cases, it was possible to obtain pure amides by long recrystallization of the primary crystals. In this way, they isolated and identified cyclopentanecarboxylic, cyclopentaneacetic, and 3-methylcyclopentaneacetic acids as the first naphthenic acids. They also identified 4-methylpentanoic and 5-methylhexanoic acids from among the fatty acids. Together, these represented only about 8% of the total acidity. Thus, it is obvious that even in their case most of the fractions did not yield amides that could be purified by recrystallization. It seems to be a safe rule of procedure that it is a waste of time to synthesize possible acids on the basis of the melting point of a derivative and that, even if an acid of known structure is available, identification should be based on mixture melting points of several derivatives of the acids.

Conversion of RCOOH to RCH_3

Since it is now generally assumed that the rings in naphthenic acids are almost exclusively either five- or six-membered, the next step will be generally an attempt to dehydrogenate the ring to the benzene ring. The dehydrogenation of the acid or the ester has been attempted [25] and would be a rather short and simple test if reliable. No thorough study of dehydrogenation of naphthenic acids of known structure appears to have been made.

More commonly, the acid is converted to the hydrocarbon by the reactions:

$RCOOH \rightarrow RCOOC_2H_5 \rightarrow RCH_2OH \rightarrow RCH_2I \rightarrow RCH_3$

The hydrogenation is carried out either by the absolute alcohol and sodium method (Bouveault and Blanc)[29, 36, 40, 53, 67, 68] or by catalytic hydrogenation.[1] The conversion of the alcohol to the iodide is usually carried out with phosphorus and iodine, instead of by hydriodic acid which would be more likely to cause rearrangements.[35] When the hydrocarbon is heated with platinum catalyst at 300° to 325°C (572° to 615°F),[36, 50, 67, 69] the elimination of hydrogen, accompanied by a marked increase in index of refraction of the hydrocarbon, should indicate that dehydrogenation has taken place. A positive test here is fairly conclusive, but a negative result might mean that there is a gem-disubstituted cyclohexane ring present which is not dehydrogenated. While such compounds have been found to rearrange by having one of the methyl groups move to another carbon, this takes place at a considerably higher temperature, and reliable procedures applicable to various types of such hydrocarbons with two substituents on the same carbon have apparently not been developed.[2, 65]

This reaction was used by a number of workers, particularly by Zelinsky who employed it to prove or disprove the presence of cyclohexane rings in naphthenic acids.[66]

Most chemists who have employed hydrogenation of the acid to its hydrocarbon have done so to demonstrate merely the similarity of the hydrocarbons obtained from petroleum to those associated with the acids in the refinery product studied. A number of workers, including von Braun and Komppa, were firmly convinced that there was a genetic relationship between acids and hydrocarbons in petroleum. Von Braun originally felt sure that naphthenic acids were, in fact, man made through oxidation during distillation or other refinery processes and, therefore, he was anxious to prove that the hydrocarbons from acids and those occurring naturally in petroleum were very similar.

After very painstaking fractionation of the acids, their esters, and their acid chlorides, Komppa and coworkers were able to obtain a C_{11} fraction of acids boiling over a 1.5° range and apparently essentially pure. When this acid fraction was converted to the hydrocarbon, it again appeared to be pure and very similar to a fraction boiling at the same temperature and obtained directly from petroleum hydrocarbons. At that time, there was no quantitative method of comparing the two hydrocarbons. While identification might or might not be possible today, there would certainly be a more detailed comparison.

Müller and Pilat [40] wanted to obtain a sample of pure high-molecular-weight hydrocarbon and wrongly decided that the best way of obtaining one was to hydrogenate an acid. It was, of course, impossible to isolate a pure acid, and so no pure hydrocarbon was obtained; however, they did obtain a low-melting hydrocarbon mixture whose physical properties were determined.

Kuwata *et al.*[36] used a modification of the hydrogenation process. They converted their acids to the phenol esters through the acid chlorides and phenol and they found that the alcohol and sodium method now gave yields of around 85% — much better than had been obtained with the ethyl esters. Rupe and Lauger reported similar high yields.[53]

Goheen [26] studied certain high-molecular-weight Gulf Coast naphthenic acids which were changed to the hydrocarbons and studied by modern type analysis methods to determine the concentration of bicyclic and polycyclic acids present in the mixture.

More recent results, obtained mainly since 1925, throw some doubt on the value of many of the earlier dehydrogenation experiments, because conditions under which they were carried out were not adequately defined, or because the conditions specified were ones under which side reactions have been reported. There are conflicting reports on the effect of platinum or other hydrogenation-dehydrogenation catalysts on cyclopentanes and cyclohexanes at 300° to 325°C (572° to 615°F).

Three different reactions have been reported in the hydrogenation and dehydrogenation of cyclopentane and cyclohexanes. One of these is the cyclization of open-chain hydrocarbons to form cyclohexanes and finally aromatic hydrocarbons; another is hydrogenolysis of cyclopentanes under hydrogenation conditions to yield open-chain hydrocarbons; and the last is the rearrangement of alkylcyclopentanes, with a dehydrogenation of these to form aromatic hydrocarbons and rearrangement of cyclohexanes to form methylcyclopentanes. Modern dual-function catalysts can, of course, be used to effect all of these conversions and results reported by early workers may be traceable to the use of acidic oxides as support for the platinum catalyst. Unfortunately early reports often do not mention such details as nature of the supporting agent for the catalyst.

Zelinsky and coworkers [70] reported that a close-cut gasoline fraction, from which aromatic hydrocarbons had been removed (presumably after an initial dehydrogenation), showed a large increase in index of refraction on reheating with platinum on charcoal at 310°C (590°F), and they attributed this to cyclization of open-chain hydrocarbons and aromatization of cyclohexanes formed in this way.

Some workers have reported rearrangements of alkylcyclopentanes to cyclohexanes, or the reverse, when these were heated over platinum at about 300°C (572°F),[34, 43, 51] as well as cleavage of cyclopentane rings.[26, 27, 68]

Yet others report that cyclopentanes can be cleaved by catalytic hydrogenation over platinum, but are not affected in the absence of hydrogen gas, i.e., under strictly dehydrogenation conditions.[27, 43, 57] However, since some workers have carried out their dehydrogenations in a slow stream of hydrogen gas and since hydrogen-donating compounds, like the cyclohexanes, could furnish the hydrogen in others, it is obvious that the index of refraction cannot be used as a measure of the concentration of hydroaromatics in the mixture. Before arriving at definite conclusions based on the older work, the conditions under which the dehydrogenation was carried out should be carefully studied.

So far, elimination tests may have shown that our acid has a cyclopentane ring and has at least one, but not more than two, methylene groups between the ring and the carboxyl group. Methods of degrading the acid to a compound with fewer carbon atoms may make it possible to eliminate a number of the many possible structures still remaining for the acid. Most of these methods have been employed in the past on fractions which were still fairly complex mixtures and the results have rarely been clear cut. When these methods are used on acids that are essentially pure, much labor should be saved.

Degradation, $RCH_2COOH \rightarrow RCH_2NH_2$

The most important degradation reaction used in naphthenic-acid chemistry so far has been the conversion of the acid to an amine of n–1 carbons. This is accomplished by three variations of the Curtius reaction, in all of which the last step is the hydrolysis of $R{-}N{=}C{=}O$ to RNH_2.

The Hofmann degradation is the oldest and it is the best known, because it is found in all textbooks, but it is now used less than the other two in naphthenic-acid work. This method includes the following series of reactions:

$$RCOOH \rightarrow RCOOCH_3(\text{or } RCOCl) \rightarrow RCONH_2 \rightarrow RNCO \rightarrow RNH_2$$

Aschan used the reaction many years ago in his petroleum-acid work [4] and von Braun used it at times, but since in some cases the yield was only 25 to 30%,[14, 18] he abandoned it in later work. Since the degradation starts with the amide, it should still be useful when a sample of the amide is available for degradation.

A study of the literature shows that the yields of amides in the Hofmann degradation are indeed low when the reaction is applied to the higher amides, using the ordinary procedure which employs an aqueous alkaline hypobromide solution. However, the modified method, using methanol as a solvent, appears to give good yields in many cases. A study of the table of Wallis and Lane [58] shows that, in some cases, acids similar to naphthenic acids have been converted to amines with good yields, while in many other cases yields are not given.

The statement is often made that the customary procedure using aqueous alkaline hypobromide gives poor yields with the higher amides, because more and more of the product consists of the nitrile obtained on dehydrogenation of the amine. In naphthenic-acid chemistry, it would often be very desirable to direct the reaction in such a way as to obtain a good yield of this nitrile, because this would make the next lower acid available by simple hydrolysis. Von Braun [15] experimented with methods of oxidizing amines to nitriles and reported fair success in early experiments, but did not publish any of his results. Berry,[10] in the author's laboratory, tried to modify the conditions of the Hofmann reaction to increase the yield of nitrile, but was rarely able to get more than 50% yield either in this way or by using various oxidizing agents to convert the amines directly to the acid with the same number of carbon atoms.

The original Curtius method consists of the following series of reactions:

$$RCOOH \rightarrow RCOOCH_3 \rightarrow RCONHNH_2 \rightarrow RCON_3 \rightarrow RNCO \rightarrow RNH_2.$$

Sometimes, the acid chloride is used in place of the ester. This procedure seems to be the method of choice when the sample available is the ester or the acid chloride. A serious drawback has been the expense involved in the use of hydrazine as a reagent.

Another modification of the Curtius method has become the most popular method of degrading acids to amines. Two variations of this method are used. In the first one, the azide is prepared by the following reactions:

$$RCOOH \rightarrow RCOCl \rightarrow RCON_3$$

using sodium azide to obtain the azide from the acid chloride. This method was studied by Nelles [41] in von Braun's laboratory, who reported that some of the technical sodium azide available to him was contaminated with some impurity which kept the azide from reacting with the acid chloride. He found that the impurity was easily removed by treatment with a little hydrazine. Whitmore [62] and Buchman [23] both used the method and reported no difficulty in using sodium azide.

Von Braun was apparently the first to use the second modification in petroleum-acid work and recommended it heartily not only for its convenience but also for its high yields (70 to 90%) when used with acids of this range. In this (the K. F. Schmidt modification), the azide is formed directly from the acid by reaction with hydrazoic acid in chloroform solution.[14, 46, 47, 55] Von Braun found the yield so uniformly high that he recommended the routine conversion of the acid, after fractionation as acid and as ester, to the amine. This was then fractionated as such or converted to a salt or solid derivative which could be recrystallized. He had so much difficulty in removing the last traces of nonacidic impurities from his acids that he recommended conversion of the amine to the oxalate, which was then to be washed with a low-boiling solvent to remove all inert impurities.

Since the amine has one less carbon than the original acid, there should be fewer isomers and, in some cases, it might be possible to identify the amine and resynthesize the acid; but von Braun did not succeed in isolating a pure compound in this way. If he had been able to obtain a pure amine and identify it, he would have had no assurance, in many cases, that the acid resynthesized from it would have been the one originally in petroleum and not a possible *cis-trans* isomer of it. Therefore, it would have been necessary to show that the original configuration had been obtained.

Failing in his attempts to obtain a pure amine by fractionation or by recrystallization of a salt or derivative, von Braun attempted to use the amine to distinguish between sterically hindered and reactive amines by determining the hydrolysis rate of the benzoyl derivatives obtained on treating the amine with benzoyl chloride. He found that for the following compounds, the amounts of benzoyl derivative hydrolyzed after 4 hours of treatment with 4 volumes of concentrated hydrochloric acid were:

	%
(1,2,2,3-tetramethylcyclopentyl)methylamine	0
(2,2,3-trimethylcyclopentyl)methylamine	15
(2,3,3-trimethylcyclopentyl)methylamine	30
cyclopentylmethylamine	55
2-cyclopentylethylamine	80

Obviously, it would be possible to achieve a definite separation into type of primary amines by this method if it were systematically repeated on a suitable mixture of amines. If a pure amine were being studied, comparison of its rate of hydrolysis with that of known naphthenylamine benzoates could perhaps be used in eliminating other possible configurations.

Conversion of an Acid to an Olefin

In a number of degradation methods, the acid is converted in some way to an olefin with double bond at the end, which can then be cleaved by ozonolysis or by vigorous oxidation to yield a lower acid or ketone.

Von Braun used two different methods of this type. In the one most commonly employed by him, he converted the acid to the amine by the K. F. Schmidt method. He then changed the amine to the olefin by the well-known Hofmann method of exhaustive methylation and pyrolysis of the resulting quaternary ammonium hydroxide to yield the olefin with the same number of carbon atoms as the amine. Subsequent ozonolysis yielded a new acid or a ketone with two carbon atoms less than the original acid.

A modification of his original decomposition of the quaternary ammonium hydroxide was used by von Braun in later work, because it was not as expensive as the method using silver oxide and because he had met with much trouble in removing colloidal silver oxide which was obtained in his work with the naphthenic acids.[13, 14]

In this modified method, the base is treated with dimethyl sulfate in the presence of alkali. The quaternary base, which usually separates as an oily liquid, is isolated from the aqueous layer, dissolved in water, and acidified with dilute sulfuric acid. The solution is heated to boiling to convert the base completely to the quaternary ammonium sulfate. Slightly less than the calculated amount of barium hydroxide solution is added, the precipitate filtered, and the filtrate concentrated by distillation. The concentrated solution is made strongly alkaline with potassium hydroxide and distilled to remove the trimethyl amine, naphthenyldimethylamine, and the olefin. The naphthenyldimethylamine, which is formed as an important by-product, can be reprocessed to increase the final yield of olefin.

The amine obtained in good yield by the Schmidt degradation can be converted to the alcohol by means of nitrous acid, but it is a well-known fact that this conversion is very often accompanied by rearrangements; therefore, its value in structure determination is slight. Von Braun reported [16] that no rearrangement takes place when this conversion is carried out in the following steps:

$$R{-}NH_2 \rightarrow R{-}NH{-}\overset{\overset{\displaystyle O}{\|}}{C}{-}C_6H_5 \rightarrow R{-}Br \rightarrow R{-}O{-}\overset{\overset{\displaystyle O}{\|}}{C}{-}CH_3 \rightarrow R{-}OH$$

Another degradation leading to lower acids or ketones over an olefin was used in early work by Aschan [6] and its modifications have been employed by others for acids having the carboxyl group at the end of a chain of at least three carbons. In this method, the acid is changed to the α-bromo acid by some modification of the Hell-Volhard-Zelinsky reaction and dehydrobrominated to the α–β unsaturated acid, as in von Braun's lactone test for chain length. The α–β unsaturated acid is cleaved at the double bond by vigorous oxidation to yield an acid or ketone with n–2 carbons.[32]

Kuwata [36] degraded petroleum acids over the α-bromo acid, which was then treated with dilute aqueous potassium hydroxide to yield the α-hydroxy acid. This was cleaved with lead peroxide to yield an aldehyde. An important by-product of the reaction is the α–β unsaturated acid formed when the bromo acid is treated with alkali. This and other complications seriously reduced the yield obtained, according to Kuwata, but Chichibabin [25] used essentially the same method and reported better yields.

Finally, there are special methods which have been tried only to a minor extent or have been found suited only for use with acids within a definite range of molecular weight.

Von Braun suggested the use of what he called the amine-phosphate method of converting an amine to an olefin.[18] This method has been used with success in work with the lower acids, but tends to yield mixtures when used with acids of ten or more carbon atoms. In this method, the amine obtained by degradation of the acid is treated with a mixture of phosphoric acid and sodium phosphate and heated in a flask. Ammonium phosphate is formed and the olefin is distilled. When this method was used with the higher acids in the author's laboratory, complex mixtures of unsaturated compounds were obtained, indicating that probably rearrangement under the influence of phosphoric acid takes place as the olefin is formed.[55] This rearrangement should be particularly extensive when the amine group is on a carbon directly attached to the ring, since unsaturated compounds of the type:

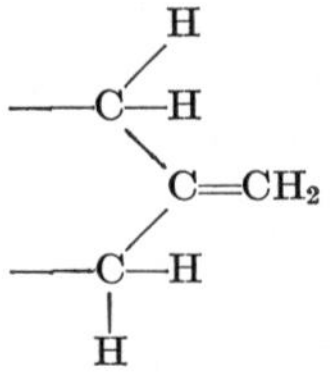

$$\begin{array}{lll} & H & \\ & / & \\ —C—H & & \\ & \backslash & \\ & & C{=}CH_2 \\ & / & \\ —C—H & & \\ \;\;\;\; | & & \\ \;\;\;\; H & & \end{array}$$

tend to rearrange to compounds having the double bond in the ring. Where applicable, the method has the advantages of convenience and single-step conversion of amine to olefin.

Another method that has been used with good results, sometimes on much larger molecules, by Wieland and coworkers [63] and many other biochemists has apparently not been employed with naphthenic acids, except by von Braun,[11, 12] and then only in an earlier form. This is the Barbier-Wieland application of the Grignard reaction to the preparation of olefins from esters. In the original method used by von Braun, the methyl naphthenate is treated with methylmagnesium bromide to yield ultimately the tertiary alcohol:

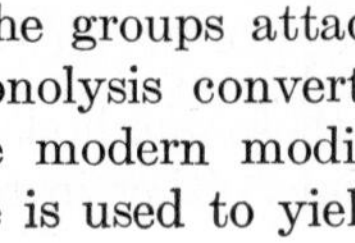

$$R{-}CH_2{-}\overset{\displaystyle CH_3}{\underset{\displaystyle CH_3}{\overset{|}{\underset{|}{C}}}}{-}OH$$

This can either be cleaved directly by oxidation or be dehydrated to yield an olefin or mixture of olefins, resulting from elimination of water toward one of the groups attached to the central atom. Vigorous oxidation or ozonolysis converts the olefin or olefins to acids or ketones. In the modern modification of this method,[28] phenylmagnesium bromide is used to yield

$$R{-}CH_2{-}\underset{\large C_6H_5}{\overset{\large C_6H_5}{\underset{|}{\overset{|}{C}}}}{-}OH$$

which can eliminate water in only one direction and so should yield only a single cleavage product. It has been found that one of the phenyl groups may be eliminated to yield a mixture of compounds after all. If steric hindrance is not too important in this reaction, it should prove to be a valuable method of degradation.

All of the methods just discussed change the acid to olefins, usually with n–1 carbon atoms, which can now be cleaved by oxidation or by ozonolysis to yield acids or ketones of lower molecular weight. Another method that was recommended by von Braun,[21, 22, 42] particularly for use with acids in which the carboxyl group is attached directly to the ring, involves the following steps:

$$\begin{matrix} {-}CH_2 & & H \\ & \diagdown\ \diagup & \\ & C{-}COOH & \\ & \diagup & \\ {-}CH_2 & & \end{matrix} \rightarrow \begin{matrix} {-}CH_2 & & COOH \\ & \diagdown\ \diagup & \\ & C & \\ & \diagup\ \diagdown & \\ {-}CH_2 & & Br \end{matrix} \xrightarrow[\text{Curtius or Schmidt}]{\text{Hofmann}} \begin{matrix} {-}CH_2 & & NH_2 \\ & \diagdown\ \diagup & \\ & C & \\ & \diagup\ \diagdown & \\ {-}CH_2 & & Br \end{matrix} \xrightarrow{HOH} \begin{matrix} {-}CH_2 & \\ & \diagdown \\ & C{=}O \\ & \diagup \\ {-}CH_2 & \end{matrix}$$

There seems to be no reason why the Curtius method could not be used with any acids that can be converted to the α-bromo acids by the Hell-Volhard-Zelinsky reaction to yield either aldehydes or ketones. Von Braun [19] found the Hofmann, and apparently also the K. F. Schmidt, method tended to eliminate HBr.

A new method that appears promising for use with acids in which the carboxyl group is at the end of a long chain has been described by Miescher and coworkers.[39] This method yields an aldehyde of n–3 carbons by the following series of reactions:

$$R{-}CH_2{-}CH_2{-}CH_2{-}COOCH_3 \rightarrow R{-}CH_2{-}CH_2{-}CH_2{-}C(OH)(C_6H_5)_2 \rightarrow$$
$$R{-}CH_2{-}CH_2{-}CH{=}C(C_6H_5)_2 \rightarrow R{-}CH_2{-}CHBr{-}CH{=}C(C_6H_5)_2 \rightarrow$$
$$RCH{=}CH{-}CH{=}C(C_6H_5)_2 \xrightarrow{Cr_2O_3} RCHO + O{=}CH{-}CH{=}(C_6H_5)_2$$

Another new method has been described by Gallagher.[32] This yields an acid of n–2 carbons by the following series of reactions:

$$RCOCl \xrightarrow{\text{diazomethane}} R{-}CH_2{-}COCHN_2 \rightarrow RCH_2{-}COCH_2Cl \rightarrow$$
$$R{-}CH_2{-}CO{-}CH_3 \rightarrow RCHBr{-}COCH_3 \rightarrow R{-}CH{=}CH{-}COCH_3 \rightarrow RCOOH$$

Apparently, neither of these methods has been tried on naphthenic or other alicyclic acids, either in regard to the yields obtainable or to the possibility of cleavage of the ring in naphthenic acids.

The Hunsdiecker reaction converts an acid to an alkyl halide of n–1 carbons by the following series of reactions:

$$RCOOH \rightarrow RCOONa \rightarrow RCOOAg \xrightarrow{Br_2^0} RBr + AgBr$$

This degradation has been used by a number of workers in other fields and should be applicable to work with naphthenic acids. The bromide can, of course, be converted to the olefin by one of the methods mentioned previously. The expense involved would not be a serious factor in work with pure acids, since only small amounts would be degraded.[3, 23, 33, 48]

Since the only naphthenic acids actually identified so far have been ones with ten or fewer carbon atoms — all of them lower in molecular weight than the typical acids — the fact should be stressed that the methods discussed in this chapter have been almost entirely methods of eliminating some of the many structural formulae fitting the molecular formula of any pure acids isolated in the separation processes.

Whenever a pure acid or a pure solid derivative is isolated, it will normally be necessary to synthesize one or more acids to establish the identity of the petroleum acids. In addition to the classical methods, infrared spectra and particularly X-ray diffraction data of the pure solids will probably be used in proving the structure of naphthenic acids.

Bibliography

1. Adkins, H., and K. Folkers, *J. Am. Chem. Soc.* **53,** 1095 (1931).
2. Adkins, H., and J. W. Davis, *Ibid.* **71,** 2955 (1949).
3. Arnold, R. T., and P. Morgan, *Ibid.* **70,** 4248 (1948).
4. Aschan, O., *Ber.* **23,** 867 (1890).
5. Aschan, O., *Ibid.* **24,** 2710 (1891).
6. Aschan, O., *Ann.* **271,** 266 (1892).
7. Aschan, O., *Ber.* **25,** 3661 (1892).
8. Aschan, O., *Ibid.* **31,** 1801 and 1803 (1898).
9. Balada, A., and J. Wegiel, *C. A.* **31,** 2398 (1937).
10. Berry, J., M. A. Thesis, The University of Texas, 1952.
11. von Braun, J., and A. Heymons, *Ber.* **61,** 2277 (1922).
12. von Braun, J., F. Jostes, and W. Munch, *Ann.* **453,** 113 (1927).
13. von Braun, J., and E. Anton, *Ber.* **64,** 2865 (1931).
14. von Braun, J., *Ann.* **490,** 100 (1931).
15. von Braun, J., Private Communication to Prof. J. R. Bailey, 1931.

16. von Braun, J., *Z. angew. Chem.* **47,** 611 (1934).
17. von Braun, J., and F. Fischer, *Ber.* **66,** 101 (1933).
18. von Braun, J., and P. Kurtz, *Ibid.* **67,** 227 (1934).
19. von Braun, J., *Z. angew. Chem.* **47,** 611 (1934).
20. von Braun, J., and P. Kurtz, *Ber.* **70,** 1224 (1937).
21. von Braun, J., E. Kamp, and J. Kopp, *Ibid.* **70,** 1750 (1937).
22. von Braun, J., and H. Ostermeyer, *Ibid.* **70,** 1004 (1937).
23. Buchman, E. R., O. A. Reims, T. Skei, and M. J. Schlatter, *J. Am. Chem. Soc.* **64,** 2696 (1942).
24. Cason, J., and H. J. Wolfhagen, *J. Org. Chem.* **14,** 155 (1949).
25. Chichibabin, A. E., *et al., Chim. et Ind.* **17,** 306, Special Number, March, 1932.
26. Denissenko, J. I., *Ber.* **69,** 1668 (1936).
27. Denissenko, J. I., and N. D. Zelinsky, *Ber.* **69,** 2184 (1936).
28. Fieser, L. F., *J. Am. Chem. Soc.* **70,** 3188 (1948).
29. Goheen, G. E., *Ind. Eng. Chem.* **32,** 503 (1940).
30. Hancock, K., and H. L. Lochte, *J. Am. Chem. Soc.* **61,** 2448 (1939).
31. Hilditch, T. P., *Chemical Constitution of Natural Fats,* Chapman and Hall, London, 1947, 464–520.
32. Hollander, V. P., and T. F. Gallagher, *J. Biol. Chem.* **162,** 549 (1946).
33. Hunsdiecker, H., and C. Hunsdiecker, *Ber.* **75,** 291 (1942).
34. Kazanski, B. A., *Ber.* **69B,** 1862 (1936).
35. Kischner, N., *J. prakt. Chem.* (**2**), **56,** 364 (1897).
36. Kuwata, T., *C. A.* **23,** 1390 (1929).
37. Markownikoff, W., and W. Oglobin, *Ber.* **16,** 1878 (1893).
38. Markownikoff, W., *Ibid.* **25,** 370 and 3355 (1892).
39. Meystre, C., L. Ehman, R. Neher, and K. Miescher, *Helv. Chim. Acta.* **28,** 1252 (1945).
40. Müller, J., and S. Pilat, *Brennstoff-Chem.* **17,** 461 (1936).
41. Nelles, J., *Ber.* **65,** 1345 (1932).
42. Nenitzescu, C. D., and G. G. Vanta, *Bull. soc. chim. France* [5] **2,** 2214 (1935).
43. Nenitzescu, C. D., and E. Cioranescu, *Ber.* **69,** 1040 (1936).
44. Nenitzescu, C. D., and I. P. Cantuniari, *Ibid.* **70,** 277 (1937).
45. Nenitzescu, C. D., D. A. Isacescu, and I. A. Volrap, *Ibid.* **71B,** 2056 (1938).
46. Ney, W. O., W. W. Crouch, C. E. Rannefeld, and H. L. Lochte, *J. Am. Chem. Soc.* **65,** 770 (1943).
47. Oesterlin, M., *Z. angew. Chem.* **45,** 536 (1932).
48. Oldham, J. W. H., *J. Chem. Soc.* **1950,** 100.
49. Pilat, S., and E. Holzman, *Brennstoff-Chem.* **14,** 263 (1933).
50. Pines, H., and V. N. Ipatieff, *J. Am. Chem. Soc.* **61,** 1076 (1939).

51. Pines, H., R. C. Oldberg, and V. N. Ipatieff, *Ibid.* **70,** 533 (1948).
52. Quebedeaux, W. A., G. Wash, W. O. Ney, W. W. Crouch, and H. L. Lochte, *Ibid.* **65,** 767 (1943).
53. Rupe, H., and P. Lauger, *Helv. Chim. Acta* **3,** 274 (1920).
54. Shidkoff, N., *J. Soc. Chem. Ind.* **1899,** 360.
55. Shive, W., J. Horeczy, G. Wash, and H. L. Lochte, *J. Am. Chem. Soc.* **64,** 385 (1942).
56. Tanaka, Y., and T. Kuwata, *C. A.* **23,** 4051 (1929).
57. Tarassowa, J. M., *Chem. Zent.* **1935,** II, 3497.
58. Wallis, E. S., and J. F. Lane, *Organic Reactions,* Vol. II, Wiley, New York, 1946, page 267.
59. Weitkamp, A. W., *J. Am. Chem. Soc.* **67,** 453 (1945).
60. Weitkamp, A. W., *Ibid.* **69,** 1936 (1947).
61. Whitmore, F. C., and H. M. Crooks, *Ibid.* **60,** 2078 (1938).
62. Whitmore, F. C., *et al., Ibid.* **63,** 2041 (1941).
63. Wieland, H., O. Schichting, and R. Jacobi, *Z. physiol. Chem.* **161,** 82 (1926).
64. Williams, M., and G. H. Richter, *J. Am. Chem. Soc.* **57,** 1686 (1935).
65. Zelinski, N., *Ber.* **56,** 1716 (1923).
66. Zelinski, N., *Ibid.* **57B,** 42 (1924).
67. Zelinski, N., and E. Pokroskaja, *Ber.* **57,** 51 (1924).
68. Zelinski, N., B. A. Kasansky, and A. F. Plate, *Ibid.* **66,** 1415 (1933).
69. Zelinski, N., and N. J. Shuykin, *Ind. Eng. Chem.* **27,** 1210 (1935).
70. Zelinski, N., I. A. Massajew, and G. D. Holpern, *Chem. Zentr.* **1939,** II, 1610; also *Universal Oil Products Co. Abstracts* **14,** 159 (1939).

CHAPTER TEN

THE HEXAHYDROBENZOIC ACID PROBLEM

During the first 15 years of naphthenic acid research, the theory that the naphthenic acids contained the hexahydrobenzene ring and so were closely related to the aromatic compounds found in coal tar influenced much of the work done on petroleum acids. Among the investigators working during this period, Aschan deserves special mention.[1, 2, 3, 4, 5, 6] In 1887, he started work on several hundred kilograms of crude acids from Baku petroleum. The acids were obtained during routine alkali washing of kerosene. Aschan decided to try to gain from them some definite information about the nature of the naphthenic acids by following one of two different lines of work, either (1) a comparison of individual naphthenic acids with the corresponding hexahydrobenzoic acids, or (2) conversion of a pure naphthenic acid into the corresponding naphthene.

He considered the first approach as the more promising, since it is normally easier to prepare pure acids and their derivatives than pure naphthenes. He also decided, quite logically, that his best chance of success should lie in work with the lower members of the homologous series of acids, i.e., with $C_6H_{11}COOH$ and $C_7H_{13}COOH$. If the theory then held had been correct, the $C_6H_{11}COOH$ acid should have been the simplest naphthenic acid and should have been easy to isolate and identify. Aschan appears to have been the only one of the early investigators who saw clearly the advantage of work with the lower members of the series, or at least he was the only one who had material enough to do so.

After tedious and painstaking fractionation of about 20 kg of acids boiling in the proper range, Aschan finally obtained 40 to 50 g of what appeared to be pure hexahydrobenzoic acid and converted this into various derivatives. At that time, the preparation of hexahydrobenzoic acid was a difficult operation, but he finally obtained some pure hexahydrobenzoic acid. He found that its boiling point was 232° to 233°C (450°F), whereas the petroleum acid boiled at 215° to 217°C. The methyl esters boiled at 179° to 180°C (355°F) and 165.5° to 167.5°C (331°F) respectively and the amides melted at 184°C and 123°C, thus proving conclusively that the small amount of acid he had isolated was not hexahydrobenzoic acid. He concluded that in this case, the naphthenic acid did not contain the cyclohexane ring.

In view of the surprising outcome of his attempts to prove the presence of hexahydrobenzoic acid, Aschan then decided to try the second of his lines of attack — conversion of the acids to the corresponding naphthenes. In this work, he used the fractions that appeared to consist mainly of $C_7H_{13}COOH$, and he tried to decarboxylate the acid by heating its sodium salt, or to degrade the amide to the amine and convert this to alcohol with nitrous acid. The alcohol could then be changed to the halide, which could be reduced to the naphthene.

About this time, Kraemer and Böttcher [14] found that, unfortunately, decarboxylation of naphthenic acids lead to extensive decomposition of the resulting hydrocarbon, so that this method could not be employed.

Hofmann's degradation to the amine goes smoothly, but he found that the reaction of the amine with nitrous acid did not yield a simple mixture. We now know that this reaction is usually accompanied by various rearrangements and, therefore, it is of little value in structure determinations of this type.

Since neither of his proposed routes led to conclusive results, Aschan decided to try Wreden's method of complete reduction, using hydroiodic acid and phosphorus. When he applied this method to a $C_8H_{14}O_2$ acid, he obtained a product analyzing for C_8H_{16} and boiling at 117° to 118°C which appeared to be identical with a naphthene isolated by Markownikoff.[18] He concluded from this work that naphthenic acids contained a saturated ring system, that the naphthene obtained from one of them was identical with a naphthene isolated from petroleum by Markownikoff, and that the naphthenes and naphthenic acids contain the same ring system.

Meanwhile, much confusion was caused by the results obtained by Wreden[28] and then by others, using the hydroiodic acid-phosphorus method of complete reduction of aromatic compounds. Wreden had treated benzene, toluene, and other simple aromatic hydrocarbons by his method and had obtained a series of saturated hydrocarbons which proved to be identical with hydrocarbons isolated by Beilstein and Kurbatov[7] from Caucasian petroleum. Since the hydrocarbons obtained from aromatic hydrocarbons were thought to be hexahydroaromatic, this appeared to prove that naphthenes from petroleum belonged to the same system and that the hydrocarbons from naphthenic acids also were of the same type. In the meantime, Freund,[9] Perkins,[20, 21, 22] and Wallach[27] had independently synthesized a series of alicyclic compounds of the type C_nH_{2n}. This appeared confusing for a time, because some of them were obviously different from the petroleum hydrocarbons which were thought to have the same structure.

The confusion was cleared up a little later by Kishner,[13] when he found that the Wreden method of reduction leads to isomerization as well as reduction, benzene yielding mainly methylcyclopentane by this method. Aschan now showed that a C_6H_{12} naphthene isolated from petroleum was identical with methylcyclopentane.[5] He also found, however, that another C_6H_{12} hydrocarbon from petroleum could be oxidized to adipic acid by nitric acid and so must have been a cyclohexane, and he then concluded that both types of rings, cyclohexane and cyclopentane, must be present in naphthenic acids. This view was soon discarded and has only recently been shown to be correct for California and some isolated other petroleum acids at least.

Soon after Aschan's work, Zelinsky[29] developed his method of catalytic dehydrogenation of hydroaromatic compounds to aromatic ones. The compound in vapor phase, often diluted with carbon dioxide or hydrogen, is passed over a platinum catalyst at 300° to 315°C (572° to 598°F) and it is found that the index of refraction rises with every pass and becomes constant when all the hydroaromatic compound has been converted. In 1924, Zelinsky[29] tried this method on some Eurasian naphthenic acids and found no benzoic acids. Kuwata[15] tried it with a number of Japanese acids and detected small amounts of benzoic acids in some and none in others. These results appeared to be conclusive and for a number of years, investigators held that naphthenic acids contained only cyclopentane rings.

Still later, conflicting reports on the behavior of cyclopentanes and hexanes under dehydrogenation conditions have appeared.[23] Kazansky [11] observed formation of aromatic hydrocarbons when alkylcyclopentanes were heated over platinized charcoal, but Tatasova [26] could observe no effect when alkylcyclopentanes were heated to 300°C over platinized charcoal in a stream of hydrogen or carbon dioxide. Zelinsky [30] observed ring cleavage when alkylcyclopentanes were treated with a platinum catalyst at 305° to 315°C. If this observation is correct, much of the evidence obtained earlier by Zelinsky, using catalytic dehydrogenation, is of doubtful value, since the decrease in index of refraction due to cleavage of cyclopentanes might well cancel the increase due to aromatization. These rearrangements may well have been due to the use of catalyst supports containing silicates or aluminates either as main ingredients or as impurities in the charcoal. Under carefully controlled conditions catalytic dehydrogenation of cyclohexanes has, of course, been in use for years.

While there is, then, no evidence for the presence in considerable concentration of naphthenic acids containing the cyclohexane ring in all naphthenic acids,[8] they are known to be present in southern California acids [19, 24, 25] and some Japanese acids, and at least one such acid has been isolated from Iranian acids.[12] The acid isolated in large amounts from California acids and from Iranian oil, 2,2,6-trimethylcyclohexanecarboxylic acid, contains a geminal dimethyl system which could not be dehydrogenated without rearrangement. Results obtained by others with similar compounds show that they are unchanged at the usual temperature of 300°C, but suffer extensive rearrangement and dehydrogenation at 375° to 400°C (708° to 752°F). It is hard to decide whether this acid is actually more plentiful than others or whether it was obtained in large amount because it is easily isolated from the residual acids left after esterification of unhindered acids.

Bibliography

1. Aschan, O., *Ber.* **23,** 867 (1890).
2. Aschan, O., *Ibid.* **24,** 1864 and 2710 (1891).
3. Aschan, O., *Ibid.* **25,** 3661 (1892).
4. Aschan, O., *Ann.* **271,** 266 (1892).
5. Aschan, O., *Ber.* **31,** 1803 (1898).
6. Aschan, O., *Ibid.* **32,** 1769 (1899).

7. Beilstein, F., and A. Kurbatov, *Ber.* **13,** 1818 and 2028 (1880).
8. Chichibabin, A. E., *et al.*, *Chim. et ind.* **17,** 306 (Special Number) (1932).
9. Freund, A., *Monatsh.* **3,** 625 (1882).
10. Hancock, K., and H. L. Lochte, *J. Am. Chem. Soc.* **61,** 2448 (1939).
11. Kazanski, B. A., *Ber.* **69B,** 1862 (1936).
12. Kennedy, T., *Nature* **144,** 832 (1939).
13. Kishner, N., *J. prakt. Chem.* (**2**), **56,** 364 (1897).
14. Kraemer, G., and W. Böttcher, *Ber.* **20,** 598 (1887).
15. Kuwata, T., *J. Fac. Eng. Tokyo Imp. Univ.* **17,** 305 (1928); *C. A.* **23,** 1390 (1929).
16. Markownikoff, W., and W. Oglobin, *Ber.* **16,** 1878 (1883).
17. Markownikoff, W., *Ber.* **25,** 370 and 3355 (1892).
18. Markownikoff, W., *Ann.* **307,** 367 (1899).
19. Ney, W., W. Crouch, C. Rannefeld, and H. L. Lochte, *J. Am. Chem. Scc.* **65,** 770 (1943).
20. Perkins, W. H., Jr., *Ber.* **16,** 208 and 2136 (1883).
21. Perkins, W. H., Jr., *Ibid.* **17,** 54 (1884).
22. Perkins, W. H., Jr., *Ibid.* **18,** 1734 and 3246 (1885).
23. Pines, H., R. C. Olberg, and N. Ipatieff, *J. Am. Chem. Soc.* **70,** 533 (1948).
24. Schutze, H., W. Shive, and H. L. Lochte, *Ind. Eng. Chem. Anal. Ed.* **12,** 262 (1940).
25. Shive, W., J. Horeczy, G. Wash, and H. L. Lochte, *J. Am. Chem. Soc.* **64,** 385 (1942).
26. Tatasova, E., *C. A.* **30,** 8173 (1936).
27. Wallach, O., *Ann.* **414,** 231 (1917).
28. Wreden, F., *Ibid.* **187,** 153 (1877).
29. Zelinski, N., *Ber.* **57B,** 42 (1924).
30. Zelinski, N., B. Kazanski, and A. Plate, *Ibid.* **68B,** 1869 (1935).

CHAPTER ELEVEN

SALTS AND DERIVATIVES OF PETROLEUM ACIDS

The naphthenic acids are typical carboxylic acids similar in practically all respects to the saturated acids of similar boiling point, but for a possible slight effect of their cyclic structure on their properties. Their strength, as far as known, is essentially the same as that of the saturated fatty acids, except a few tertiary acids which are known to have a slightly lower ionization constant, i.e., they have K_a values of 10^{-6} and 10^{-7} instead of 10^{-5}. Aschan, in very early work, reported that the acids could not be dried over calcium chloride because they reacted with it to liberate hydrogen chloride. Presumably this was due to the evolution of insoluble hydrogen chloride gas rather than to the strength of the naphthenic acids. The sodium salts of fairly high molecular weight acids behave like sodium phenolates of much lower molecular weight. Treatment of an aqueous solution of sodium salts with an excess of carbon dioxide liberates part of the naphthenic acids.[21] Goodman [13] found that approximately 40% of the C_{10-12} California petroleum acids can be liberated by treating a 5% sodium-salt solution with dry ice or carbon dioxide gas. Since most elementary textbooks state that phenols are liberated by carbon dioxide while the sodium salts of carboxylic acids are not, this may seem surprising. Actually, this behavior is shown also by aliphatic acids of equal molecular weight.[16]

As typical carboxylic acids, the naphthenic acids form various salts, esters, amides, and other derivatives analogous to those of the fatty acids which have been covered thoroughly in recent monographs.[15, 18, 20, 24, 28]

The sodium and potassium salts are very soluble in water, yielding clear to milky solutions similar to ordinary soaps (to which they are, of course, closely related). Many attempts have been made to utilize the naphthenic acids by converting them to sodium soaps, but their generally objectionable odor (due probably to slight hydrolysis to the evil-smelling acids) appears to have blocked their use except for special purposes. At present, their high cost prohibits their use as ordinary soaps, even if it were possible to completely deodorize them. Since the sodium and potassium salts and many of the others tend to be amorphous or at least only partially crystalline, they are not well suited for use in the purification of naphthenic acids.

Chichibabin [9] and a number of the other early workers attempted to use the difference in solubility of the cadmium, lead, and zinc salts for the separation of aliphatic from naphthenic acids. Apparently, the difference in solubility depends on whether the acids are primary, secondary, or tertiary rather than whether they are cyclic or not. Successes reported by Chichibabin were not impressive, but a carefully worked out systematic procedure may be found useful in separating primary from secondary acids.

Some of the heavy-metal salts, particularly the copper salts, are soluble in nonpolar solvents and may be used in solution. The Charitschkoff [7, 8] qualitative test for naphthenic acids depends on the green color of the solution obtained when copper naphthenates are dissolved in petroleum ether.

Insoluble heavy-metal salts which are capable of forming soluble ammono salts may be precipitated from their concentrated ammonium hydroxide solution by neutralization or evaporation of the ammonia, thus leaving the insoluble salt. Such a solution of copper salts is the basis of some methods of rotproofing sails, ropes, and sandbags.[1, 2, 29, 30] Their chief value may lie in the gradual formation of copper ions on the surface of the fibers treated.

The aluminum salts, probably a mixture of basic salts, deserve special mention because of their use in greases and in Napalm.[3, 23] They are insoluble in water and in alcohol, but dispersible in nonpolar solvents, like gasoline, turpentine, and benzene. The gel may be used as such or concentrated by evaporating part of the solvent from the solution.

The acids can be esterified by any of the standard methods. In the most commonly used method for large batches of acids, the acids are refluxed with a large excess of methanol containing dry hydrogen chloride, a small amount of concentrated sulfuric acid, toluenesulfonic acid, or some other catalyst. Yields of 80 to 95%, based on the acids, are commonly obtained.

Various schemes of improving the yield or reducing the time required have been reported. Several workers added benzene or petroleum ether boiling at 80° to 100°C and slowly distilled the solvent, the water formed, and part of the excess methanol.[19] Goheen [12] used the method of Corson, Adams, and Scott [10] and passed the vapors of ethanol through the naphthenic acids containing the mineral-acid catalyst heated at 115° to 120°C (239° to 248°F). The higher temperature reduced the time required and the yield was satisfactory.

The esters or the acid chlorides are readily converted to amides, anilides, or toluidides, which are generally well crystallizing solids. Unfortunately, the amides in particular tend to form mixed crystals or solid solutions from which pure compounds either cannot be obtained by recrystallization or can be prepared only with excessive recrystallization losses. While workers have attempted to isolate pure acids through recrystallization of the amides, their results have been very disappointing.[9, 25]

The esters can be reduced with good yield by either the sodium and alcohol method of Bouveault and Blanc [5, 14, 17, 22] or by catalytic hydrogenation under high pressure.[11] The alcohols do not have the usual objectionable odor of the acids and the acetates are said to have a very pleasant odor.[6] The alcohols can, in turn, be converted in fair yield to the iodides, bromides, or chlorides, which can be used in a number of synthetic methods, such as the Wurtz, Wurtz-Fittig, Fridel-Craft, and Grignard reactions. Unfortunately, the products are all mixtures, and little work has been done in this direction.

Acids having at least one hydrogen on the α-carbon atom can be converted to the α-bromo acids by the Hell-Volhard-Zelinsky procedure and the bromo acids can be converted to α–β unsaturated acids or even to the amino acids.[9] The unsaturated acids have been used in degradation experiments.

Kraemer and Böttcher [24] found that the calcium salts of naphthenic acids cannot be converted to ketones, except in poor yield, because of extensive pyrolysis during the heating operation. Later work has shown, however, that good yields of ketones can be obtained when the vapors of acids that can be distilled without decomposition are passed over hot manganese oxide or thorium oxide. Zelinsky and Rjachina [31] reported good yields of mixtures of acetone, methyl-naphthenyl, and dinaphthenyl ketones, when mixtures of acetic acid and naphthenic acid vapors were passed over thorium or manganese oxide. This should make it possible to prepare a whole series of methyl ketones and dinaphthenyl ketones. Unfortunately, as in the case of other compounds prepared from the mixtures known as naphthenic acids, the ketones would all be mixtures.

Finally, what may prove to be a very important chemical reaction (aside from the use von Braun made of it in the determination of structure through degradation) is the conversion of the acids to amines by one of the degradation methods, preferably the Schmidt or the Curtius reaction.[6, 26] The amines can be converted to amides of the lower acids or of naphthenic acids, or they can be converted to quaternary ammonium salts and used as cationic detergents or disinfectants.

As might be predicted from their structure, the naphthenic acids are fairly stable toward dilute oxidizing agents, but more vigorous conditions, such as refluxing with permanganate or dichromate, lead to deep-seated decomposition. Treatment with dilute cold permanganate has been used to remove unsaturated and sulfur-containing compounds, but even this treatment is not generally employed. A different type of reaction takes place, however, when oxygen is blown through hot naphthenic acids; they appear to be polymerized in the process. Nothing has been published about the exact nature of this reaction.[24]

Bibliography

1. *Anon, Chem. Ind.* **45,** 661 (1939).
2. *Anon, Met. and Chem. Eng.* **51,** 88 (1944).
3. *Anon, Chem. and Met. Eng.* **58,** 162 (1951).
4. Aschan, O., *Ber.* **24,** 2710 (1891).
5. Bouveault, L., and G. Blanc, *Bull. soc. chim. France* **31,** 666 (1904).
6. von Braun, J., *Ann.* **490,** 100 (1931).
7. Charitschkoff, K. W., *Seifensieder Ztg.* **1907,** No. 22, 509.
8. Charitschkoff, K. W., *Chem. Ztg.* **34,** 479 (1910).
9. Chichibabin, A. E., *et al., Chim. ét ind.* **17,** 306 (Special Number, March, 1932).

10. Corson, B. A., E. Adams, and R. W. Scott, *Org. Synth.* **10,** 48 (1930).
11. Folker, K., and H. Adkins, *J. Am. Chem. Soc.* **54,** 1145 (1932).
12. Goheen, G. E., *Ind. Eng. Chem.* **32,** 503 (1940).
13. Goodman, H. H., Ph. D. Thesis, The University of Texas, 1951.
14. Hansley, V. L., *Ind. Eng. Chem.* **39,** 61 (1947).
15. Jezl, J. L., *Pet. Processing* **8,** 89 (1953).
16. Kamm, O., *Qualitative Organic Analysis,* Wiley, New York, 1929, page 55.
17. Kastens, M. L., and H. Peddicord, *Ind. Eng. Chem.* **41,** 438 (1949).
18. Komppa, G., *Ber.* **62,** 1562 (1929).
19. Kozicki, G., and S. Pilat, *Petroleum Z.* **11,** 310 (1915).
20. Markley, K. S., *Fatty Acids, Their Chemistry and Physical Properties,* Interscience, New York, 1947.
21. Markownikoff, W., and W. Oglobin, *Ber.* **16,** 1878 (1883).
22. Martens, K. H., *Seifen, Oele, Fette, u. Wachse* **78,** 601 (1952).
23. Mysels, K. J., *Ind. Eng. Chem.* **41,** 1437 (1949).
24. Naphthali, M., *Chemie, Technologie und Analyse der Naphthensäuren,* Wissenschaftliche Verlagsgesellshaft, Stuttgart, 1927; see also, Supplement 1927–1933.
25. Nenitzescu, C. D., D. A. Isacescu, and T. A. Volrap, *Ber.* **71,** 2062 (1938).
26. Oesterlin, M., *Z. angew. Chem.* **45,** 536 (1932).
27. Pyhala, E., *Kolloid Z.* **9,** 212 (1911).
28. Ralston, A. W., *Fatty Acids and Their Derivatives,* Wiley, New York, 1948.
29. Smith, P. I., *Soap,* No. 11, 1938, page 86.
30. Wise, L. E., *Ind. Eng. Chem., News Edition* **13,** 371 (1935).
31. Zelinsky, N., and E. Rjachina, *Ber.* **57B,** 1932 (1924).

CHAPTER TWELVE

PHENOLS

Since the commercial isolation of the mixture of phenols known as cresylic acids is expanding rapidly and detailed information on the petroleum phenols will probably be given some day in a separate monograph on cresylic acids, no attempt will be made in this volume to include any discussion of the commercial isolation, refining, and uses of phenols.[1, 2, 3, 6, 8] However, in view of the practically universal occurrence of phenols in crude mixtures of acidic compounds from petroleum fractions, a brief chapter will be devoted to phenols.[15]

Apparently, phenols are present in all petroleum fractions of the proper boiling-point range — they have even been found in considerable concentrations in casing-head gasoline obtained from high-pressure gas wells.[16] It is generally believed that the phenol content of a petroleum fraction is increased during cracking operations. Certainly, phenols are more easily isolated from cracking-process products. Since the amount of phenols that can be isolated from a particular boiling range of products varies widely from field to field, it would be necessary to determine the phenol concentration in both straight-run and cracking-process mixtures from a number of different fields before this question could be settled definitely. Available information appears to indicate that the cracking-process fractions contain more phenols than corresponding straight-run fractions.

Many investigators reported failure to isolate phenol itself, although they had no difficulty in obtaining the higher alkylated phenols.[13, 17, 20, 21] Story and Snow [20] pointed out that the water-soluble phenols would probably have been extracted from crude oil by underground water. Another probable reason for the common failure to isolate phenol lies in the fact that the acidic compounds are usually isolated by extraction with alkali which, for reasons of economy and to avoid emulsions, is kept at a rather low concentration where only a fraction of the phenol content would be converted to sodium salt. The free phenol would, of course, be distributed between the organic and the water layers. The relatively large volume of the organic layer should mean that much of the phenol remains in it. Another, and perhaps more serious, loss of phenol occurs at the refinery when the solution of sodium salts is treated with sulfuric acid to liberate the acids. The phenols will again be distributed between aqueous and organic layers and, in view of the low acidity of the final aqueous layer, phenol, which is quite soluble in water as well as in the main layer of organic acids, should again be partially lost to the wrong phase, with the result that there is not sufficient phenol left to permit its isolation in the ordinary routine study of petroleum acids. A similar fate, in all probability, befalls pyridine which is usually not isolated from petroleum bases.

Phenols Isolated and Identified

While the lower alkylated phenols have been found in many types of petroleum fractions from many different fields the exact structure of these phenols has been determined in a few cases only. This is not surprising in view of the fact that the phenols are toxic, have quite disagreeable odor, and are so similar in properties that isolation and positive identification is tedious. Even in the case of the coal-tar phenols, the number identified is surprisingly small in comparison with the number of bases and hydrocarbons from this source that have been identified.[14]

Phenols can be concentrated through fractional countercurrent neutralization by alkali at the refinery, as was done with the California acids studied at the University of Texas. In this case, two towers were used in the first of which practically no phenols were extracted while in the last only very low concentrations of carboxylic acids were obtained. Considerable amounts of neutral substances, mostly hydrocarbons, were mixed in with the phenols obtained in the second tower.[5]

More commonly, the preliminary isolation of acids at the refinery yields a mixture of carboxylic acids and phenols, from which the phenols may be concentrated by fractional neutralization or liberation. The Spitz and Honig method, or one of its modifications, removes most of the carboxylic acids. The remaining mixture of phenols and hydrocarbons may contain only 70 to 80% of phenols and the neutral substances can be removed from these only by tedious fractionation.

Careful fractional distillation at this stage yields mixtures of phenols of nearly the same molecular weight, along with neutral compounds which are also within a narrow, but much higher, molecular-weight range. Probably, the best method of fractionating kilograms of such mixtures is that of systematic fractional neutralization, using some modern countercurrent method embodying the principles set forth by Golumbic.[10, 11] In molecular-weight ranges in which emulsions are very troublesome, large-scale separation of phenols from hydrocarbons by chromatographic analysis on silica gel or carbon columns should be feasible. After the hydrocarbons have been almost completely removed, the remaining phenols can be separated by the Craig apparatus and the methods of Golumbic [11] or by a careful chromatographic procedure on tall, slender columns.

With the exception of the work of Golumbic, none of the refined modern methods of separation appear to have been used and the few alkylated phenols which have been identified have been isolated and purified by the special methods used in the study of coal-tar phenols.

Unfortunately, most of the phenols which have been identified were obtained from cracking-process fractions. Therefore, they may or may not have been present as such (or even as some other phenolic compound) in the crude oil. We know as little about the origin of phenols as about that of acids in petroleum, but there can be no doubt that a considerable fraction of the phenols in petroleum products was present in the crude oil from which the fraction was obtained, since phenols are found not only in many condensates from gas wells, but also in water produced by many oil and gas wells. The isolation of the very weakly acidic phenols from crude oil is a difficult problem and since crudes contain usually less than 0.1% of phenols, the quantitative extraction from crude oil would be a tedious operation at best.

As far as the naphthenic acids are concerned, the phenols represent one of the important separation problems. Since the phenols are not esterified by the usual method and since the methyl esters boil about 50°C lower than the acids (and phenols) from which they are derived, careful fractionation of the acids followed by esterification and very careful fractionation of the resulting ester-phenol mixture should remove the naphthenic acids as methyl esters and leave the phenols in the still pot, provided the original acid was available as a cut of sufficiently narrow boiling range. Actually, this method has been used, e.g., by Nenitzescu, Isacescu, and Volrap,[17] in the separation of carboxylic acids from phenols.

Phenols Identified

Catlin [6] studied an alkali wash obtained in refining cracked distillates, presumably from mid-continent crudes. He liberated the acidic compounds by treating the alkaline solutions with an excess of hydrogen sulfide obtained in the refinery. The liberated mixture, boiling between 195° and 227°C (383° and 440°F), was fractionated into ten cuts, all of which gave positive tests for cresylic acids. He reported a yield of about 0.0068% of cresylic acids from this distillate, but did not identify any individual compound.

More important early work was done by Story and Snow.[20] They studied both straight-run and cracking-process gasoline and distillate, presumably of Oklahoma origin.

The cracking-process distillate, boiling up to 300°C (572°F), was treated with alkali and the emulsified or dissolved hydrocarbons removed by steam distillation. The acids were then liberated, dried, and fractionated under water-pump vacuum, using several distillations through a simple Hempel column. Most of the fractions still appear to have been too complex to permit preparation of solid derivatives. *p*-Cresol was identified by the boiling point and density of the acetate; *m*-cresol did not yield a solid derivative and was not definitely identified. Zeisel tests did not indicate the presence of alkoxyphenols. Bromination yielded only oils. Ferric chloride gave colors and the phthalein-fusion test was positive for all fractions. Finally, solid phenylurethanes were obtained and *p*- and *m*-cresol identified by these. No urethanes of xylenols could be isolated.

In the study of another series of phenols, low-boiling, cracking-process gasoline was washed with 20% sodium hydroxide and the acidic compounds were liberated from the aqueous layer by sulfuric acid and extracted with benzene. The carboxylic acids were removed by repeated washing with sodium bicarbonate solution.

The phenols were extracted from the benzene solution by strong alkali, and the strongly alkaline solution was finally steam distilled to remove hydrocarbons. The phenols were liberated by sulfuric acid and fractionated by four distillations through a twenty-disk Young column. They were found to boil between 196° and 235°C (385° and 454°F). Apparently, no attempt was made to identify the phenolic compounds in this relatively simple mixture.

To determine whether the phenols were formed in the cracking process or were present in the crude oil, or at least in straight-run fractions, samples of straight-run gasoline and distillate were extracted as in the previous case, but the yield of phenols was very low and no attempt was made to identify any compound. However, qualitative tests indicated that the compounds were very similar to the cracking-process cresylic acids.

Tanaka and Kabayashi [21] studied phenolic compounds derived from Japanese distillates. They liberated acidic compounds from concentrated sodium-salt solution by means of dilute sulfuric acid and then carefully neutralized the carboxylic acids, leaving the phenols which were extracted by ether. This process of fractional neutralization was repeated several times so as to remove carboxylic acids as completely as possible. Petroleum ether boiling below 70°C was finally used in the extraction and this solution, containing probably some hydrocarbons, was fractionated into a series of narrow cuts boiling at 190° to 220°C (374° to 428°F). Bromine yielded white precipitates expected for monohydric phenols. Finally, the careful fractionation was repeated and the main phenol content was found to be in the 193° to 196°C (379° to 385°F) and 199.3° to 203.0°C (390° to 397°F) fractions. Carbolic acid was not isolated, but *o*- and *p*-cresol were isolated as cresoxyacetic acids; *m*-cresol was isolated as the trinitro derivative; 1,3,4-xylenol and 1,2,3-xylenol were, by means of their oxyacetic acid derivatives, shown probably to be present; there was also indication of the presence of diethyl- and triethylphenol, although these were not identified.

Pilat and Holzmann [13] reported the first careful and systematic study of Polish petroleum phenols. They studied crude naphthenic acids as they were routinely obtained in refining Boryslaw distillate. They did not indicate whether their product was obtained from cracking process or straight-run material, but it was probably straight-run. To avoid further work with gums and resins, they steam distilled their crude acids at 210°C (410°F) and worked mainly with the distillate which was then treated with a 6% solution of "soda" (apparently sodium bicarbonate). They removed neutrals and phenols by extraction with ether. The ether extract was made alkaline and extracted repeatedly with ether to remove hydrocarbons and other neutral compounds, thus leaving only the phenolic compounds. The phenols were finally liberated by dilute sulfuric acid. They report a yield of:

2,320 g naphthenic acids
1,985 g phenols
735 g neutrals
2,990 g residue and losses

from an original total of 8,030 g of crude acids.

Since carbolic acid and other lower phenols are quite soluble in water, they saturated the water condensed in their steam distillation with sodium chloride and extracted with ether, but obtained only 45 g of phenolic mixture, of which only 8.8 g boiled at 180° to 200°C (356° to 392°F). This mixture was insoluble in ammonium hydroxide; therefore, they decided that carbolic acid was absent. This was confirmed when they failed to obtain any aryloxyacetic acid derivative of carbolic acid. They identified the three cresols in this water-soluble mixture.

The main batch of 1,880 g of phenols was distilled at 40 mm pressure to yield 1725 g of distillate and a residue. Zeisel tests on the fractions collected were negative for methoxy groups which could have been present. They next carefully fractionated the phenols at 6 mm pressure into five cuts which were then individually fractionated at 6 mm, except the first cut which was fractionated at atmospheric pressure. Each fraction was then studied separately in attempts to identify individual phenols and obtain an estimate of the total amount of each present. *m*-Cresol was identified by means of its trinitro derivative. The aryloxyacetic acid derivatives were used in identifying *o*- and *p*-cresol and 1,2,4-, 1,4,5-, and 1,3,4-xylenols, using mainly the difference in solubility of their sodium salts in separating these phenols from their mixtures in various fractions.

They attempted to use the Brückner splitting method of separating phenols by converting them to their sulfonic acid and then hydrolyzing them with steam. According to Brückner, this proceeds at a definite temperature for each phenol; therefore, if the temperature is held at some fixed point until all of the least stable sulfonic acid has been decomposed, all of that individual phenol should have been steam distilled.[4] By raising the temperature and holding it constant at another, somewhat higher, value, the next phenol should be collected, until the whole series for which it can be used has been separated.

Pilat and Holzmann reported that the separation was not satisfactory, probably because the decomposition temperatures are not the same for mixtures of sulfonic acids as for single pure compounds; Givens [9] had similar difficulties, but Field *et al.*[7] reported satisfactory separations.

Since they were not able to get satisfactory separations by Brückner's method, they decided to test various reagents on small aliquots of the higher fractions and identified 1,3,5-xylenol by means of its tribromo derivative and 1,3,4-xylenol by means of the potassium salt of its sulfonic acid which is insoluble in concentrated potassium chloride solution. This phenol had previously been identified by means of its aryloxyacetic acid derivative.

They found that the higher-boiling xylenols and other higher phenols could not be converted satisfactorily to solid derivatives from the complex mixtures found in the fractions; therefore, they fractionated these very carefully through a Widmer column and collected thirteen fractions. They tested for long side chains in each of the higher cuts, but could get no evidence for such structure.

The first fraction again yielded 1,3,4-xylenol, and 1,2,4-xylenol was indicated (but not proved) by the fact that its aryloxyacetic acid showed no depression when mixed with a known sample of this xylenol, but did give a large depression when mixed with other known xylenol derivatives. The novelty in the use of this method was that the derivative isolated melted many degrees below the melting point of pure 1,3,4-xylenol, but its melting point was not further depressed when mixed with the 1,3,4-xylenol derivative. It was depressed, however, when mixed with others. In the second Widmer cut, they identified 1,2,4-xylenol in the same way.

The third fraction yielded another xylenol derivative, but they did not identify it. The cuts from this fraction to the eleventh were found to be complex mixtures which were not identified.

The twelfth and thirteenth fractions had given a typical blue color test for naphthol when treated with potassium hydroxide and chloroform. On standing, solid β-naphthol separated out and was found to be identical with the known β-naphthol. No test for α-naphthol could be obtained.

Their semiquantitative analyses for individual phenols showed that their crude naphthenic acids contained:

	%
naphthenic acids	28.9
phenols	24.7
neutral compounds	9.1
losses and residues	37.2

Of the phenols obtained they estimated that they had:

	%
o-cresol	1.91
p-cresol	0.78
m-cresol	2.40
1,3,5-xylenol	1.68
1,3,4-xylenol	4.76
naphthol	0.17

They reported that, contrary to their earlier negative results on isolation of naphthenic acids from crude oil, they were able to extract phenolic compounds from both Boryslaw and Bitkow crudes. They extracted the phenols by boiling crude oil with alcoholic potassium hydroxide.

Williams and Richter [22] studied a Dubbs-process distillate from West Texas crudes from which the acids were extracted by dilute sodium hydroxide. This aqueous solution was distilled with steam to remove neutral compounds. The residue was acidified and again steam distilled to remove the more volatile acids. The steam distilled acids were then converted to methyl esters which were extracted with cold alkali to remove phenols. The acids which were not steam volatile were also converted to methyl esters and the phenols remaining unesterified were combined with those from the steam distillable matter. The combined phenols were fractionated to yield fractions boiling at 175° to 181°C (346° to 357°F), 181° to 185°C (357° to 364°F), 190° to 192°C (374° to 378°F), 201° to 202°C (394° to 396°F), and 212° to 228°C (414° to 442°F). Valeric acid, but no phenol, was identified in the first cut; the second cut yielded no solid derivative. Consequently, it was extracted with large volumes of water and, on treatment with bromine, yielded tribromophenol, that is, carbolic acid was identified in this case. The third fraction yielded the oxyacetic acid of *o*-cresol, while the fourth yielded *p*-cresol as the oxyacetic acid and *m*-cresol as the trinitro derivative. Finally, 1,2,3-xylenol was identified as the aryloxyacetic acid in the fraction boiling between 212° and 228°C (414° to 442°F).

Nenitzescu and coworkers, in 1938,[17] isolated a number of phenols from technical naphthenic acids from Ploesti crude oil and found a high concentration of phenols, even though the acids had been obtained from straight-run gasoline from this asphaltic field.

They removed most of the neutral compounds by dissolving the acids in sodium carbonate and extracting this solution with petroleum ether. The acids were next liberated by dilute sulfuric acid, the resulting layer of free acidic compounds removed, and the bottom layer extracted with petroleum ether to remove dissolved and emulsified acidic compounds. All acidic material was combined and again treated with sulfuric acid to decompose sodium salts which had remained in the original acid layer. The organic layer was distilled to remove hydrocarbons used in extractions. The acidic compounds remaining in the still amounted to approximately 83% of the original batch of technical acids and consisted of a mixture of phenols and carboxylic acids.

They decided that the usual methods of separating phenols from acids were ineffective and they converted the acids to methyl esters from which the phenols were extracted with dilute alkali. The alkali extract was acidified with mineral acid and again esterified to obtain another batch of methyl esters. A third treatment yielded no additional esters, and it was assumed that the remaining acidic material was phenolic in nature. The first esterification yielded 4,800 g of ester and the second only 200 g; the phenols separated from the hydrocarbons amounted to 157 g and those from the ester layer, to 478 g.

They could find no carbolic acid but isolated "in the usual manner" *o*-, *m*-, and *p*-cresol and 1,2,3- and 1,3,5-xylenol. A straight-run mixture was used for these separations. No yields were reported except the fact that with 5,000 g of methyl esters, they isolated 635 g of phenols.

Field and coworkers, in 1940, studied three grades of California acids with boiling ranges of 196° to 222°C (385° to 432°F), 225° to 235°C (437° to 455°F), and 228° to 258°C (442° to 496°F).[7] They removed the carboxylic acids and neutral compounds in the usual manner, carefully fractionated the phenols into three degree cuts, and then repeated this operation. The lowest-boiling cut yielded small amounts of phenol and the next higher cuts were separately nitrated and found to yield the trinitro derivatives of phenol and *m*-cresol. The aryloxyacetic acids were used to identify *o*- and *p*-cresol, each cut being studied separately, so that they were finally able to make an approximate estimate of the amount of each phenol isolated. The xylenols and higher phenols are not easily separated through oxyacetic acids or other simple derivatives. Thus they were sulfonated and separated by Brückner's [4] acid-splitting method in which the phenolic sulfonic acids are heated to various definite temperatures and steam distilled. Each of the lower phenols has a characteristic decomposition temperature at which it is hydrolyzed to sulfuric acid and the parent phenol and they can, therefore, be separated and identified by collecting the individual Brückner cuts. In this manner, they identified five of the xylenols and isopseudocuminol from each phenolic compound in each sample. A coal-tar cresylic acid, which was studied at the same time, gave very similar results.

Table 19. **Fractionation of Texas Petroleum Acids**

	Boiling Point 35 mm		*Approximate Volume*	n_D^{20} *of*	*Acids*	n_D^{20} *of*
Cut	*°C*	*°F*	*Liters*	*Cut*	*%*	*Acids*
1	100–110	212–230	3.80	1.4751	10.80	1.5279
2	110–120	230–248	9.50	1.4807	15.40	1.5275
3	120–130	248–266	45.40	1.4903	20.30	1.5258
4	130–140	266–284	34.10	1.4949	24.10	1.5230
5	140–150	284–302	26.50	1.5008	33.80	1.5159
6	150–160	302–320	34.10	1.5003	37.00	1.5095
7	160–170	320–338	26.50	1.4965	42.60	1.4935
8	170–180	338–356	11.35	1.4872	43.40	1.4795
9	180–190	356–374	7.60	1.4852	54.30	1.4765
10	190–200	374–392	1.90	1.4825	60.20	1.4753
Residue			15.20	—	—	—

A group at the University of Texas [9, 12, 19] separated phenols, lower aliphatic acids, and lower naphthenic acids from a three-barrel batch of acids obtained at the refinery from a mixture of straight-run and cracking-process fractions. They distilled off the water and other low-boiling compounds (including probably some of the carbolic acid and other lower phenols) at atmospheric pressure in a stream of natural gas. After the water had been removed and quiet boiling had been attained, distillation was continued at about 35 mm pressure, using water pumps. Rough fractionation of the mixture obtained in the initial distillation yielded the fractions shown in Table 19.

The third fraction obviously contained a considerable amount of phenolic matter and it was decided to study this large cut in detail to determine the nature of all the acids in it. They attempted to separate the neutral compounds from the acids by making strongly alkaline a solution of sodium salts and steam distilling it to remove the hydrocarbons, but since it was soon found that an excessive amount of steam was required, they used this on only a small aliquot. Most of the mixture was diluted with petroleum ether and treated with 1/6 equivalent of 1N potassium hydroxide, stirred vigorously for an hour, settled, separated, and again treated with 1/6 equivalent of base until all acids had been extracted as a series of potassium salt mixtures. Each of these was then steam distilled until the distillate was clear and the acids in each cut were liberated by dilute sulfuric acid. Each fraction obtained in this way was then fractionally neutralized in petroleum ether solution by addition of 1/6 equivalents of 0.5N sodium carbonate solution to yield six series of fractions of acidic mixtures. Properties of a typical series are shown in Table 20.

Table 20. **Properties of Neutralization Fractions**

Cut	*Volume, cc*	n_D^{20}
1	120	1.4478
2	93	1.4515
3	72	1.4636
4	54	1.4868
5	55	1.5035
6	49	1.5000

The last two or three cuts of each series were studied by Henson [12] and Givens [9] who isolated and identified a series of phenols from them. The first cuts of several series were studied by Schutze [19] who isolated a series of aliphatic acids and carbolic acid. The method of isolation of the simplest phenol, which has usually been missed, illustrates the fact that its water solubility may result in its loss to water solutions at the refinery — or even in the laboratory. In this case, a distillation fraction with $n_D^{20} = 1.4490$, which had been obtained on fractionation of the first cut of a series, like that of Table 19, was converted to silver salts by a slight excess of silver nitrate solution; these salts were dissolved in concentrated ammonium hydroxide solution and the silver salts of aliphatic acids were finally liberated from the ammonia complex by careful addition of dilute nitric acid under the surface of the well-stirred solution. As precipitates of silver salts appeared, they were filtered off until a series of fractions of silver salts had been obtained. The final combined filtrates were strongly acidified with hydrochloric acid and distilled to recover the acids which had not been obtained as insoluble silver salts. It was observed that a yellow precipitate formed in the distillate so obtained. This precipitate proved to be *o*-nitrophenol resulting from the slight excess of nitric acid and the phenol in these aqueous filtrates.

Henson,[12] working with the same series of cuts, and Givens [9] studied the neutralization fractions containing high concentrations of phenols as shown by high index of refraction and high density. They carried out an elaborate series of fractionations through efficient columns, operated at high reflux, and collected constant-volume fractions whose boiling points were determined later by noting the temperature at which 40% of a 10 cc fraction had been distilled from a small insulated distilling flask. In many cases, several fractions boiled within one degree, but few of these plateaus coincided with the boiling point of a known phenol, and even then it was not always possible to isolate that phenol. The boiling points were obviously those of mixtures. It was possible, however, to isolate a number of phenols from the mixtures. Starting with 2 cc of one of a series of cuts boiling at 222°C (432°F), they found its density to be 0.9948 at 20°C (78°F) and they were able to prepare an aryloxyacetic acid melting at 124° to 125°C (256°F) which was not depressed when mixed with synthetic 2-methyl-4-ethylphenol. Another phenol in the same fraction yielded a much more soluble oxyacetic acid and could not be obtained in pure form.

The next fraction, boiling at 222°C (428°F), was converted to the 3,5-dinitrobenzoate ester and on recrystallization melted at 193.5° to 194.0°C (381°F), corresponding to the melting point of 1,3,5-xylenol. This had been previously isolated from a lower fraction. It was easy to separate because of the extreme insolubility of the 3,5-dinitrobenzoate compared with other phenols in this boiling-point range.

From one of the fractions boiling at 226°C, 2 cc was converted to the aryloxyacetic acid melting at 160°C (320°F) and to the 3,5-

dinitrobenzoate melting at 179° to 180°C (356°F); the corresponding derivatives of 1,2,4-xylenol melt at 161° to 162°C (322°F) and 181.6°C (360°F) and it boils at 225.0°C (437°F). A mixed melting point with the aryloxyacetic acid derivative of an Eastman 1,2,4-xylenol was not depressed. This xylenol constituted the major portion of the fraction.

A series of seven fractions, boiling between 219.4° and 219.6°C (427°F), was studied because this appeared to be a boiling point plateau and also had a nearly constant index of refraction. One cubic centimeter of the third fraction was converted to the aryloxyacetic acid melting at 130° to 131°C (267°F). The portion not converted at this time solidified on cooling and was found to be the main constituent of this fraction. Further purification showed the melting point of this phenol to be 70° to 73°C (158° to 163°F). 2,4,6-Trimethylphenol boils at 219.5°C (426°F) and has a reported melting point of 69°C (156°F). The melting point of the aryloxyacetic acid has been reported as 131°C (268°F) and 143° to 144°C (290°F). In view of conflicting reported melting points, this trimethylphenol was synthesized and found to melt sharply at 72°C (162°F), whereas its aryloxyacetic acid was found to melt at 149° to 150°C (300° to 302°F) and its phenylurethane at 141.5°C (286°F). The phenol isolated was not mesitol (2,4,6-trimethylphenol).

Another sample of the before-mentioned third cut was treated with 1 equivalent of dilute alkali and steam distilled in the hope that if 2,4,6-trimethylphenol were present, it could be separated from others by this method. The steam-distilled phenol now showed $n_D^{20} = 1.5330$, whereas that of the phenol which did not hydrolyze was only 1.5273; but there was 13 cc of the first and only 2 cc of the second. The distilled portion gave no aryloxyacetic acid derivative melting near 130°C, but did yield one melting at 86° to 87°C which is the reported melting point of that derivative of 1,3,5-xylenol. There was only a low yield of this phenol. The filtrate from this preparation yielded another oxyacetic acid derivative, melting at 74° to 75°C which was shown by mixed melting point with a synthetic sample to be *m*-ethylphenol. This phenol was further identified by melting point and mixed melting point of the phenylurethane melting at 138°C.

Steam distillation of the rest of this series of fractions boiling between 219.4° and 219.6°C (426°F), after they had been treated with 3 equivalents of sodium hydroxide, resulted in slight separation, which was sufficient, however, to permit isolation of 2-ethyl-4-methylphenol and its identification through the melting point and mixed melting points of its aryloxyacetic acid and urethane derivatives with those of synthetic 2-ethyl-4-methylphenol.

An attempt was made to effect a separation of the other xylenols and higher-boiling phenols by using the Brückner scheme of preparing the sulfonic acid of the phenols and decomposing these by steam distillation at definite progressively higher temperatures to obtain a series of phenols; but the separation obtained was not satisfactory and no further attempt has been made to identify other phenols in the large volume of higher-boiling phenols available.

A different type of source material was studied by Henson, in 1936,[12] when he investigated the acidic contents of a 5 gal sample of "spent doctor solution" obtained in refining of Texas mixed cracking and straight-run gasoline. He removed hydrocarbons by steam distillation of the basic solution, then acidified the residual liquid, and again steam distilled to remove steam volatile acids. The steam distillate was found to be heavier than water. The portion which was not volatile in steam represented only about 1% of the total volume and was not investigated.

The distillate was made strongly alkaline and again steam distilled to remove remaining hydrocarbons, but only a very small volume of distillate was obtained. The acids were again liberated and steam distilled to remove salts. The yield of volatile acids at this stage was 250 cc from the original 5 gal sample.

Attempted separation of carboxylic acids by careful fractional neutralization showed that all fractions were heavier than water and had the high index of refraction and density expected of phenols. Careful fractional distillation at atmospheric pressure led to the data of Table 21.

Table 21. **Properties of Henson's Phenolic Fractions**

Cut	*Boiling Point at 760 mm* °*C*	°*F*	*Volume* *cc*	n_D^{25}	d_4^{25}
1	185–195	372–383	96	1.5422	1.0448
2	195–200	383–392	32	1.5413	1.0381
3	200–205	392–401	27	1.5395	1.0294
4	205–220	401–428	27	1.5349	1.0191

Fraction 1 was found to contain no carbolic acid, but at least

40% of *o*-cresol identified by its cineol and by its α-naphthylurethane derivative. Fraction 3 was found to contain *m*-cresol, identified through a 12% yield of trinitroderivative, and *p*-cresol which was sulfonated and gave an 11% yield of barium salt. Fraction 4 probably consisted of xylenols, but these were not isolated and identified. Apparently, this is the first time that phenols have been isolated from this refinery by-product.

Potts and Morrow [18] studied an alkaline sodium salt solution obtained in treating cracking-process products at Pt. Isabel, Texas. They acidified the solution, extracted with ether, and separated the resulting layers. Evaporation of the ether left a peculiar black viscous liquid which was fractionated at 25 mm pressure in an atmosphere of nitrogen. The distillate obtained at this pressure at 162° to 182°C (324° to 360°F) hardened to a white solid melting, after two recrystallizations, at 170° to 171°C (338°F). This solid was shown by its melting point, as well as by those of its acetate and benzoate, to be trimethylhydroquinone. This appears to be the only dihydric phenol isolated from petroleum products and since no such solid has been reported in other studies, it does not seem to occur generally.

Obviously, a whole series of phenols, probably with no long side chains, and possibly also naphthols, can be isolated from straight-run and especially from cracking-process distillates. At least, they have been extracted from crude Polish petroleum and they have been found in water produced with some American crude oil. Until there will be an urgent reason for isolating and identifying additional higher-boiling phenols and naphthols, it seems unlikely that such work will be done, in this country at least.

Bibliography

1. *Anon, Chem. Ind.* **60,** 48 (1947).
2. *Anon, Met. and Chem. Eng.* **55,** 312 (1948).
3. Aries, R. S., and S. A. Savitt, *Chem. Eng. News* **28,** 316 (1950).
4. Brückner, H., *Erdöl* und *Teer* **4,** 580 (1928).
5. Campbell, S. E., *Refiner* **14,** 381 (1935).
6. Catlin, L. J., *Ind. Eng. Chem.* **18,** 743 (1926).
7. Field, E., F. H. Dempster, and G. E. Tilson, *Ibid.* **32,** 489 (1940).
8. Gallo, S. G., C. S. Carlson, and F. A. Biribauer, *Ibid.* **44,** 2610 (1952).
9. Givens, R. C., M. A. Thesis, The University of Texas, 1941.
10. Golumbic, C., *J. Am. Chem. Soc.* **71,** 2627 (1949).

11. Golumbic, C., *Anal. Chem.* **23,** 1210 (1951).
12. Henson, D.D., M.A. Thesis, The University of Texas, 1936.
13. Holzman, E., and S. Pilat, *Brennstoff-Chem.* **11,** 409 (1930).
14. Kruber, O., and W. Schmieden, *Ber.* **72,** 653 (1939).
15. Mabery, C. F., *Proc. Am. Phil. Soc.* **1903,** 36.
16. Natural Gasoline Assn. of America, Proceedings Corrosion Committee, Numerous reports on phenols in condensate well products (1945–1950).
17. Nenitzescu, C. D., D. A. Isacescu, and T. A. Volrap, *Ber.* **71B,** 2056 (1938).
18. Potts, W. M., and H. N. Morrow, *Ind. Eng. Chem.* **31,** 1270 (1939).
19. Schutze, H., W. Shive, and H. L. Lochte, *Ind. Eng. Chem. Anal. Ed.* **12,** 262 (1940).
20. Story, L. G., and R. D. Snow, *Ind. Eng. Chem.* **20,** 359 (1928).
21. Tanaka, Y., and Kabayashi, R., *C. A.* **22,** 1032 (1928).
22. Williams, M., and G. H. Richter, *J. Am. Chem. Soc.* **57,** 1686 (1935).

CHAPTER THIRTEEN

THE ALIPHATIC OR FATTY ACIDS

If we accept the theory that fatty acids are in some manner decarboxylated underground to produce at least part of the petroleum hydrocarbons,[6, 18] or the theory of Brooks [4] and the older theory of Tanaka [22, 23] that a main source of cyclic compounds in petroleum are the unsaturated fatty acids, we should expect fatty acids to be of quite general occurrence in petroleum and to constitute perhaps the main type of acids in some field in which these transformations have not been carried as far as in others.

As a matter of fact, the aliphatic acids have been reported by a large number of workers studying acids from widely separated fields, such as in Eurasian, Japanese, California, and Texas petroleum and have been found in condensates obtained from high-pressure gas wells.

Apparently, all workers who have studied petroleum acids with less than ten carbons have reported the occurrence of the lower fatty acids, but since the commercial naphthenic acids have a considerably higher molecular weight and contain very small amounts of acids with less than ten carbons, relatively few studies of the lower acids have been undertaken. In the common naphthenic acid range of ten to twenty carbon atoms, the aliphatic and naphthenic acids should be present side by side and their separation is still largely an unsolved problem. Certainly, fractional distillation alone, which has been almost universally used in separating the naphthenic acids, cannot be expected to separate the aliphatic from the naphthenic acids. Consequently, very few of the workers would have been able to separate the aliphatic from the naphthenic acids if they had both been present.

In spite of this situation, we find a number of reports — especially ones dealing with Russian and Roumanian acids — in which refractive index determinations show values typical of naphthenic acids and the claim is made that the acids are pure.[1,2,9,14,15,16,24,25] Many of these acids might contain low concentrations of phenols, which could bring up the average index of refraction of a mixture of aliphatic, naphthenic, and phenolic compounds to the typical naphthenic acid range. If, however, we assume that some phenols are present, we should also assume that at least small amounts of hydrocarbons are also present, and their low index of refraction counteracts the effect of the high values for the phenols. Furthermore, we find that a number of these workers report carbon and hydrogen analyses that agree well with the expected values for pure naphthenic acids — this again could be due to an error, since combustion analyses of naphthenic acids require considerable care and experience to obtain really accurate results (see Chapter 6), or more probably it could represent average results on mixtures of aliphatic and naphthenic acids or phenols. However, in view of the number of similar reports, we must assume that, in some fields at least, the higher acids consist essentially of naphthenic acids alone. We know that in some cases, especially in some California and Japanese acids, considerable concentrations of aliphatic acids occur even in the higher-boiling range of more than twelve carbons.[22, 23, 24, 25]

In one of the first studies of petroleum acids, Markownikoff and Oglobin,[14] in 1883, mentioned isolation of acetic acid and the presence of some higher fatty acids in fractions obtained on distillation of Caucasian crude oil. Shidkoff,[21] in 1899, found lower aliphatic acids in the study of Grozny acids. Most of the reports on identification of aliphatic acids have appeared since 1925, especially during the decade of 1930 to 1940.[3, 10, 17]

Tanaka and Kuwata,[26] in 1928, culminated the results obtained in a series of earlier papers by Tanaka and others [24, 25] by announcing the isolation and identification of palmitic, stearic, myristic, and arachidic acids, which separated as solid acids from fractions of Ishikari petroleum acids and determined that, in this case, the total concentration of aliphatic acids in this boiling-point range is 7.7% of the total acidity. Pilat and Holzmann,[9] in 1933, reported the identification of the same solid acids in Boryslaw acids, but Tyutyunikoff and Pervukina [27] failed to isolate any of them from Baku acids; Lapkin [11] could not find them either, this time in Grozny acids. Von Braun, in his study of Roumanian acids, identified 3-ethylpentanoic and 3-methylhexanoic acids and decided that all acidity below the C_7 acids was due to aliphatic acids which were present, mixed with naphthenic acids from C_7 to C_{10} and were absent above C_{10} acids.

Chichibabin,[5] after a prolonged and elaborate study of Baku petroleum acids boiling below 215°C (419°F), decided that the primary aliphatic acids predominated in the Baku region while above 215°C (419°F), the aliphatic acids rapidly decreased in concentration until, above 260°C (500°F), the acidity was due entirely to naphthenic acids. He was convinced, after a study of density and index of refraction data reported much earlier by Aschan [1, 2] and by Markownikoff,[14, 15] that the acids studied by them also had high concentrations of fatty acids.

Richter and coworkers, in 1935,[28] reported the isolation and identification of *n*-heptanoic, *n*-octanoic, and *n*-nonanoic acids from West Texas cracking-process material. A number of later workers also studied cracking-process acids, perhaps because it was very hard to obtain samples of acids not derived from cracking-process. Since, however, fatty acids, in this case, may well have been formed by cracking of higher acids, these studies will not be discussed.

Lapkin, in 1932,[10] and later workers determined the total aliphatic acid content of petroleum acids, but did not attempt to identify any of them. Therefore, their results are significant only because they show that aliphatic acids are present and sometimes indicate what percentage of the total acidity is due to aliphatic acids. Lapkin decided that Grozny acids contained about 3.6% of fatty acids in the boiling range of 100°to 200°C (212° to 392°F) at 20 mm pressure, or roughly in the range 200° to 300°C (392° to 572°F) at atmosphermic pressure.

Pilat and Holzmann [9] found fractionation of the methyl esters of the higher petroleum acids obtained from spindle oil boiling above 350°C (662°F) led to aliphatic acids with sixteen to twenty carbon atoms, but these acids could not be purified as such. Consequently, they were fractionally converted to their magnesium salts by adding $Mg(Ac)_2$ solution in small increments and filtering off the precipitate formed. Four systematic fractionations yielded the acids which had been reported by the Japanese.

Nenitzescu *et al.* (1938) [17] identified 4-methylpentanoic and 5-methylhexanoic acids from Ploesti naphthenic acid over the methyl esters and the amides in the same project in which they identified a series of simple phenols and simple naphthenic acids. In this project, they deliberately worked with the lower-boiling and simpler acids to be able to isolate and identify individual acids from petroleum.

A group studying petroleum acids at the University of Texas developed methods of separating aliphatic and mixtures of aliphatic and naphthenic acids and isolated and identified a series of fatty acids between 1938 and 1942. The simpler ones were identified by careful fractionation of the acids and then of the methyl esters of these fractions, and the more complex mixtures were identified by a combination of fractional distillation and systematic fractional neutralization and liberation.

With the exception of the work of Schutze,[20] concerned mainly with development of methods, the study was all done on a large sample of acids extracted from gasoline and kerosene from Signal Hill, California, straight-run fractions. Hancock [7] worked with the lowest boiling 3.6 liters of acidic material obtained on fractionating 70.2 liters of crude acids. The water layer of the 3.6-liter cut amounted to 760 ml and consisted of a water solution of soluble fatty acids. Careful and repeated fractionation of this layer yielded 20 ml of organic acids which were combined with the large organic layer of the original cut, while the water layer (740 ml) was made alkaline and distilled to dryness. Half of the resulting powder was heated with syrupy phosphoric acid and distilled to yield 4.5 g of low-boiling acids. Tests of a few drops of this were positive for formic acid, as indicated by reduction of silver nitrate and of mercuric chloride. The rest of the 4.5 ml portion was converted to methyl esters by transesterification with methyl hexanoate in the presence of a small amount of concentrated sulfuric acid. This yielded a 3-g batch of methyl esters which was fractionated through a 70 cm semimicro column and yielded five fractions found to contain mainly acetic acid.

The main batch of lower acids was separated into five neutralization fractions by fractional neutralization of a petroleum ether solution of the acids. This separation actually achieved a separation largely on the basis of molecular weight, as seen by the fact that the equivalent weight increased regularly from 104.5 to 157. When this operation was repeated separately on the first two fractions, the first fraction of an equivalent weight of 104.6 yielded three fractions of 90.6, 98, and 109 equivalent weights, and the second one, with an equivalent weight of 115.6, gave similar cuts of 107, 113.1, and 126.5 equivalent weights. When these fractions were separately esterified, part of the unesterified material crystallized and was found to be dimethylmaleic anhydride — the first unsaturated anhydride isolated from petroleum. This was also found previously by Schutze in Texas petroleum acids.[20] The methyl esters, on careful fractionation, yielded propionic, butyric, *n*-butyric, isovaleric, and *n*-valeric acids — all identified by physical properties and melting points and, usually, by mixed melting points of various derivatives. Methylethylacetic and trimethylacetic acids were not found, even though a search for them was conducted.

A naphthenic acid isolated at this stage out of the unesterified material was camphonanic (1,2,2-trimethylcyclopentane carboxylic) acid, a solid melting at 194° to 195°C (382°F), which was isolated in larger amounts than the other acids — probably because it could not be esterified and because it was a solid and thus easy to isolate.

Quebedeaux, Wash, Ney, and Crouch [19] worked with the higher-boiling aliphatic acids of the batch studied by Hancock. They esterified all the acids that could be esterified in the usual manner in three or four separate treatments and then fractionated the methyl esters very carefully, first through a 12-ft Berl saddle-packed adiabatic column under high reflux, after a preliminary distillation through still heads giving only a few plates. The preliminary fractions were added to the still pot of the 12-ft column as the boiling point at the bottom of the column approached that of the preliminary cut. The curve in Figure 8 shows the points at which additional esters were introduced into the still pot by sharp breaks at fractions 220, 355, etc. The peaks in the index of refraction correspond to the naphthenic acids.

Another search was made for trimethylacetic and methylethylacetic acids to determine whether all of the isomeric pentanoic acids were present (perhaps in equilibrium) or not, but no evidence for their presence could be found. Isolation and identification of 2- and 3-methylpentanoic and 2-, 3-, 4-, and 5-methylhexanoic acids were achieved, but no other isomeric seven-carbon acid was found, except the normal heptanoic acid which was present in a large amount (2 liters of methyl ester). The normal octanoic and nonanoic acids were located from the curve of Figure 8 and again were present in much larger amounts than the isomeric acids and thus were fairly easy to separate by fractional neutralization. No effort was made to identify other eight-carbon or nine-carbon fatty acids, as their separation from each other would have been a tedious task with no important effect on the possible results. Separation efforts from this stage on were directed to identification of the naphthenic acids which were indicated by the peaks in the index of refraction in Figure 8.

Ridgway and coworkers [8] studied Texas acids and separated them through fractionation of the methyl esters. They identified formic, acetic, propionic, isobutyric and isovaleric acids, and thiophenol and thiocresol.

An interesting development, which is of importance in connection with the origin of petroleum acids and which will be discussed in greater detail, is the isolation and identification of a series of fatty acids in the liquid phases produced by certain high-pressure gas wells. As might perhaps be expected, the main acids produced are acetic, propionic, and butyric. Since production is usually at over 2,000 lb per sq in. pressure and at bottom-hole temperatures of up to 85°C (185°F), aliphatic acids of eight or nine carbon atoms are also produced in small amounts, but these have not been identified.[12, 13]

While, then, a number of the lower liquid and a few higher solid aliphatic acids have been identified, several of them in widely separated fields and laboratories, there is as yet no definite evidence that fatty acids with ten or more carbons exist in all petroleums, or that the C_{10} to C_{14} aliphatic acids exist in any typical naphthenic acid. The solid acids had fourteen or more carbons; the liquid ones had nine or less carbons. There are reports that most of the Eurasian naphthenic acids contain only naphthenic acids in that range. It is true that separation of aliphatic from naphthenic acids by fractional distillation cannot be expected. Therefore, earlier methods could not have separated any aliphatic acids that might have been in the Eurasian naphthenic acids; but the index of refraction and, particularly, the index of refraction times density ($n \times d$) value of many of these acids falls at least partially in the naphthenic acid range. This could be due to the presence of phenols which are not easily removed or to bicyclic or polycyclic acids, which would raise the average index of refraction to normal values even if aliphatic acids were present. Analytical results, along with density or index of refraction (usually only one is given), lead to the conclusion that some of the Roumanian and Polish acids do not contain appreciable amounts of aliphatic acids with ten or more carbon atoms.

This question could be settled only by careful and tedious work on a number of different samples of acids in the ten-to-fourteen-

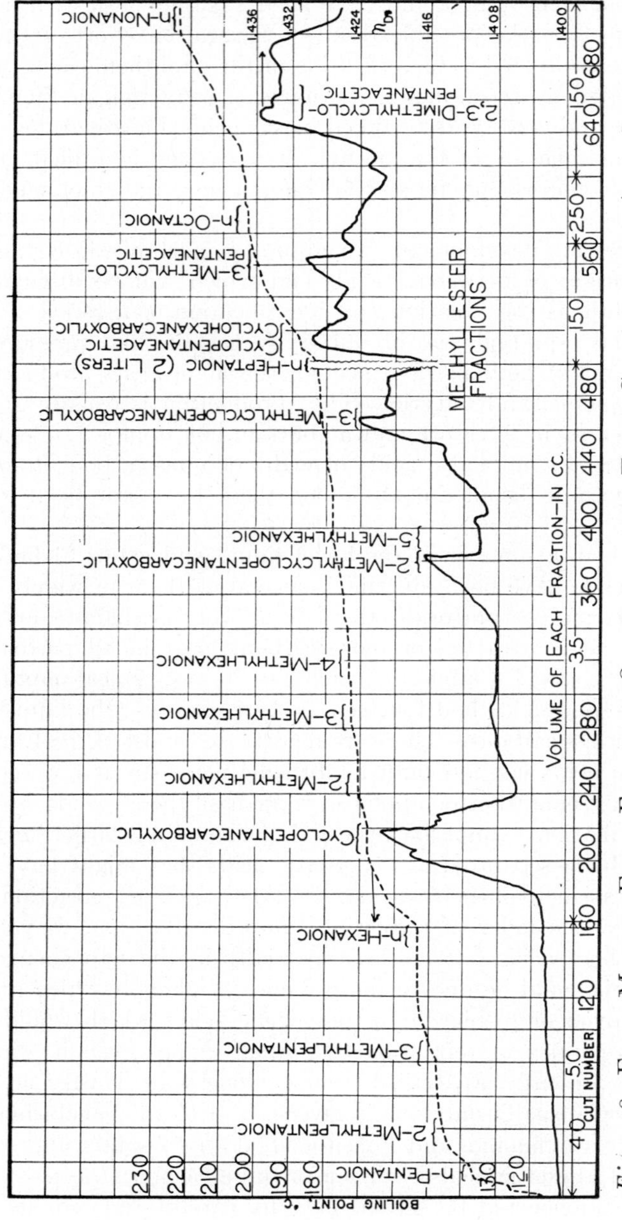

Figure 8. Final Methyl Ester Fractions Obtained by the Texas Group and the Acids Isolated

[Reprinted by permission from *J. Am. Chem. Soc.*, **65**, 768 (1943)]

carbon range. This would determine once and for all whether isolatable amounts of aliphatic acids of this range are present in all or any of these acids.

Bibliography

1. Aschan, O., *Ber.* **23,** 871 (1890).
2. Aschan, O., *Ibid.* **24,** 2711 (1891).
3. von Braun, J., *Ann.* **490,** 100 (1931).
4. Brooks, B. T., *Science* **111,** 650 (1950) and 114, 240 (1951).
5. Chichibabin, A. E., *et al.*, *Chim. et ind.* **17,** 306 (Special Number, March 1932).
6. Engler, C., *Die Chemie u. Physik des Erdöls,* Volume I, S. Hirzel, Leipzig, 1913, pages 427–455.
7. Hancock, K., and H. L. Lochte, *J. Am. Chem. Soc.* **61,** 2448 (1939).
8. Henderson, L. M., G. W. Ayers, and C. M. Ridgway, *Oil and Gas J.* **38,** 114, 118, 121 (1940).
9. Holzmann, E., and S. Pilat, *Brennstoff-Chem.* **14,** 263 (1933).
10. Lapkin, I. I., *C. A.* **26,** 6106 (1932).
11. Lapkin, I. I., *Ibid.* **34,** 611 (1940).
12. Lochte, H. L., C. W. Burnam, and H. W. H. Meyer, *Pet. Eng.* Aug. 1949, page 225.4.
13. Lochte, H. L., *Condensate Well Corrosion,* N. G. A. A. Condensate Well Corr. Committee, Natural Gasoline Assn. of America, Tulsa, Okla., 1953.
14. Markownikoff, W., and W. Oglobin, *Ber.* **16,** 1876 (1883).
15. Markownikoff, W., *Ann.* **307,** 367 (1899).
16. Müller, J., and S. Pilat, *Brennstoff-Chem.* **17,** 461 (1936).
17. Nenitzescu, C. D., D. A. Isacescu, and T. A. Volrap, *Ber.* **71B,** 2062 (1938).
18. Petrov, A. D., and I. Z. Ivanov, *J. Am. Chem. Soc.* **54,** 240 (1932).
19. Quebedeaux, W. A., G. Wash, W. O. Ney, W. W. Crouch, and H. L. Lochte, *Ibid.* **65,** 767 (1943).
20. Schutze, H. G., W. Shive, and H. L. Lochte, *Ind. Eng. Chem. Anal. Ed.* **12,** 262 (1940).
21. Shidkoff, N., *J. Soc. Chem. Ind.* **1899,** 360.
22. Tanaka, Y., and S. Nagai, *J. Am. Chem. Soc.* **45,** 754 (1923).
23. Tanaka, Y., and S. Nagai, *C. A.* **19,** 1135 (1925).
24. Tanaka, Y., and S. Nagai, *C. A.* **20,** 583 (1926).
25. Tanaka, Y., *J. Soc. Chem. Ind. Japan* **29,** 1 and 58 (1926).
26. Tanaka, Y., and T. Kuwata, *C. A.* **23,** 4051 (1929).
27. Tyutunikov, B., and J. Pervukhina, *Ibid.* **27,** 5956 (1933).
28. Williams, M., and G. H. Richter, *J. Am. Chem. Soc.* **57,** 1686 (1935).

CHAPTER FOURTEEN

VON BRAUN'S STUDY OF THE PROPERTIES AND REACTIONS OF NAPHTHENIC ACIDS

In 1927, Julius von Braun undertook a long-range ambitious program of research on naphthenic acids. He decided to enter what had been a rather discouraging field of research for a number of prominent chemists of whom we only name Markownikoff, Aschan, Komppa, Tanaka, Kuwata, Pyhala, Zelinsky, and Pilat. In spite of an enormous amount of tedious and time-consuming work, these and many others had succeeded only in establishing that the naphthenic acids are (1) saturated toward potassium permanganate and bromine, (2) monobasic cyclic carboxylic acids, and (3) mainly cyclopentane derivatives.

Apparently, he was attracted by this field because he felt especially qualified to carry out both the synthetic and analytical work involved, since he had had very extensive experience in both. He felt, further, that results obtained would be of interest and value to the rapidly expanding petroleum industry — partly because of their interest in the naphthenic acids as such, but mainly because he felt that the hydrocarbons and the acids would be found to be genetically related and that, aside from the carboxyl group, they should be similar in structure.[8, 19] There was at that time an awakening of interest in the nature of the hydrocarbons which make up petroleum oils (establishment of the Rockefeller funds for fundamental petroleum research and the start of American Petroleum Institute projects, for instance), and von Braun felt it should be easier to determine what is in petroleum by isolating and identifying the chemically active acids and then decarboxylating them or converting them to esters and these to the hydrocarbons over the alcohols and iodides, than it would be to separate, isolate, and identify the unreactive individual hydrocarbons of petroleum. We know now that American petroleum chemists chose what von Braun considered the hard road and proceeded to separate and identify the hydrocarbons.

Unfortunately, both in his appraisal of the acid problem and in his estimate of the difficulty of separating the hydrocarbons, he underestimated the power of modern methods of separation (and so made his own acid problem inordinately difficult by not using such methods extensively) and overestimated the difficulty of separating the hydrocarbons. It should be pointed out, however, that for some reason, he felt that the only results that would be of interest would be on compounds with at least ten carbons, and in this range, even the best methods of separation have still not been successful after a quarter of a century of feverish development of such methods.

As a result of this situation, which was partly brought on by von Braun's inability to obtain the funds required to carry on modern research of this type, we shall find it best to treat his naphthenic acid studies in two chapters — one dealing with his very important and many-sided experiments with natural and synthetic naphthenic acids in which these were characterized by various methods, were degraded to compounds of fewer carbon atoms, were converted to various derivatives, and were tested to determine the number of carbons between the carboxyl group and the ring; and the other, less important one, dealing with his attempts to identify individual bicyclic, as well as monocyclic, acids.

In the more important phase of his work — a study of properties and reactions of natural and synthetic alicyclic acids — von Braun used methods of degrading the acids to amines of lower molecular weight, of converting the amines to acids or ketones of still lower molecular weight, or of converting the amines after further fractional distillation into solid salts which might be purified by recrystallization.

These operations on natural naphthenic acid mixtures seemed to indicate that there was usually at least one methylene group between the ring and the carboxyl group, but it would take an inordinate amount of time to carry out a sufficient number of such degradations by this method to arrive at any general conclusions on this part of the structure of the acids; therefore, he made use of the imide-chloride method that had been developed in his laboratory [3, 6, 16] to determine in a relatively short time whether naphthenic acids generally had two hydrogens on the α-carbon atom. To obtain still more information regarding the length of the chain carrying the carboxyl group, he used his lactone method [6] which should tell whether there are at least three carbons between the carboxyl group and the ring. He wanted to use these tests to determine if the carboxyl group was attached directly to the ring or if there was at least one carbon between the carboxyl group and the ring. If he found at least one carbon, then he tried to determine whether there were as many as three methylene carbons. In addition, he studied the preparation and properties of the naphthenyl amines obtainable by the K. F. Schmidt degradation of the acids.[6, 10, 14, 17] These main studies, and a number of minor investigations that may or may not have practical importance, should probably be considered von Braun's basic contributions to our knowledge of the naphthenic acids.

Since he realized very soon that the fractionation methods which he employed were not able to separate the acids or any of their derivatives to the point where they could be identified, he studied available methods of degrading acids to compounds with a lower number of carbons.

Among the methods studied briefly to determine their suitability for degradation of naphthenic acids was what, in a modified form, is now known as the Barbier-Wieland method of degradation.[4, 6, 25] This reaction, as used by him, involved the following steps:

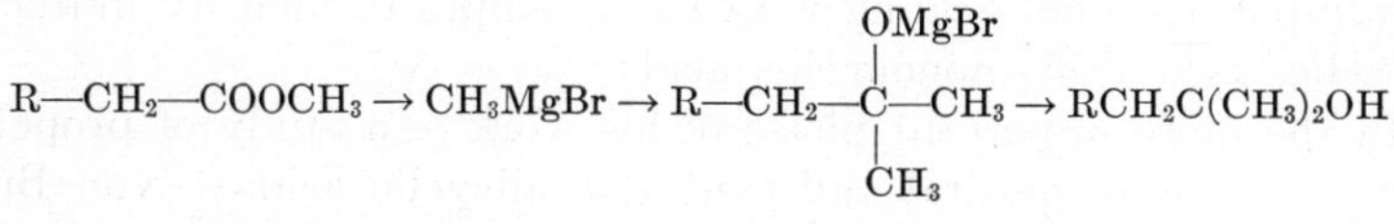

$$R{-}CH_2{-}COOCH_3 \rightarrow CH_3MgBr \rightarrow R{-}CH_2{-}\overset{\displaystyle OMgBr}{\underset{\displaystyle CH_3}{\overset{|}{\underset{|}{C}}}}{-}CH_3 \rightarrow RCH_2C(CH_3)_2OH$$

The tertiary alcohol formed yielded, on dehydration, not only the olefin:

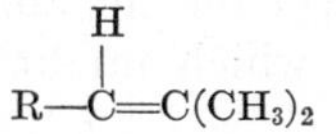

$$R{-}\overset{\displaystyle H}{\overset{|}{C}}{=}C(CH_3)_2$$

which was wanted, but also:

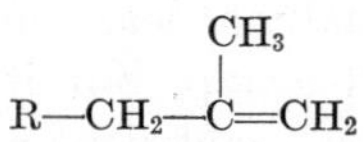

$$R{-}CH_2{-}\overset{\displaystyle CH_3}{\overset{|}{C}}{=}CH_2$$

On oxidation or ozonolysis, this would yield a mixture of an acid and a methyl ketone. If the more modern method using phenyl-Mg-Br had been employed, he might have been more favorably impressed, since acids with the carboxyl on the ring would have yielded ketones while others would have given acids with n-1 carbons which could have been carried through the same operations as before after separation from the ketones.

The only degradations that he considered valuable were those in which the acid or amide was degraded to the next lower amine. Of the various methods available for this,[1] he selected the then new K. F. Schmidt method [6, 10, 11, 14, 15, 17, 21, 23] because of its simplicity and because the yield of amine from naphthenic acids was 70 to 90%, while the old degradation of the amides with sodium hypobromide gave less than 50% yield and was not as simple as the Schmidt method.

The reactions involved in the Schmidt method are:

$$R\text{—}COOH + HN_3 \xrightarrow{\text{conc. } H_2SO_4} RNH_2 + CO_2 + N_2$$

Presumably the isocyanate $R\text{—}CH_2\text{—}N{=}C{=}O$ is an intermediate here as in the other degradations in which an acid is degraded to an amine with n-1 carbon atoms. Aside from danger of inhaling the very poisonous hydrazoic acid (which means that the reaction must be carried out under a good hood), this degradation is very simple. The hydrazoic acid is always used in a solvent, usually chloroform or benzene. It may be generated and used in the same step by adding about 2 moles of sodium azide to the mixture of chloroform or benzene, concentrated sulfuric acid, and the acid to be degraded or it may be taken up in chloroform or benzene, titrated to determine the strength of the hydrazoic acid which is then added to a solution of the acid to be identified in the same solvent. A large excess of concentrated sulfuric acid is used and the unknown acid is present in dilute solution. The resulting amine is isolated, dried, and distilled.

He found that the boiling range of the amine from a fraction of acids was much larger than that of the original acids.[6] Therefore, he decided that it should be possible to obtain additional separation by distillation. His attempts in this direction failed to yield any pure amine, however.

Von Braun then attempted to obtain a pure amine by recrystallization of the amine oxalate or the salt of naphthalene disulfonic acid, but again he was not able to obtain a pure compound, although he felt he had achieved a considerable amount of purification in this manner.[6]

Another method which he tested, but apparently did not use, was the separation of the amines obtained from mixtures of acids by difference in the rate of hydrolysis of the substituted benzamides, $R\text{-}NH\text{-}CO\text{-}C_6H_5$, obtained in the reaction of benzoyl chloride with the amines. If the amine were a hindered one, hydrolysis to RCl and $C_6H_5CONH_2$ would be slow, while unhindered benzamides would be hydrolyzed rapidly. Trial on typical compounds showed that there is a large difference in reaction rate in extreme cases, but the method obviously separated only these two types and there did not appear to be many hindered amines in these mixtures.[11]

Since he did not succeed in isolating a pure amine, he concluded that additional degradation of the amines should lead to mixtures of lower-molecular-weight compounds which, being less complex mixtures, might be separable. If ammonia could be eliminated from the amines as hydrogen chloride and water are eliminated from alkyl chlorides or alcohols, the resulting olefin could be converted to the next lower acid or ketone, depending on the structure of the original acid.

$$RCH_2CH_2COOH \rightarrow RCH_2CH_2NH_2 \rightarrow RCH{=}CH_2 \rightarrow RCOOH \qquad (1)$$

or,

$$\begin{matrix} -CH_2 & & -CH_2 & & -CH_2 & & -CH_2 \\ \diagdown & & \diagdown & & \diagdown & & \diagdown \\ CH{-}CH_2COOH & \rightarrow & CHCH_2NH_2 & \rightarrow & C{=}CH_2 & \rightarrow & C{=}O \\ \diagup & & \diagup & & \diagup & & \diagup \\ -CH_2 & & -CH_2 & & -CH_2 & & -CH_2 \end{matrix}$$

The resulting acids could be separated from the ketones readily, the ketones identified, and the acids further degraded.

A method that gets essentially this change from acid to amine to olefin is the amine-phosphate method which was studied and used by von Braun and has seen use since.[15, 20, 22, 24] He apparently concluded that the method (which consists of heating a mixture of concentrated phosphoric acid, sodium phosphate, and the amine and finally distilling the olefin which is formed) was fairly satisfactory but experience in the author's laboratory on pure synthetic naphthenic acids and on close-cut fractions in the C_6 to C_{10} range indicated that olefin formation was accompanied by rearrangements. Thus the method seems to be of doubtful value in establishing the structure of a compound.[20]

He finally turned to the Hofmann degradation, one of his favorite methods of cleaving nitrogen compounds — which had been found very valuable in much of his earlier work with alkaloids — exhaustive methylation and decomposition of the quaternary ammonium base to yield an olefin:

$$RCH_2CH_2NH_2 \rightarrow RCH_2CH_2N(CH_3)_3I \rightarrow RCH_2CH_2N(CH_3)_3OH \rightarrow RCH{=}CH_2 + N(CH_3)_3$$

Usually, the hydroxide is obtained by heating the iodide or bromide with AgOH, but von Braun not only obtained colloidal silver suspensions which were difficult to break, but also found that the higher naphthenic acids showed a strong tendency to eliminate methanol, when the quaternary ammonium base was heated, in addition to the desired decomposition to yield an olefin.[5, 6, 9] He decided that it should be possible to shift the reaction in favor of the olefin by using an excess of concentrated sodium or potassium hydroxide and found, on trial, that the yield was improved. Even then, however, it was necessary to treat again the $RCH_2CH_2N(CH_3)_2$ formed as by-product once or twice to improve the yield of olefin. Goodman, in the author's laboratory, found that two repetitions of the methylation and pyrolysis, using the concentrated-base technique, still gave a total yield of only 20 to 35%. Probably, with their enormous amount of experience with this reaction, the workers in von Braun's laboratory were able to get considerably better yields, although here as elsewhere few specific yields are given.

He later used another modification of this degradation to obtain an olefin with the same number of carbons as the original acid and, therefore, n − 1 carbons in the final acid or ketone, but apparently did not make use of the method extensively. In this method, he converted the acid to the ester, reduced this to the alcohol, RCH_2OH, changed this to the bromide, RCH_2Br, which was treated with trimethyl amine to yield $RCH_2C(CH_3)B_3r$. This was then converted to hydroxide, treated with concentrated base, and decomposed in the usual manner to yield the olefin.[6]

The olefins obtained by von Braun by either method were subjected to ozonolysis to yield the lower acid or ketone. Since the amount of olefin obtained from the fractionated amine was not large and the amount of olefin ozonized probably was never over 10 to 20 g, the fact that ozonolysis was carried out on small batches was no serious objection — particularly since, at this stage of the work, any fraction of olefin would represent such a great expenditure of time and effort that no worker would want to risk the loss of a large portion of expensive olefin in any one experiment.

Since the main product obtained on ozonolysis was usually a mixture of ketones, he concluded that the original acids must have been substituted acetic acids:

```
—CH2
    \
     CH—CH2—COOH
    /
—CH2
```

which would yield ketones. Although he admitted that acids having the carboxyl group attached directly to the ring were present in naphthenic acids, he apparently ignored them in most of his work and made no serious attempt to isolate any of the acids which should have been obtained from these on ozonolysis. Probably he recognized that most of such acids would yield mixtures of dicarboxylic acids and ketoacids whose identification he did not wish to attempt. Since they must have been present, nevertheless, they complicated his separation problems and this was a serious weakness in his degradation scheme involving elimination of two carbons in each series of reactions.

Von Braun's conclusion that the original naphthenic acids must have been substituted acetic acids was arrived at after a very tedious series of operations on his petroleum acids and there were not enough data to warrant any general conclusion concerning the structure of naphthenic acids. In the hope of finding a shorter method which would make it possible to examine a large number of different acids, he tested and used a procedure which had been developed in his laboratory in 1927.[3, 6, 16] In this method, the acid is changed to its acid chloride which is reacted with ethyl amine to yield the ethylamide, $RCH_2CONHC_2H_5$. With phosphorus pentachloride, this will give first the imide chloride, $RCH_2CCl{=}NC_2H_5$, then the compound $R{-}CCl_2CCl{=}N{-}C_2H_5$, which is finally hydrolyzed to the dichloro compound, $RCCl_2CONHC_2H_5$, which on analysis shows two chlorines.

Secondary acids of the type:

```
—CH2
    \
     CHCOOH
    /
—CH2
```

would be changed to:

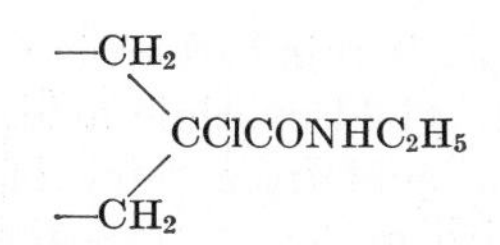

with one chlorine atom and tertiary acids containing no α-hydrogen would not be converted to a chlorinated compound and so would show zero chlorine. Von Braun multiplied the apparent number of chlorines by 100 and called this the chlorine number, thus the primary acids would have a chlorine number of 200, the secondary acids one of 100, and the tertiary acids, of course, zero. An equimolecular mixture of secondary and primary acids would show a chlorine number of 150. Most of his fractions gave chlorine numbers above 150, often near 200, but in some cases, the treatment with PCl_5 had to be repeated to get such high values. He concluded that most of the fractions contained largely acids with two α-hydrogens, i.e., at least one methylene group between ring and carboxyl group. In the absence of data on yields based on original acid fractions, it is difficult to decide whether his conclusions are valid or whether perhaps the yield of chlorine compounds of primary acids is much better than that of secondary acids. In view of the fact that most of the acid fractions readily esterified in the ordinary manner, the presence of any considerable concentration of tertiary acids is unlikely and these may be ignored, at this stage at least. The chlorine number has been used not only by von Braun but by others and appears to be a valuable tool in the study of naphthenic acids.[2, 6, 22, 24]

After von Braun had found that, in the few cases studied by the tedious degradation procedure, some of the acids had one and only one methylene group on the carboxyl side chain and in a number of other cases, the acids had at least one methylene group in the chain, he attempted to determine by some short method if generally such acids have only one methylene group. The best method he could find was a rather long one giving low yields, but he did use it in a few cases and decided that there was some acid present with at least three methylene groups between ring and carboxyl group.[6]

The method makes use of the following series of reactions:

$$RCH_2CH_2CH_2COOH \rightarrow RCH_2CH_2CHBrCOBr \rightarrow RCH_2CH_2CHBrCOOH$$

$$\rightarrow RCH_2CH{=}CHCOOH \rightarrow RCH{=}CHCH_2COOH \rightarrow R{-}\underbrace{CH{-}CH_2{-}CH_2{-}C}_{O}{=}O$$

to yield a lactone. If the carbon atom next to the R group still has a hydrogen atom on it, i.e., if this carbon is not part of the ring, the lactone can be oxidized to yield succinic acid. Von Braun obtained small amounts of succinic acid from some of his acids and so concluded that molecules with at least three methylene groups in the chain were present. He found that the higher naphthenic acids, unfortunately, in some way reverted back to the saturated acid from the bromoacid when an effort was made to form the α–β unsaturated acid by elimination of HBr. This peculiar reaction has apparently not been studied by others, but whatever the mechanism may be, the yield of unsaturated acid was considerably reduced — a serious matter when highly fractionated acids serve as samples to be tested. Consequently, he used the method for only a few cases and preferred to get the same type of information by degrading the acids, two carbons at a time, via the amines and olefins until ozonolysis would yield a ketone, indicating that the ring had been reached. Although he said, after some experience, that this series of degradations was not difficult to carry out, workers in the author's laboratory have not found them very satisfactory, either in regard to yields or to time consumed.

Von Braun studied,[14, 17] but apparently made no serious effort to apply, a method by which acids could have been degraded, one carbon at a time, to mixtures of acids and ketones with n-l carbon atoms, thus avoiding the loss of ketones that would be obtained after elimination of an odd number of carbon atoms as in the amine-olefin method regularly used in his laboratory. This method, when applied to a substituted acetic acid, proceeds over the steps:

$$RCH_2COOH \rightarrow RCHBrCOOH \rightarrow RCHBrNH_2 \rightarrow RCHO$$

The aldehyde can be oxidized to the acid.

When applied to an acid with the carboxyl group attached directly to the ring, the reactions are:

```
—CH2              —CH2               —CH2              —CH2
    \                 \                  \                 \
     CHCOOH →          CBrCOOH →          CBrNH2 →          C=O
    /                 /                  /                 /
—CH2              —CH2               —CH2              —CH2
```

The acids obtained from the first type could be separated from the ketones obtained from the second; the ketones could be identified and the acids again degraded, until finally no acids were obtained on further degradation.

As a result of the development and testing of the methods just outlined, von Braun felt capable of starting with a well-fractionated and refined naphthenic acid sample and determine (a) whether or not the carboxyl group is attached directly to the ring, (b) whether there is at least one methylene group between ring and carboxyl group, and (c) whether there are at least three groups in the side chain carrying the carboxyl group. In addition, he felt able to degrade a given acid finally to a ketone which could perhaps be identified if the amine obtained in a previous step could not be. He felt sure that, if a mixture were degraded two carbons at a time, it would be possible to separate any cyclic ketone formed from the acids, identify the ketones, and degrade the acids through another series of reactions.

With these tools available for use, he was confident that he could now make progress in the identification of the acids where little or none had been made before. How adequate these methods proved to be when applied to poorly fractionated samples, such as those studied by von Braun and workers, we shall see in the next chapter. Meanwhile von Braun's painstaking work in the study of the fundamental reactions of the naphthenic acids gave us more useful information on the chemistry of these compounds than had accumulated from all previous work, even though some of his work should be repeated to confirm his results or to obtain information on yields that can be expected in some of his reactions.

Running parallel with his attempts to isolate, or at least to identify, a petroleum naphthenic acid, we find von Braun carried out a number of studies on methods of degrading acids,[5, 6, 13, 15, 16] methods of synthesizing monocyclic and bicyclic naphthenic acids,[5, 7, 14, 17, 18] and on the validity of methods of characterizing naphthenic acids.[11, 12, 13, 17] Syntheses were needed because few bicyclic acids of various types were known and a knowledge of the properties of such acids was almost essential if any bicyclic acid that might be isolated were to be identified. All of these contributions added to our knowledge of acids — possibly of naphthenic acids in petroleum — even though he never was able to make use of the information he gained.

Bibliography

1. Adams, R. *Organic Reactions*, Volume III, Chapters 7, 8, and 9, Wiley, New York, 1946.

2. Balada, A., and J. Wegiel, *C. A.* **31,** 2399 (1937).
3. von Braun, J., F. Jostes, and W. Munch, *Ann.* **453,** 114 (1927).
4. von Braun, J., and A. Heymons, *Ber.* **61,** 2277 (1928).
5. von Braun, J., *Ann.* **472,** 1 and 121 (1929).
6. von Braun, J., *Ibid.* **490,** 100 (1931).
7. von Braun, J., W. Keller, and K. Weissbach, *Ann.* **490,** 179 (1931).
8. von Braun, J., *Z. angew. Chem.* **44,** 661 (1931).
9. von Braun, J., and E. Anton, *Ber.* **64,** 2865 (1931).
10. von Braun, J., *Ibid.* **66,** 684 (1933).
11. von Braun, J., *Ibid.* **66,** 1373 (1933).
12. von Braun, J., *Ibid.* **66,** 1464 (1933).
13. von Braun, J., *Ibid.* **66,** 1499 (1933).
14. von Braun, J., *Ibid.* **67,** 218 (1934).
15. von Braun, J., and P. Kurtz, *Ibid.* **67,** 227 (1934).
16. von Braun, J., and H. Ostermayer, *Ibid.* **70,** 1004 (1937).
17. von Braun, J., E. Kamp, and J. Kopp, *Ibid.* **70,** 1751 (1937).
18. von Braun, J., *Oel u. Kohle* **13,** 799 (1937).
19. von Braun, J., *Ibid.* **14,** 283 (1938).
20. Goodman, H., Ph.D. Thesis, The University of Texas, 1951.
21. Nelles, J., *Ber.* **65,** 1345 (1932).
22. Ney, W. O., W. W. Crouch, C. E. Rannefeld, and H. L. Lochte, *J. Am. Chem. Soc.* **65,** 770 (1943).
23. Oesterlin, M., *Z. angew. Chem.* **45,** 536 (1932).
24. Shive, W., J. Horeczy, G. Wash, and H. L. Lochte, *J. Am. Chem. Soc.* **64,** 385 (1942).
25. Skraup, S., and E. Schwamberger, *Ann.* **462,** 141 (1948).

CHAPTER FIFTEEN

VON BRAUN'S STUDY OF ROUMANIAN PETROLEUM ACIDS

Von Braun's laboratory was actively engaged in the study of natural and synthetic naphthenic acids for about 10 years. During that time, a number of different samples of petroleum acids were investigated. While brief references to results obtained with some of the other samples will be made from time to time, only the main project — which was concerned with Roumanian acids — will be covered in this chapter, because practically all important results were obtained in the study of these acids. He was hoping to be able to have a chance to study acids isolated directly from different crude oils, as well as those obtained from distillation fractions of the same crudes, but in spite of strenuous efforts in Europe as well as in the United States where his friend, J. R. Bailey, tried to obtain such samples, von Braun received such duplicate samples from only one field, the Roumanian oil studied in this project. Contrary to his earlier theory that the naphthenic acids are man made during refinery operations, he found that the acids isolated from Roumanian crude oil were very similar to those from distillation fractions, except the aliphatic acids which he considered somewhat altered, probably by cracking of higher-nolecular-weight acids to lower ones. In the absence of detailed data on steps involved in the isolation of acids from crude oil, it is not possible to state definitely that the lower acids were not lost to the water phase in some isolation step.

The main sample (I) of Roumanian acids was a commercial acid isolated from kerosene; another (II) was from the same source but from a different supplier and was a gas and lubricating-oil acid; the third (III) was isolated directly from crude oil of the same Roumanian field.[1] They were all dark-brown acids with a pungent odor. Sample I had a low viscosity, but the others were viscous oils. I and II were purified in an identical manner and yielded 10 and 60% of phenols and nonacidic matter, while III was treated with 3 volumes of glacial acetic acid, resulting in the separation of an upper nonacid phase and a lower carboxylic acid layer. This lower layer was extracted several times with petroleum ether to remove the remaining phenols and hydrocarbons, while the upper phase was extracted with sodium carbonate solution to remove acids. The yield of acids, in this case, was 50%.

The purification of samples I and II was carried out separately and consisted of extraction with an excess of sodium carbonate solution and repeated ether extraction of the aqueous layer to remove the remaining hydrocarbons. Ether dissolved in the water layer was removed on a steam bath with a stream of air blowing through the solution. In both cases, troublesome foaming occurred at this stage.

Oxidizable impurities were discarded by treatment with about 0.1 mole of 3% potassium permanganate solution. The manganese dioxide was removed by filtration, the acids liberated from their sodium salts by sulfuric acid and separated, dried over anhydrous sodium sulfate and distilled at 12 mm pressure. Sample I boiled at 120° to 230°C (248° to 446°F), II at 160° to 270°C (320° to 518°F), and III at 150° to 290°C (302° to 554°F).

Table 22. **Properties of von Braun's Roumanian Samples**

Sample I *9 kg*			Sample II *0.8 kg*			Sample III *1 kg*		
Boiling Range at 12 mm		*Weight g*	*Boiling Range at 12 mm*		*Weight g*	*Boiling Range at 12 mm*		*Weight g*
°C	*°F*		*°C*	*°F*		*°C*	*°F*	
57–125	135–257	78	170–200	338–392	61	145–170	293–338	95
125–146	257–295	1703	200–220	392–428	86	170–237	338–458	245
146–170	295–338	3339	220–240	428–464	85	190–280*	374–536	98
170–215	338–419	2290	240–280	464–536	58			
Above 215		140						
% Accounted for by Fractions		84.3		36			44	

* 0.2 mm

The three samples were separately esterified by ethanol or methanol, apparently to remove phenols and other impurities, and the esters were washed with cold alkali to remove unesterified matter. Any tertiary acids present and not esterified were presumably lost at this stage.

Each batch of esters was saponified, separated, the acids liberated by sulfuric acid, dried, and fractionated through a column to yield the data shown in Table 22.

Apparently this fractionation was carried out only to determine the boiling range of each sample and the fractions from each sample were recombined and again, esterified separately, and the esters then fractionated. Although he did not say so, von Braun apparently saponified the ester fractions separately and then fractionated the acids obtained from each ester fraction to get a series of narrow-boiling cuts of acids which were combined on the basis of boiling point and studied as individual samples. Since there are gaps in the boiling points of the fractions studied, he apparently selected only fractions that were large, i.e., those which represented plateaus in the boiling-point curve.

The fractions obtained from sample I (Table 22) were studied most carefully and the results obtained are presented in the next few pages.

Table 23. **Data on Petroleum Fractions Studied**

Cut	*Boiling Point Range* °C	°F	*Pressure* mm	*Nature of the Acids*
1	About 190	374	760	C_6 and C_7; mainly aliphatic
2	About 210	410	760	Similar to 1
3	118–127	244–261	12	C_8, monocyclic and C_7, aliphatic
4	130–137	266–279	12	C_8, monocyclic
5	139–141	282–286	12	C_9, monocyclic; studied in detail
6	140–148	284–298	12	C_9 and C_{10}, monocyclic; not studied
7	148–155	298–311	12	$C_{10}H_{18}O_2$; largest sample; studied most
8	165–179	329–354	12	C_{12}, monocyclic and C_{13}, bicyclic
9	179–194	354–381	12	Mainly bicyclic
10	190–210	374–410	12	Mainly bicyclic

Detailed Study of Cut 7

A fraction boiling in the range of 148° to 155°C (300° to 311°F) at 12 mm was obtained and studied by von Braun's men in working with a number of different samples of acids and von Braun reported that the analysis always approached that of a $C_{10}H_{18}O_2$ acid. However, really careful analyses and degradation work appear to have been done on Cut 7 of Roumanian sample I. Typical analyses:

C, found: 70.56; calculated for $C_{10}H_{18}O_2$:70.80
H, found: 10.66; calculated for $C_{10}H_{18}O_2$:10.81
d, 0.9718; n_D^{15}, 1.4607; $n \times d = 1.419$
M_r, 47.79; calculated: 47.72

Thionyl chloride gave a good yield of the acid chloride of boiling point 100°C at 13 mm; not sharp. The acid chloride with ammonia gave, in quantitative yield, the amide of melting point 76° to 83°C (169° to 181°F) and boiling point 150° to 165°C (302° to 329°F) at 12 mm. The distilled amide melted at 91° to 94°C (196° to 201°F). When this was recrystallized, the melting point was raised above 110°C (230°F) without becoming constant or sharp. The ethylamide from the acid chloride was an oil boiling at 170° to 180°C (338° to 356°F); the anilide in the same way boiled at 225° to 240°C (436° to 464°F) at 13 mm. The ethyl ester made from the acid in the usual manner boiled at 110° to 130°C (230° to 266°F) at 12 mm, but analyzed correctly. Reduction of the ester by sodium and absolute alcohol gave, in 80% yield, the naphthenyl alcohol which had an agreeable odor. The acetate and the propionate of this alcohol had even more pleasant odors.

In attempts to learn something about the structure of the main acid or acids in the cut, the chlorine-number test was run and yielded 200 after the third treatment of the same sample. It was difficult to remove the chlorine of the acid-chloride portion of the molecule while leaving the chlorines on the α-carbon atom, and von Braun had to reflux with water for an hour to decompose the acid chloride. He repeated this test several times with similar results, but admitted that, in some cases, he could not get a chlorine number greater than 180. On the basis of the chlorine number of 200, he concluded that the acids consisted almost entirely of a mixture of C_8H_{15}—CH_2—COOH isomers. He attempted to saponify the dichloroethylamide and thus obtain the acid, which could be converted to the next lower acid, but found he could not carry out the saponification step without loss of some of the chlorine on the α-carbon atom. He confirmed this result with stearic acid.

While he now considered it certain that there was at least one methylene group between the ring and the carboxyl group, he decided to try to determine whether there were additional methylene groups in that chain. He, therefore, tried his lactone method. He converted a sample of the C_{10} acid to the α-bromoacid without difficulty by the Hell-Volhard-Zelinsky method. Hydrogen bromide was not easily removed from this bromoacid, but a 65% yield of the α–β unsaturated acid was obtained after heating the bromoacid with diethylamine for 24 hours at 230°C (446°F). Rearrangement of this to the β–γ unsaturated acid gave $C_6H_{11}CH{=}CH{-}CH_2{-}COOH$. Ozonolysis of this acid to malonic acid and $C_6H_{11}COOH$ was not satisfactory. Thus the unsaturated acid was converted to the lactone $C_6H_{11}{-}\overset{\lceil\quad O\quad\rceil}{CH{-}CH_2CH_2CO}$ from which a small amount of succinic acid was obtained on oxidation with concentrated nitric acid. His conclusion was that at least a low concentration of $C_6H_{11}{-}CH_2CH_2CH_2COOH$ was present in this mixture of C_{10} acids.

From the properties of all of the cuts, it was evident that none was even approximately a pure compound. Consequently, von Braun decided to attempt to degrade the acids of cut 7 to compounds of lower molecular weight which could perhaps be identified (Flowsheet 1). A hypobromide degradation of the amide, using 4 moles of potassium hydroxide, 15 volumes of water, and 1 mole of bromine and heating on a water bath for 2 hours gave, on steam distillation, only a 25% yield of C_9 amine boiling at 170° to 180°C (338° to 356°F) at 12 mm. It proved to be identical with that obtained by the next method.

After testing the K. F. Schmidt hydrazoic acid degradation method on campholic and fencholic acids,[2] he carried out this degradation on 270 g of the C_{10} acid and obtained 190 g of C_9 amine boiling at 170° to 180°C (338° to 356°F). It was a typical primary nonaromatic amine. Its picrate melted at 170° to 178°C (338° to 352°F) and the benzamide was an oil boiling at 215° to 230°C (419° to 446°F) at 12 mm and analyzing correctly.

While he had recommended fractionation of the amines or preparation and recrystallization of the oxalate, he did not mention these methods in this paper. Probably separate trials had shown that no great concentration of any individual amine would result. Instead, he prepared (in 40% yield) the carboxylic acid by converting the amine, through his phosphorus pentabromide degradation of the benzamide of the amine, to the bromide and this to the alcohol,

$C_8H_{15}CH_2OH$, which was oxidized by Cr_2O_3 in glacial acetic acid to $C_8H_{15}COOH$. This acid boiled at 136° to 139°C (277° to 282°F) at 12 mm, had a density d_4^{22} of 0.9819 and n_D^{22} of 1.4516. (The n·d value calculated is 1.42 which is at the lower limit of naphthenic

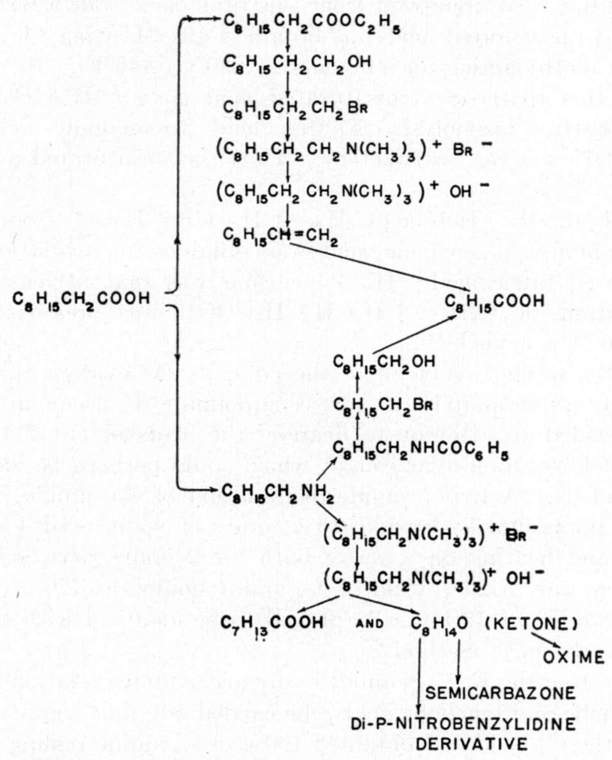

Flowsheet 1. MAIN OPERATIONS IN VON BRAUN'S STUDY OF FRACTION 7

acids of this molecular weight.) A chlorine number determination gave 130. Thus he decided that most of the acids in this mixture had the carboxyl group attached directly to the ring. C_9 acids isolated from adjacent fractions were not identical with this one and, therefore, the C_{10} acid which he degraded was not simply a homologue of the C_9 acid in other cuts. Finally, he obtained the same C_9 acid on degrading his C_{10} acid through the reactions:

Acid → ethyl ester → naphthenyl alcohol → bromide → quaternary ammonium base → olefin → acid

The olefin was converted to the acid by ozonization in glacial acetic acid solution, with subsequent decomposition of the ozonide by treating with water, ether, and zinc dust. The resulting aldehyde-ketone mixture could not be separated satisfactorily. Therefore, the aldehydes were oxidized to the acids by warming for 24 hours with 10% sodium carbonate solution to which perhydrol was added. The product consisted of about 2 parts of acids and 1 part of ketones. The acids were a complex mixture with the main portion boiling at 127° to 135°C (261° to 275°F). He considered this identical with the C_9 acid obtained by the first method. The reported density of 0.9810 and index of refraction, n_D^{22}, of 1.4516 yields an n·d value of 1.42. Since both were mixtures, there is probably not much significance in either set of properties.

The ketone mixture yielded a $C_8H_{14}O$ compound which analyzed correctly and gave a semicarbazone and a di-*p*-nitrophenylhydrazine derivative which could be purified by recrystallization. It was this ketone which was obtained in fairly large volume from cut 7 of the Roumanian acids and was studied very carefully. It was this ketone, also, which he reported to be a degradation product from fractions of corresponding boiling-point range from California, Texas, Galician and other petroleum acid samples, although admittedly in very low yield from Galician acids. From these results, he concluded that the ring and substituents involved in this $C_8H_{14}O$ ketone must be of some fundamental importance somewhat like the benzene ring of aromatic compounds. A separate section will deal with the structure of this ketone.

EXAMINATION OF THE OTHER FRACTIONS OF SAMPLE I

Cut 5, Boiling Range 139° to 141°C (282° to 286°F) at 12 mm

Fractionation of sample I acids repeatedly yielded a plateau, with a density range of 0.9704 to 0.9714 and an index of refraction range of 1.4505 to 1.4497. Chemical examination indicated the presence of C_8 and C_{10} as well as C_9 acids, in spite of the narrow boiling range and other physical properties. Degradations yielded mixtures of $C_7H_{13}COOH$, $C_7H_{12}{=}CH_2$, $C_7H_{12}{=}O$, and C_7H_{12} olefins. A chlorine number of 192 indicated that over 90% of the molecules were of the type, RCH_2COOH. Degradation over the alcohol, bromide, Hofmann degradation, and ozonolysis yielded 72% of a mixture of acids and only 4% of ketones, from which, however, the $C_8H_{14}O$ ketone of cut 7 was easily isolated as the semicarbazone melting at 162° to 163°C (324° to 325°F).

The mixture of acids obtained in this degradation was studied by chlorine-number determination (which was found to be 163) and by degradation over the amine, but these studies failed to yield results of interest except that here again von Braun reported a small amount of the same $C_8H_{14}O$ ketone.

Fraction 4, Boiling Range 130° to 137°C (266° to 279°F) at 12 mm

This mixture analyzed as $C_8H_{14}O_2$, had a density of 0.9694, was different from C_8 acids obtained by degradation, and since it had a chlorine number of 200, it must have contained only acids of the type $C_6H_{11}CH_2COOH$. Degradation attempts on this acid mixture failed to identify any compound.

Fraction 3, Boiling Range 118° to 127°C (244° to 261°F) at 12 mm

Analysis indicated that this cut was a mixture of C_7 aliphatic and C_8 naphthenic acids. The chlorine number again approached 200, but von Braun failed to identify any fatty acid definitely, although he did conclude that one of them probably was either 3-methylpentanoic or 3-methylhexanoic acid.

Fractions 1, Boiling about 190°C (374°F), and 2, Boiling about 210°C (410°F)

These fractions were found to contain hydrocarbons and thus were esterified again separately, washed well, saponified, liberated, and fractionated. Fraction 2 was now found to boil over the range of 205° to 220°C (401° to 428°F) and its weight was only 1.5 g. A similar situation in regard to fraction 21 made it impossible to identify any compound.

Fraction 8, Boiling at 165° to 179°C (329° to 354°F) at 12 mm

Analyses and molecular-weight determination of fraction 8 indicated that this contained a mixture of mono- and bicyclic acids. A mixture of amines boiling, after purification over the oxalate, at 95° to 123°C (203° to 253°F) at 15 mm was obtained by the hydrazoic acid method of degradation.

Fraction 9, Boiling Range 179° to 194°C (354° to 381°F) at 12 mm

Fraction 9 appeared, from analyses and molecular-weight determination, to be almost entirely bicyclic. The amine mixture obtained from it by degradation boiled at 110° to 135°C (230° to 275°F) at 15 mm. The chlorine number of the acids was found to be near 200, thus indicating that the acids were of a bicyclic, $C_{11}H_{19}CH_2COOH$, type possibly with additional methylene groups between the carboxyl group and the bicyclic portion of the molecule.

Fraction 10, Boiling Range 190° to 200°C (374° to 392°F) at 12 mm

This fraction, with the unusually high density of 0.9924 and a molecular weight of 226, seemed to consist mainly of $C_{14}H_{24}O_2$ bicyclic acids. The amine obtained by the Schmidt degradation also analyzed as a bicyclic base as expected. The chlorine number of the acids again was near 200, but an attempt to prepare a lactone, indicating that there were four carbons in the chain carrying the carboxyl group, failed. An attempt to obtain a pure compound by a reaction series for which von Braun had held high hopes also failed. In this scheme, the acid was esterified, reduced to the alcohol, and converted to the bromide which was then condensed with malonic ester and converted to the barbiturate. When the scheme was tried on this mixture, most of the sample was lost in the long series of reactions and the rest did not yield a pure barbiturate through recrystallization. Even if a pure barbiturate had been obtained, any attempt to reverse this long series of steps seems doomed.

Study of Roumanian Acid Sample II

Von Braun's study of this higher-boiling sample by the methods used on sample I led to little of interest but for the fact that the dividing line between mono- and bicyclic acids again appeared at C_{12} and that the C_{11}—C_{12} amine mixture finally led again to the same $C_8H_{14}O$ ketone after degradation. Fraction 5 of this series, boiling over the range of 240° to 270°C (464° to 518°F) at 12 mm, gave no indication of the presence of tricyclic or more complex acids. Therefore, von Braun concluded that the Roumanian acids above C_{13} were bicyclic.

Study of Acids from Roumanian Crude Oil, Sample III

A partial study of the acids isolated from crude Roumanian oil showed that the break between mono- and bicyclic acids again was at C_{12} and that the low-boiling aliphatic acids appeared to differ materially from those derived from distillation fractions, samples

I and II. Whether this difference was due to acids formed by cracking and oxidation during distillation, as von Braun thought, or to different losses through solubility in the different isolation procedures cannot be decided definitely.

Study of California Stove-Oil Acids

Stove-oil acids from California were subjected to essentially the same purification and distillation operations as employed in the study of the Roumanian acids and the cuts obtained were those listed in Table 24.

Table 24. **California Stove-Oil Acids**

Cut	*Boiling Point* °C	*Boiling Point* °F	*Pressure* mm	*Grams*
1	125–142	257–288	12	7
2	140–150	284–302	12	41
3	162–173	324–343	11	164
4	180–200	356–392	16	176
5	200–220	392–428	16	138
6	220–240	428–464	16	86
7	240–265	464–509	16	36
8	220–260	428–500	0.4	18

It is apparent that cut 2 corresponded roughly to cut 7 of sample I of the Roumanian acids and this was studied most carefully. Its density of 0.9739 approached that of the Roumanian fraction, but analysis indicated this was a mixture of C_9 and C_{10} acids instead of the C_{10} mixture of the Roumanian acids. Three treatments brought the chlorine number only to 188. The amine obtained by the hydrazoic acid degradation boiled at 72° to 82°C (162° to 180°F) at 16 mm, while the Roumanian amine boiled at 170° to 180°C (338° to 356°F), presumably at atmospheric pressure. Degradation of the amine yielded a ketone whose semicarbazone melted at 110°–137°–150°–163°–163°C (230°–279°–302°–325°–325°F) on recrystallization and the last melting point was not depressed when the semicarbazone was mixed with the $C_8H_{14}O$ ketone obtained from Roumanian acids.

Fraction 3 of the California acids also yielded a ketone whose semicarbazone melted at 162°C, but this melting point was depressed when the semicarbazone was mixed with the old $C_8H_{14}O$ ketone; therefore, it must have been a different one. Cuts 4 to 8 appeared to be mainly bicyclic in nature. Except for the isolation of the old

$C_8H_{14}O$ ketone, the California acids yielded little of interest or novelty and the same was true of von Braun's study of the Galician acids whose only interesting difference was that up to C_{20}, the mixture appeared to contain no bicyclic acids, i.e., it was monocyclic throughout and seemed to have a higher aliphatic acid content than the others studied.

Bibliography

1. von Braun, J., *Ann.* **490,** 100 (1931).
2. von Braun, J., *Ber.* **66,** 684 (1933).

CHAPTER SIXTEEN

THE STRUCTURE OF VON BRAUN'S KETONE

We have seen that when von Braun found that purification and fractional distillation of the naphthenic acids and their esters failed to yield fractions from which pure compounds or derivatives could be obtained, he decided to simplify the mixture by degrading it, hoping that the new lower-molecular-weight compounds (amines) could be more easily separated and purified than the parent acids. When the amines obtained by the hydrazoic acid method of degradation of acids could not be separated by distillation or by recrystallization of their oxalates, he decided to remove another carbon atom by exhaustive methylation and ozonolysis of the olefin obtained in the Hofmann degradation. A C_{10} acid mixture yielded some $C_8H_{14}O$ ketones from which an individual ketone was isolated, purified over the semicarbazone, and characterized in a number of ways.

The only clue in regard to the structure of the ketone at this stage of the investigations was the fact that it formed a dibenzylidine derivative and thus must have had both of the α positions unsubstituted. Assuming that the ketone was a cycloalkanone, i.e., had the carbonyl oxygen attached to a ring carbon, the structure of the ketone was known to be of type I, and the di-p-nitrobenzylidine derivative must have been of type II.

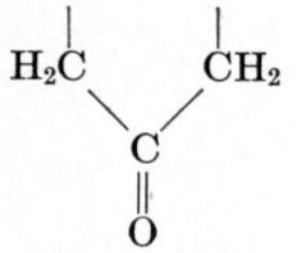

```
   |       |                   |        |
  H2C     CH2       C6H5CH=C        C=CHC6H5
     \   /                    \    /
      C                         C
      ‖                         ‖
      O                         O

      I                         II
```

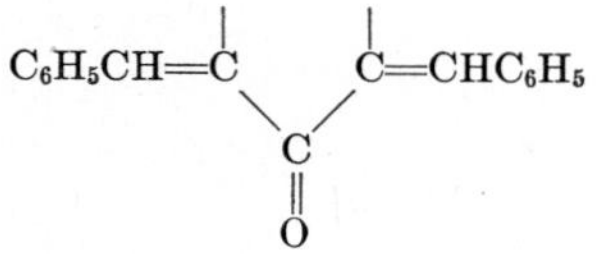

Assuming that acids containing the cyclobutane or cycloheptane rings could be excluded because such ring compounds are rare in nature, von Braun had only to decide between cyclohexane and cyclopentane derivatives. In common with most other workers in this field, he held that cyclohexane acids are not found in high concentration in petroleum acids and he, therefore, felt that this ketone which was formed from a number of different acid samples (and was reported to be a major ingredient in two of them) could not be a cyclohexanone but must have been a substituted cyclopentanone carrying three carbons as alkyls on carbon atoms 3 and/or 4. The possible structures, then, would be:

```
HC——CHCH2CH2CH3     HC——CHCH(CH3)2     HC——C(C2H5)CH3
|    |              |    |             |    |
H2C  CH2            H2C  CH2           H2C  CH2
  \ /                 \ /                \ /
   C                   C                  C
   ||                  ||                 ||
   O                   O                  O

CH3HC——CH(C2H5)       CH3HC——C(CH3)2
   |    |                |    |
   H2C  CH2              H2C  CH2
     \ /                   \ /
      C                     C
      ||                    ||
      O                     O
```

and as a remote possibility, cyclopentylacetone.

The physical properties of the petroleum-derived ketone did not help much in determining its structure because only a few of the isomers were known. Von Braun decided to synthesize some of the possible isomers which either were not known or were not well characterized. He prepared 3-ethyl-3-methylcyclopentanone and 3-ethyl-4-methylcyclopentanone and attempted to prepare 3,3,4-trimethylcyclopentanone. Unfortunately, in the case of 3-ethyl-4-methylcyclopentanone, he either overlooked the fact that *cis-trans* isomers are possible or, more probably, found that his synthesis yielded only one form and decided that the other one was not stable (or at least not accessible to synthesis) and dismissed it from his mind. At any rate, he found only one form which was not identical with his petroleum ketone and so considered 3-ethyl-4-methylcyclopentanone eliminated.

Table 25 lists the substituted cyclopentanones and cyclohexanones, one of which could have been identical with his ketone.

Table 25. **Properties of Ketones Possibly Identical with von Braun's Ketone**

Ketone	Boiling Point °C	°F	n_D^t	t °C	d_4^t	t °C	Semicarbazone Melting Point °C	Di-p-Nitrobenzylidine Derivative Melting Point °C	Reference
von Braun's	172–174	343	1.4390	20	0.8945	20	162–163	188–190	14
CYCLOPENTANONES									
3-Ethyl-3-Methyl	174	345	—	—	0.9074	18	170	180	14
3-Ethyl-4-Methyl	180	356	—	—	0.9058	18	208	192	2
3-Ethyl-4-Methyl	180	356	—	—	—	—	208	192	14, 37
3-Propyl	190	374	1.4456	12	0.9041	20	178–179	222	14
3-Isopropyl	78 (13 mm)	172	—	—	0.921	—	198	—	11
3-Isopropyl (*levo*)	183	361	1.4443	18	—	—	191–192	149–150	20
3-Isopropyl (*dextro*)	183	361	1.4438	17	—	—	191	149–150	20
CYCLOHEXANONES									
3,3-Dimethyl	173–174	345	—	—	—	—	195	—	28
3,3-Dimethyl	—	—	—	—	—	—	195–198	—	46
3,3-Dimethyl	—	—	—	—	—	—	203	—	33
3,3-Dimethyl	—	—	—	—	—	—	195–196	—	42
4,4-Dimethyl (Solid melting 40–42°C)	73 (14 mm)	163	1.454	24	0.9282	24	203	—	3
3,4-Dimethyl	187	369	—	—	—	—	175	—	43
3,4-Dimethyl	—	—	1.4510	12.5	—	—	189	—	45
3,4-Dimethyl	186–188	369	1.4513	20	—	—	184–185	—	6
3,5-Dimethyl	181–182	360	1.4450	—	0.8994	12	190–196	—	31
3,5-Dimethyl (*cis*)	182–183	361	1.4407	15	0.8942	15	202–203	—	12, 15
3,5-Dimethyl (DL, *trans*)	180–181	358	1.4475	15	0.9032	15	193–194	—	12, 15
3,5-Dimethyl (D, *trans*)	—	—	—	—	—	—	187–189	—	12, 15
3,5-Dimethyl (L, *trans*)	—	—	—	—	—	—	189	—	12, 15
3-Ethyl	192–194	381	1.4543	20	0.9196	18	184	176	17
3-Ethyl	81 (12 mm)	178	1.4511	20	0.9145	20	166–167	—	1
3-Ethyl	189–191	376	1.4511	20	—	—	181–182	—	45
4-Ethyl	192–194	381	1.4532	20	0.9214	18	174	156	17, 45

Some of these ketones are eliminated on the basis of physical properties, although here the situation is unfortunate because von Braun did not report the index of refraction of the ketones he synthesized and we, therefore, have to depend on the melting points of two key derivatives, the semicarbazones and the di-*p*-nitrobenzylidine derivatives. The semicarbazones, in turn, may usually exist in *syn* and *anti* geometrical forms and the rather poor agreement in melting points reported may be due to mixtures of such isomers. Since this happens to be the derivative which von Braun found easiest to isolate and purify, it is the one on which most reliance has to be placed in comparing synthetic ketones with his compound.

As stated before, he apparently did not seriously consider the substituted cyclohexanones as possibilities. Thus he decided, on the basis of the data shown in Table 25, that 3,3,4-trimethylcyclopentanone was the only possible ketone which could be the one identical with his ketone. Table 26 lists the properties of 3,3,4-trimethylcyclopentanone prepared by different workers and those of von Braun's petroleum ketone.

Table 26. **Properties of Synthetic 3,3,4-Trimethylcyclopentanones and von Braun's Ketone**

Source	*Boiling Point* °C	°F	n_D^t	*t*	d_4^t	*t*	*Semicarbazone Melting Point* °C	*Di-p-Nitrobenzylidine Derivative Melting Point* °C	*Form*
Buchman and Sargent [19]	172–173	343	1.4386	25	0.892	25	213–214	204.7–205.1 202.0–202.5	α β
Ruzicka and Seidel [41]	94.6 (90 mm)	203	1.4395	18	0.8974	18	221–222	207–208	
Mukherji [35]	174	345	—	—	—	—	172	190–191	
Baumgarten and Gleason [5]	172–173	343	1.4378	25	0.882	25	213	205 203	α β
von Braun's Ketone [13]	172–174	345	1.4390	20	0.8955	20	162–163	188–190	

While boiling points, densities, and indices of refraction are not seriously different, the melting points of both the semicarbazones and di-*p*-nitrobenzylidene derivatives are so far from von Braun's (in all cases except that of Miss Mukherji) that there can be no doubt that von Braun had a ketone which was not 3,3,4-trimethylcyclopentanone and thus his substituted acid could not have had the structure he assigned to it.

While von Braun's structure of the ketone and the parent acid appeared to be generally accepted, there seemed to be no valid reason for assigning to it the fundamental importance in naphthenic acid chemistry which he did in his lectures and papers, because there seemed nothing about the 3,3,4-trimethylcyclopentane structure that would set it aside either structurally or genetically. Meanwhile, Cosciug [26] attempted to obtain von Braun's acid and ketone from a Texas naphthenic acid, but in spite of active correspondence with von Braun, who advised him in regard to details of steps involved, he was able to isolate only a mixture of ketones from which none of the von Braun ketone could be obtained in the form of the semicarbazone melting at 162° to 163°C (324° to 325°F).

Since von Braun claimed [13] that the C_{10} acid from which the ketone was derived was the major constituent of both California and Roumanian fractions of acids boiling between 148° and 155°C (298° to 311°F) at 12 mm, Goodman in the author's laboratory decided to attempt the isolation of von Braun's acid from California acids available at the time. He thought if von Braun, with relatively rough fractionation by distillation alone, was able to obtain a fraction in which the acid was the main component, it should be possible to obtain evidence of very high concentration of one acid, von Braun's, in some one fraction or some adjacent fractions by using a combination of efficient fractional distillation followed by countercurrent fractional neutralization or liberation. The plateau obtained in this way should then be due to von Braun's acid, even if it should have properties different from the one he synthesized from the petroleum-derived ketone, since we have seen that his synthesis may have yielded the *cis-trans* isomer not found in oil. It might then be possible to purify this acid by recrystallization of some derivative, regenerate the acid from the derivative, and characterize the pure acid. It was realized that the naphthenic acids studied by von Braun came from central California,[18] while those available now were from the Signal Hill field and might differ materially from the acids studied by von Braun. Results obtained in this attempt indicate this may be true.

Starting with 825 cc of methyl esters, which had been carefully fractionated through a 225 cone Stedman still at 20:1 reflux, collected in narrow-boiling fractions and then combined to make up the 825 cc batch boiling at 218°–221°C (424° to 429°F) at atmospheric pressure, Goodman [29] saponified the esters, liberated the acids, and fractionally neutralized them in a countercurrent extraction column with a throughput of 2,500 cc an hour, but an efficiency of only five to six stages. Ten fractions were collected in this fractionation, but a study of physical properties showed no indication of a plateau in the density or index-of-refraction curve; each property changed steadily from cut to cut and it was found that the index of refraction of von Braun's synthetic acid lay between cuts 4 and 5, while the density of the synthetic acid lay between cuts 2 and 3. Throughout this attempt at fractionating the acid to locate the von Braun acid, it was always found that the molecular-weight, density, and index-of-refraction values of von Braun's acid obtained from his ketone never came all in the same fraction, or even two adjacent fractions. Even this might be comprehensible if we assume that the acid he obtained by synthesis was the other *cis-trans* isomer of the natural acid and if there were any evidence of accumulation of any acid in *any* of the ten fractions. Analyses indicated that the California acids studied were indeed C_{10} acids and the melting-point range of the ureide and amide mixtures was about the same as that of the pure synthetic derivatives, i.e., 117° to 121°C (247° to 250°F) for the amides against 122° to 123°C (252° to 253°F) for von Braun's synthetic amide, and 189° to 195°C (372° to 383°F) for the ureide against 197°C (387°F) for the synthetic derivative; the amides of fractions 1 and 5 showed no depression in a mixed melting-point determination.

Another 725 cc batch of highly fractionated California esters, boiling over the range 218° to 225°C (424° to 437°F) was treated like the first one, except that a systematic batch scheme of countercurrent fractional neutralization was used and fourteen fractions were collected. There was no evidence of accumulation of any one acid, and the properties of von Braun's synthetic acid now came between cuts 10 and 11 for the index of refraction and 7 and 8 for density. When cuts 7 to 11 were combined and carefully fractionally distilled through a three-foot spinning-band column at high reflux, the index of refraction lay at cut 1 while the density of the synthetic acid came at cut 7 out of nine fractions, i.e., these properties were further apart than ever and all properties changed regularly from cut to cut.

This led to the conclusion that the California acid, this particular one at least, did not contain a high concentration of any one acid in this boiling range and that, if an acid with the properties of von Braun's synthetic acid was present, its properties were hidden by those of other acids in the mixture.

The Study of Texas Gulf Coast Acids

Since a sample of Texas Gulf Coast acids of the correct boiling range was available by this time, Goodman decided to attempt once more the direct isolation of von Braun's acid by fractionation. Three liters of carefully fractionated methyl esters of C_9–C_{10} acids were available for study and the whole batch was saponified and the acids liberated from the resulting sodium salts. Fractional neutralization of such a large batch would have been very tedious. Therefore, it was decided to attempt fractional liberation of the acids by treatment of a 5% sodium salt solution with dry ice or carbon dioxide gas. This procedure had been found in test-tube trials to liberate about 45% of the acids in one treatment. The salts from 925 g of methyl esters were diluted to 5% concentration, cooled to 10°C (50°F), and treated with carbon dioxide gas until

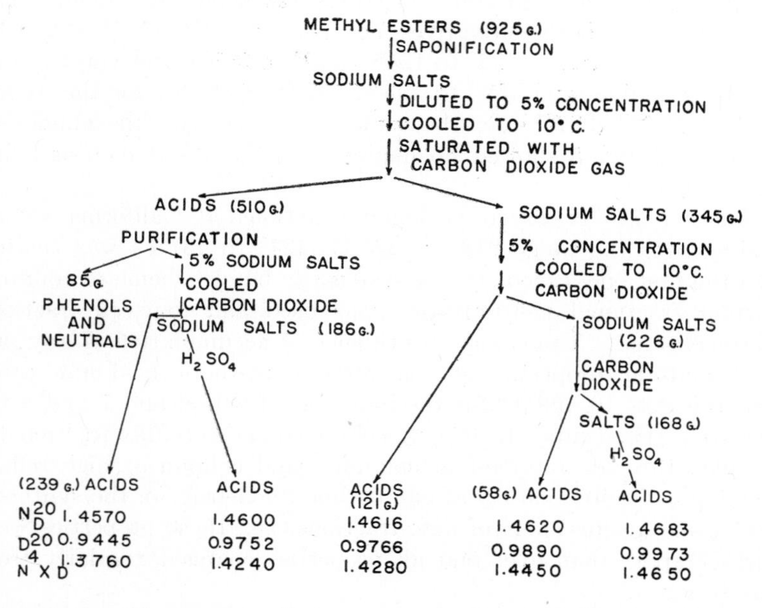

Flowsheet 2. Fractional Liberation of Acids by Carbon Dioxide

no additional organic layer seemed to form. The top layer was separated and the aqueous layer extracted with petroleum ether to remove emulsified acids. The combined organic layer was then stirred for 2 hours with 4 liters of 5% sodium bicarbonate solution to convert all carboxylic acids to the sodium salts. The bicarbonate solution, in turn, was extracted with ether to remove phenols and neutral compounds. A total of 85 g of such material was obtained from the 925 g batch of esters. The bicarbonate solution was concentrated to about 5% concentration, cooled to 10°C (50°F), and treated anew with carbon dioxide. The acids liberated were treated as before and finally another batch of acids was liberated by carbon dioxide.

The aqueous layer from the initial treatment gave 345 g of acids which were converted to sodium salts, adjusted to 5% concentration, cooled to 10°C (50°F), and treated with carbon dioxide to yield another 121 g of acids. Flowsheet 2 shows how these operations yielded, finally, five fractions of acids with a wide range of physical properties. This method of separation, which at first sight seems peculiar, because elementary texts usually state that only phenols would be liberated by this treatment, could probably be developed into a systematic method of fractionation by fractional liberation of acids.

Table 27. **Fractional Neutralization of Texas Acids**

Cut	n_D^{20}	d_4^{20}	$n \times d$	*Comment*
1	1.4562	0.9517	1.386	High concentration aliphatic
2	1.4572	0.9581	1.395	
3	1.4583	0.9643	1.406	Lower limit naphthenic
4	1.4597	0.9676	1.412	
5	1.4599	0.9707	1.417	
6	1.4602	0.9733	1.421	
7	1.4605	0.9764	1.427	
8	1.4613	0.9772	1.430	
9	1.4625	0.9722	1.439	
10	1.4638	0.9890	1.449	
11	1.4648	0.9927	1.453	
12	1.4654	0.9958	1.460	Upper limit naphthenics

From these fractions and similar ones obtained from additional batches of methyl esters, 306 g of acids, with an index of refraction range of n_D^{20} = 1.4600 to 1.4616 and a density range of d_4^{20} = 0.9768 to 0.9802, was selected for study. Von Braun's synthetic acid had n_D^{20} 1.4587 and d_4^{20} 0.9783. Thus it appears again that the American acids and von Braun's acid did not correspond in properties, for the range selected for study had an average n_D^{20} range that was too high throughout while the density range was too low.

Using a sixteen-bottle extraction apparatus,[34] this batch of acids was fractionally neutralized to yield twelve cuts of the properties shown in Table 27.

The same series of operations on 305 g of another batch of acids with an n_D^{20} range of 1.4598 to 1.4625 and a d_4^{20} range of 0.9736 to 0.9822 yielded a very similar series of neutralization fractions with fraction properties ranging from $n_D^{20} = 1.4583$ to 1.4652 and $d_4^{20} = 0.9638$ to 0.9958 — almost the identical range if the first two fractions of the first series are discarded. There was, then, still no evidence of accumulation in any one fraction of an acid with the properties of the von Braun acid or, indeed, of any acid, since there was only the usual slight flattening out of the curve near the middle of the series of fractions normal for the series of fractions obtained from any batch of petroleum compounds.

Degradation Studies

Since it was obvious that the combination of fractional distillation and of fractional neutralization did not lead to any appreciable concentration of any acid present as a major constituent in the C_{10} acid range, it was decided to attempt to degrade acid fractions to the amines and then to the ketone with n-2 carbon atoms, in the hope that recrystallization of the semicarbazone might lead to the isolation of the pure ketone described by von Braun.[13] To this end, degradation experiments were at first made with neutralization fractions which, from the physical properties, might contain an appreciable concentration of the parent acid and thus should yield a semicarb azone that could be purified without high loss during recrystallization.

The degradation steps described by von Braun [13] were followed carefully, except as will be described or in connection with experiments with specific samples.

The hydrazoic acid (K. F. Schmidt) degradation of the acid to the amines with n-1 carbons proceeded smoothly in all cases, with yields of 60 to 80% or higher.

The Hofmann degradation of the amines was carried out as modified by von Braun and Anton.[15] In this method, the amine is treated with about 3 moles of 25% sodium hydroxide and the mixture converted to the quaternary ammonium hydroxide by gradual addition of 1.5 moles of dimethyl sulfate. The quaternary ammonium sulfate layer was removed in a separatory funnel and refluxed for 2 hours with 0.52 mole of 20% sulfuric acid; this was next neutralized with a very slight excess of hot saturated barium hydroxide. The precipitated barium sulfate was digested and filtered. The filtrate was distilled under water-pump vacuum on a steam cone until only a clear viscous liquid remained. These steps avoided the preparation of the iodide and its decomposition by silver hydroxide which was accompanied by formation of colloidal silver precipitate that could be removed only by tedious centrifugation or filtration. No attempt was made to determine the yield of the quaternary base as such, but the over-all yield — amine to olefin — was somewhat lower than that reported by von Braun in his work.

The next step, decomposition of the quaternary base to the olefin, followed von Braun's procedure which yielded somewhat less of the troublesome tertiary amine by-product, but still required double or triple treatment of the base to obtain a satisfactory yield of olefin. The modification used by von Braun involves distillation of the quaternary ammonium base in the presence of 3 moles of 50% potassium hydroxide to which water is added as fast as removed by distillation, until finally no organic matter remains in the distillation flask. This required about a day of distillation and if the methylation and this decomposition has to be repeated once or twice, it is obvious that this step in von Braun's procedure is quite time consuming. The unsaturated hydrocarbon obtained was extracted with ether, dried, and distilled. The yield of olefin, after several treatments, ranged from 20 to 35%.

Ozonolysis was carried out by passing air containing about 8% ozone through a cold solution of the olefin in carbon tetrachloride at a rate of 4 liters per hour for several hours, after which the ozonide was decomposed by pouring it into ice water containing an excess of powdered zinc. After thorough shaking, the organic matter was extracted with carbon tetrachloride, separated in a separatory funnel, and then poured into water containing hydrogen peroxide to oxidize any aldehydes present. After standing at room temperature for a day with occasional shaking, the acids were separated from the ketones by extraction with dilute alkali. The ketones were converted to semicarbazones and, in most cases, the acids were not studied. The yield from olefin to ketone again was lower than desirable, usually less than 25%. The semicarbazone was, in all cases, highly impure and could be purified only with such loss that no ketone having the melting point of von Braun's semicarbazone could be obtained with certainty that the melting point could not be raised higher by continued recrystallization.

To check the yields to be obtained in this series of operations, 100 g of cyclohexaneacetic acid was converted to 73.5 g of cyclohexylmethylamine (88.6%), from which 37.8 g was then converted to N-trimethylcyclohexanemethylammonium hydroxide — a viscous liquid — in undetermined yield. The whole batch was degraded to methylenecyclohexane in 8 g (25%) yield. The second trial gave only 19% yield, from which 6.5 g was then ozonized and the ozonide decomposed to give a 3.7 g (39%) yield of semicarbazone and 1 g (11%) of adipic acid obtained by cleavage of the cyclohexane ring during ozonolysis. This indicates that von Braun's acid, if present, could probably be converted in about 7% over-all yield to the semicarbazone, which would, however, be mixed with a number of others and could not be purified without much loss during recrystallizations.

Since von Braun obtained a sufficient quantity of pure semicarbazone to permit isolation and characterization of the pure ketone from the purified semicarbazone and the conversion, in turn, of the ketone to the acetic acid which was characterized, it is obvious that, even though von Braun's yields were perhaps double those obtained at the University of Texas, he must have degraded a total of several hundred grams of the parent acid present in a complex mixture of isomers — even if it was the major constituent in the fraction. He complained that very large samples had to be processed to get conclusive results, and he was not able to make a complete study of most of his samples except that from California and the original Roumanian samples.

Degradation of the First California Acid Fraction

A sample of acids boiling at 130° to 132°C (266° to 270°F) at 2 mm pressure and having a neutralization equivalent of 171, an n_D^{20} of 1.4596 to 1.4600, and a d_4^{20} of 0.9756 to 0.9787 was selected for study. The amide prepared over the acid chloride melted at 110° to 112°C (230° to 234°F) from 70% methanol. Degradation of 18 g of this sample yielded 11.5 g (77%) of C_9 amine boiling at 69° to 72°C (156 to 162°F) at 12 mm pressure, with an n_D^{20} of 1.4590, and a d_4^{20} of 0.8193. Its picrate melted at 168° to 174°C (331° to 345°F) recrystallized from methanol. It was analyzed and agreed with the formula $C_{15}H_{21}N_4O_6$. Exhaustive methylation of 11 g of this amine gave, after double treatment, a 3.17 g (35%) yield of olefin boiling at 139° to 146°C (282° to 295°F) with an n_D^{20} of 1.4415 to 1.4456 and a d_4^{20} of 0.7911. On ozonolysis with 8.5% ozone, 2.5 g of the olefin gave 0.5 g of a neutral product which yielded a semicarbazone melting at 127° to 129°C (261° to 264°F) with a yield of only 10 mg so that further purification was not possible. The acid portion consisted of 0.8 cc of an acid boiling at 100° to 105°C (212° to 221°F) at 8 mm pressure. It had a neutralization equivalent of 167, an index of refraction of $n_D^{20} = 1.4510$, and gave an amide melting from 50% methanol at 119° to 123°C (246° to 253°F). The analysis showed that it was the amide of a C_8 acid. Consequently, the high neutral equivalent probably indicated the presence of inert impurities which were not removed from the 0.8 cc product.

Degradation of the Second California Acid Fraction

Degradation of 25 g of another sample, boiling at 148° to 150°C (298° to 302°F) at 12 mm pressure gave results similar to those of the first sample in the preparation of the amine. At this stage, it was decided to try the amine-phosphate method of degradation, in spite of the fact that in some earlier work, rearrangements had obviously taken place as the amine was heated with phosphoric acid. A 10 g portion of the amine obtained from the second sample was placed in a distilling flask connected to a dropping funnel and a condenser and 9 g of 85% phosphoric acid was added dropwise with shaking. A white solid amine-phosphate formed which was washed with absolute alcohol and then heated on a sand bath to 300°C (572°F) until no more distillate came over. The olefin was washed and distilled. It weighed 9 g (67% yield) and was boiling at 138° to 152°C (280° to 306°F), but the index of refraction (1.4504 to 1.4605) was much higher than that of the previous fraction, indicating extensive changes. Therefore, the work on this sample was discontinued in spite of the fact that the yield here was much higher than by the more tedious exhaustive methylation procedure.

As another check on the nature of the products obtained in the amine-phosphate method of degradation, 10 g of cyclohexanemethylamine was treated in the same manner as the second California sample. The olefin obtained boiled at 101° to 106°C (214° to 223°F) and had an index of refraction of 1.4420 to 1.4468, while the Hofmann degradation gave an olefin that boiled at 102.5° to 103.5°C (216° to 218°F) and had an index of refraction of 1.4493. Zelinsky had reported [47] a boiling point of 101° to 102°C (215°F) and an index of refraction, $n^{18} = 1.4506$. Thus it is obvious that the amine-phosphate method leads to extensive rearrangements and cannot be used in degradations of this type.

Degradation of the Third California Acid Sample

A large fraction of acids, boiling at 149° to 151°C (300° to 304°F) at 12 mm and having a neutral equivalent of 173 to 175, was converted in 93% yield to 84.5 g of the amine which, on exhaustive methylation, gave 39 g (54%) of olefin boiling at 147° to 167°C (297° to 313°F) — mostly at 147° to 151°C (297° to 304°F). Ozonolysis of 4.8 g of the olefin yielded 3 g (55%) of acids, with a neutral equivalent of 143 and practically no ketone. The acid was proved to be caprylic (*n*-octanoic) by converting it to the acid chloride and the amide and anilide which proved to be those of this fatty acid. Evidently, the major acid in this fraction was, then, a straight-chain aliphatic acid with ten carbon atoms (*n*-decanoic acid).

Degradation of the First Texas Acid Fraction

A 68 g sample of acid, boiling at 148° to 150°C (298° to 302°F) at 12 mm pressure, with $n_D^{20} = 1.4606$ to 1.4628, $d_4^{20} = 0.9768$ to 0.9802 and a neutral equivalent of 169, obtained as described previously from Texas Gulf Coast acids, was degraded by the hydrazoic acid method to yield 47 g (85%) of amines. Exhaustive methylation yielded 10 g of olefin, which, on second treatment, was increased by 4 g. Ozonolysis in the usual manner of 8 g of this olefin gave 3 g of neutral products and 2 g of acids. The neutral fraction could not be distilled at atmospheric pressure; at 4 mm pressure, 0.5 cc distilled at about 56°C (133°F), with an additional 1.5 cc distilling at 70° to 115°C (158° to 239°F). Only the low-boiling fraction yielded a solid semicarbazone. This was recrystallized until the melting point finally reached 160°C (320°F) for the last 5 mg. The acids were not studied.

Another Texas sample of the same size and similar properties was degraded, in exactly the same manner as the first, up to the degradation by ozonolysis. Here 4.5 g of olefin was dissolved in 300 cc of ethylacetate and ozonized in the usual manner until no more ozone was absorbed. Decomposition of the ozonide was carried out by its catalytic hydrogenation over Adams catalyst, followed by extraction of acidic matter and isolation of the neutral portion which amounted to 2 cc. The aldehydes in this portion were oxidized by a slight excess of dilute permanganate in acetone solution; the excess permanganate was eliminated by a few drops of formalin. After removal of the acetone in vacuum, the residue was dissolved in ether and washed with 10% sodium carbonate solution. In this step, 0.5 ml of water-insoluble acids was removed. The neutral portion, measuring 0.5 ml, was distilled to give 0.3 cc of a substance which yielded a semicarbazone. The melting point of this semicarbazone was raised from 149° to 156°C (300° to 313°F) by recrystallization from 25% methanol. The final yield was 10 mg.

While the work on Texas acids indicated that a low concentration of the von Braun acid might be present, the yields obtained were so low that a sufficient amount of ketone to permit not only recrystallization to final melting point but also isolation of the pure ketone from the semicarbazone and final characterization of the ketone and substituted acetic acid which von Braun prepared from the ketone would have required the degradation of batches of acids of kilogram size. However, such operations would certainly not show that the major compound in the fraction of acids was the von Braun acid.

By this time, it was obvious that if a ketone similar to von Brauns' were isolated from an American acid, it would not be the degradation product of a major constituent of the acid mixture and it would be necessary to completely characterize the ketone unless it would be possible to obtain an authentic sample of von Braun's ketone from Germany, which seemed problematic. To obtain a sufficient amount of the ketone to prepare various derivatives and to prepare the acid which von Braun had synthesized from it would require such a large batch of ketone that a very large fraction of the acids would have to be processed. If this were the requirement of identification of a certain $C_8H_{14}O$ ketone, it would be probably far simpler to synthesize this ketone and show that all of its properties, those of its derivatives, and those of the acetic acid obtainable from it agreed with the data reported by von Braun.

Von Braun's own choice, 3,3,4-trimethylcyclopentanone, had by now been shown to be definitely different from his ketone, so that it could be eliminated. Cyclooctanone and the methylcycloheptanones were eliminated by their much higher index of refraction, as well as by the fact that such cyclic compounds are rare in nature and that this ketone was obtained from every naphthenic acid studied by von Braun. The parent acid was reported as the main C_{10} acid in the fraction boiling at 148° to 155°C (298° to 311°F) at 12 mm pressure. The substituted cyclobutanones were, like cycloheptanones, rare in nature and were eliminated on that basis. This left the cyclohexanones and cyclopentanones. Of these, only those with both α-carbons unsubstituted had to be considered, because the ketone gave a di-*p*-nitrobenzylidene derivative on treatment with benzaldehyde. Von Braun gave only slight consideration to the cyclohexanones, because he held that cyclohexane acids were so rare that the major parent acid could not be a cyclohexane acid.

In view of the number of cyclohexane acids isolated since von Braun's day, this assumption can no longer be held, but the index of refraction appears to be higher in every case for the known cyclohexanones with both α-carbons unsubstituted than that reported by von Braun for his ketone. Nevertheless, there remained a slight possibility that one of the four possible cyclohexanones, namely, 3- or 4-ethylcyclohexanone and 3,4- or 3,5-dimethylcyclohexanone, might prove to be the correct one.

Cyclopentylacetone was prepared by Goodman, even though the fact that the petroleum ketone does not give a positive iodoform test and yields degradation products different from those expected from cyclopentylacetone seemed to eliminate it. The synthetic compound was found to be different.

There remained also five substituted cyclopentanones, 3-propyl- and 3-isopropylcyclopentanone, 3-ethyl-3-methylcyclopentanone, and the *cis*- and *trans*- isomers of 3-ethyl-4-methylcyclopentanone. All of these had been prepared at least once and the first two seemed to be definitely different. The 3-ethyl-3-methyl ketone had been prepared by von Braun and found to be different, but as his method of synthesis would be expected to lead to mixtures of isomers at least, there was some doubt about it; and in the case of 3-ethyl-4-methyl-cyclopentanone, it was found that von Braun prepared one form which was different, but did not mention the other one. Since he had mentioned isomers earlier in the synthesis, it was assumed that he probably obtained only one form and failed to mention the other one.

Goodman, accordingly, proceeded to prepare both forms of 3-

ethyl-4-methylcyclopentanone. He used the following series of reactions:

$$CH_3—CO—CH(Et)—COOEt \xrightarrow[NaOEt]{Cl—CH_2—COOEt} CH_3—CO—C(Et)(CH_2—COOEt)—COOEt \quad (I)$$

$$\xrightarrow[HCl]{5\%\ NaOH,\ then} CH_3—CO—CH(Et)CH_2COOH \xrightarrow[HCl]{CH_3OH} \quad (II)$$

$$CH_3—CO—CH(Et)CH_2COOCH_3 \xrightarrow[NH_4Ac\ +\ HAc]{NC—CH_2—COOEt} \quad (III)$$

$$EtOOC—C(CN)═C(CH_3)—CH(Et)—CH_2—COOCH_3 \quad (IV)$$

$$\xrightarrow[Pt]{H_2} EtOOC—CH(CN)—CH(CH_3)—CH(Et)—CH_2—COOCH_3 \xrightarrow{conc.\ HCl} \quad (V)$$

$$HOOC—CH_2—CH(CH_3)—CH(Et)—CH_2—COOH \ (VI) + CH_3HC—CH(Et),\ H_2C—CH_2,\ C{=}O \ (VII)$$

$$(VI) \xrightarrow{Ba(OH)_2} CH_3HC—CH(Et),\ H_2C—CH_2,\ C{=}O \ (VIII)$$

Product I was obtained by the procedure of Blaise [10] in poor yield, due to its ketonic cleavage. Berry later found that the yield was better if the bromoester was used, as suggested by Goodman. As II appeared to decompose during distillation, it was esterified and purified as the ester. IV was prepared by the procedure of Cope[25] as modified by Cragoe [27] and gave a single product, but only in 50% yield. Catalytic hydrogenation to produce V was an erratic reaction. Goodman found that a second batch of Adam's catalyst was usually needed and that the operation required 24 hours at room temperature and about 20 lb pressure, with 4 volumes of methanol as solvent. Berry later observed that a very active catalyst completed the hydrogenation in about half that time, but yielded only one isomer of the final ketone. Goodman and Berry both found that a viscous residue of 15 to 20% was always obtained when V was distilled at 5 mm pressure, but the final total yield of V was nearly 80%. Treatment of V with concentrated HCl yielded a small amount of the ketone VII, along with the expected product VI. When VI was heated with barium hydroxide, the main yield of ketone was obtained. This was found to boil over the range of 173° to 175°C (343° to 347°F) and, on careful fractionation, yielded two fractions. The alpha (low-boiling) form boiled at 171°C (340°F) and showed values of n_D^{20} = 1.4384 and d_4^{20} = 0.8930. The semicarbazone recrystallized from 50% methanol melted at 163° to 165°C (325° to 329°F) and the 2,5-bis-*p*-nitrobenzylidine derivative, recrystallized from benzene and petroleum ether, melted at 185° to 186°C (365° to 367°F). The other ketone had the properties of the one synthesized by von Braun and thus was not the form he had obtained on degrading a naphthenic acid. Unfortunately, however, when the corresponding acetic acid was synthesized by the better-yielding method used by Mukherji,[35] it had the following properties:

Boiling point at 12 mm = 144° to 145°C (291° to 293°F); n_D^{20} = 1.4557; d_4^{20} = 0.9682. Amide, melting point = 122° to 123°C (252° to 253°F). Ureide, melting point = 190° to 191°C (374° to 376°F). Anilide melting point = 82° to 85°C (180° to 185°F).

The corresponding values for von Braun's synthetic acid from the natural ketone were:

144° to 145°C (291° to 293°F), 1.4609; 0.9783; 122° to 123°C (252° to 253°F); 197° to 198°C (387° to 388°F); and 55° to 58°C (131° to 136°F).

These data showed definitely that the two acids were different. Since the method of synthesis had unfortunately not been the one used by von Braun, it was felt that Goodman's synthesis might have yielded the *cis-trans* isomer of von Braun's acid. Therefore, the synthesis of the ketone was repeated by what appeared to be a shorter and better method.

The same synthesis was used as by von Braun [14] and later by Asahina and Okazaki,[2] but the Japanese workers had not recorded any properties, merely stating that their synthesis gave the same results as von Braun's. It was hoped that perhaps von Braun did not fractionate his ketone carefully enough to notice the lower-boiling ketone. Von Braun's synthesis was followed closely up to the reduction of the diester, in which Goodman used the modern lithiumaluminumhydride, and the next step, in which the conversion of the glycol to the dibromide was accomplished by the method of Adams and Noller.[36] Finally, the cyclization was done by the usual $Ba(OH)_2$ reaction, instead of using ferrous sulfate as von Braun. Unfortunately, no trace of the low-boiling ketone could be detected in the product and work on the problem was temporarily discontinued.

Two years later, Berry, in the author's laboratory, again took up the synthesis of the low-boiling ketone, after preliminary attempts to obtain samples of von Braun's compounds failed. He tried to repeat the synthesis of the low-boiling ketone by using Goodman's first procedure, but was unable to isolate any of the form which had been obtained as a minor product in only one of Goodman's experiments. It is not known definitely at what stage in Goodman's synthesis the mixture was accidentally obtained, but it seems probable that it was in the catalytic hydrogenation in which two different *dl* or racemic mixtures should be obtainable. It is known that, while catalytic hydrogenation usually yields the *cis* isomer when *cis* and *trans* products are possible, the use of old or slightly active catalyst sometimes yields a mixture of *cis* and *trans* isomers.[38] The fact that Goodman found it difficult to hydrogenate the compound with his catalyst and had to shake the mixture for 24 hours and add an extra quantity of catalyst may indicate that he accidentally had a catalyst of the correct activity. Berry's experiments. in which very active catalyst yielded no trace of the low-boiling ketone, seems to agree with this theory. However, a series of trials in which old or used catalyst was employed also failed to yield any of von Braun's petroleum-derived ketone.

Berry used ethyl bromoacetate, as had been suggested by Goodman, and obtained a 70% yield where Goodman's had been 30%. All other steps gave the yields and products reported by Goodman, until the catalytic hydrogenation was reached. At this stage, the yields were always poorer than those reported by Goodman and the yield of a viscous by-product remaining in the still pot at the end of the distillation was sometimes as high as 50% instead of the 20 to 30% reported by Goodman. Another difference was observed by Berry. While in Goodman's case, the amount of cyclopentanone obtained directly when the diester was saponified by concentrated hydrochloric acid was the minor product, it was the main product in Berry's work. Berry's VI solidified on standing, as had von Braun's, while Goodman's had remained a viscous liquid in spite of efforts to make it solidify. Possibly, the stereoisomeric form, which yielded the low-boiling ketone in one of Goodman's runs, was the one which was lost to form the viscous by-product observed by both Goodman and Berry — in much larger amounts by Berry.

Since the high-boiling ketone obtained synthetically by von Braun and also obtained as the usual product by both Goodman and Berry was now available in considerable amounts, it seemed worth while to attempt to determine the configuration of the high-boiling form by converting it to 1-ethyl-2-methylcyclopentane and determining its structure by comparison with the known hydrocarbons. The removal of the carbonyl oxygen by the Huang-Minlen [30] modification of the Wolff-Kishner reduction was readily accomplished and the substituted cyclopentane obtained was found to have the properties reported fro the *trans* hydrocarbon.[23, 24] From this, it is apparent that the high-boiling ketone (which is not the one obtained on degradation of von Braun's acid) must be the *trans* form; the low-boiling one should then be the *cis* ketone.

While Goodman's results, as far as the low-boiling ketone was concerned, were definite and clear cut, his failure to obtain the acid reported by von Braun and the final failure to obtain any of von Braun's samples from Germany (Professor Karl Ziegler undertook to ascertain whether any of von Braun's preparations with which we were concerned were still available and reported that there was none to be found) made it desirable to attempt to find some other route that might lead to the desired ketone — perhaps as a major product.

The first new route attempted was patterned after the syntheses of Best and Thorpe [9] and Carpenter and Perkin,[21] in which they prepared cyclic compounds by reaction of the sodio derivative of ethyl malonate or of ethyl cyanoacetate with ethylene dibromide and had obtained good yields of cyclopropane and cyclopentane derivatives. It seemed doubtful that 2,3-dibromopentane would react smoothly to yield 3-ethyl-4-methylcyclopentanone directly, but as trial runs could be made in a short time, this scheme was tried. The yield consisted mainly of gum, but there was a low yield of liquid which was converted directly to the semicarbazone melting at 210° to 211°C (410° to 412°F), i.e., the ketone was obviously again the wrong one.

On the suggestion of Professor Ettlinger of Rice Institute, a Diels-Alder synthesis was next attempted:

$$CH_2{=}CH{-}CH{=}CH_2 + CH(COCH_3){=}CHCH_3 \xrightarrow{\text{I}} \text{(II)} \xrightarrow{\text{II}}$$

(II): ring of CH_2, $HC{=}HC$, $CCOCH_3$, $CHCH_3$, CH_2

(III): ring of CH_2, $HC{=}HC$, $CHCH_2CH_3$, $CHCH_3$, CH_2

$$\text{(III)} \xrightarrow{\text{III}} \text{HOOC–}CH_2\text{–}CHCH_2CH_3\text{–}CHCH_3\text{–}CH_2\text{–COOH}$$

$$\xrightarrow{\text{IV}} O{=}C\langle(H_2C{-}CHCH_2CH_3)(H_2C{-}CHCH_3)\rangle$$

There was no reliable method of predicting the configuration of II to be expected from the Diels-Alder reaction, but it was hoped that at least a mixture of the *cis* and *trans* forms would be obtained and could be converted without change in configuration to III. Of the three common methods of eliminating the carbonyl oxygen to obtain III, that of reduction to the alcohol, dehydration to the unsaturated compound, and hydrogenation to the saturated compound was immediately eliminated, because the hydrogenation would also convert the cyclohexene ring to the saturated ring — if it did not also change the configuration. The Clemmensen reduction was considered the less desirable, because it was feared that addition of HCl to the cyclohexene double bond might occur and lead to extensive rearrangements in the ring. The Wolff-Kishner type of reduction was tried, in spite of the fact that, under the strongly alkaline conditions required, enolization toward the ring might result in the elimination of one of the *cis-trans* isomers and so lead to only the stable form — probably the *trans*. The ethylmethyl cyclohexene obtained as III was found to boil over a very narrow range, indicating that probably only one isomer was present. Oxidation by a dilute acetone solution of $KMnO_4$ was easily carried out and gave a good yield of adipic acid which yielded again the wrong ketone — the *trans* form.

After an elaborate series of experimental studies in which Berry changed to the use of ethyl-2-ethylideneacetoacetate in the Diels-Alder synthesis, he found that he could obtain a mixture of the *cis* and *trans* forms of II and that this mixture would again lead only to the wrong form of the ketone. From this, it was obvious that the Wolff-Kishner reduction had been accompanied by rearrangement to the stable form. When the Clemmensen reduction with Zn and concentrated HCl was finally employed for removing the carbonyl oxygen to obtain III, it was found that both geometrical isomers were produced and that the final ketone was a mixture of the *cis* and the *trans* ketones, each having properties like those reported initially by Goodman. In all probability, the low-boiling ketone is the one obtained from petroleum acids by von Braun and Berry's very careful work has shown definitely that the higher-boiling ketone is the *trans* form and the lower one the *cis*. Unfortunately, an attempt to synthesize the acetic acid reported by von Braun from the natural ketone again yielded a substituted acetic acid which appears to be different from von Braun's. Whether, for some reason, the acid obtained in the author's laboratory is the pure *cis-trans* isomer of the one obtained by von Braun, or whether von Braun's ketone used in his synthesis was impure and yielded an acetic acid different from the expected one or mixed with a different one, or, finally, whether von Braun's ketone was different in spite of the many points of agreement in properties of the ketone and its derivatives with the Goodman-Berry ketone, will apparently have to await the possible reisolation and characterization of the ketone from petroleum since Richter and Baumgarten [39] report the isolation, in very low yield, of a ketone which gives a semicarbazone melting at 162° to 163°C, or the possible synthesis of yet another ketone which not only has the properties of the von Braun petroleum ketone, but also yields the same acetic acid.

The conclusion is that the von Braun ketone was:

```
                CH3
                 |
   H3C           C————————CH2
    |           /|          \
   H2C         / H           C=O
    |         /             /
    C——————————————CH2————
    |
    H
```

and the acetic acid prepared by him was either;

```
                CH3       H              COOH
                 |        |             /
   H3C           C————————C          CH2
    |           /|        |\        /
   H2C         / H    H   H  \    /
    |         /       |        C
    |        /        |       / \
    C————————————————C———————     H
    |                 |
    H                 H
```

or

```
                CH3       H
                 |        |
   H3C           C————————C         H
    |           /|        |\       /
   H2C         / H    H   H  \   /
    |         /       |        C
    |        /        |       / \
    C————————————————C———————    CH2
    |                 |             \
    H                 H              COOH
```

so that the acid from which his ketone was obtained was one of these geometric isomers or their mixture. Since there seems to be no reason for supposing that there is any fundamental importance to this particular acid, the author plans no further study of the problem.

Bibliography

1. Adams, R., S. E. Loewe, and C. M. Smith, *J. Am. Chem. Soc.* **64,** 2653 (1942).
2. Asahina, Y., and K. Okazaki, *C. A.* **41,** 6235 (1947).
3. von Auwers, K., and E. Lange, *Ann.* **401,** 303 (1913).
4. von Auwers, K., R. Hinterseber, and W. Treppman, *Ann.* **410,** 257 (1915).
5. Baumgarten, H. E., and D. C. Gleason, *J. Org. Chem.* **16,** 1658 (1951).
6. Baumgarten, H. E., and R. L. Eifert, *Ibid.* **18,** 1180 (1953).
7. Berry, J., M.A. Thesis, The University of Texas, 1950.
8. Berry, J., Ph.D. Thesis, The University of Texas, 1953.

9. Best, S. R., and J. F. Thorpe, *J. Chem. Soc.* **95,** 685 (1909).
10. Blaise, E. E., *Bull. soc. chim., France* (**3**), **23,** 918 (1900).
11. Bouveault, L., and G. Blanc, *Compt. rend.* **146,** 235 (1908).
12. von Braun, J., and W. Haensel, *Ber.* **59,** 2008 (1926).
13. von Braun, J., *Ann.* **490,** 100 (1931).
14. von Braun, J., W. Keller, and K. Weissbach, *Ibid.* **490,** 179 (1931).
15. von Braun, J., and E. Anton, *Ber.* **64,** 2865 (1931).
16. von Braun, J., *Ibid.* **66,** 684 (1933).
17. von Braun, J., *Ibid.* **66,** 1499 (1933).
18. von Braun, J., and H. Wittmeyer, *Ibid.* **67*B*,** 1739 (1934).
19. Buchman, E. R., and H. Sargent, *J. Org. Chem.* **7,** 148 (1942).
20. Burger, G., and A. K. Macbeth, *J. Chem. Soc.* **1946,** 145.
21. Carpenter, H., and W. H. Perkin, Jr., *Ibid.* **75,** 921 (1899).
22. Cason, J., *J. Am. Chem. Soc.* **69,** 1548 (1947).
23. Chiurdoglu, G., *Bull. soc. chim. Belg.* **44,** 527 (1935).
24. Chiurdoglu, G., *Ibid.* **53,** 45 (1944).
25. Cope, A. C., C. M. Hofman, C. Wyckoff, and E. Hardenburgh, *J. Am. Chem. Soc.* **63,** 3452 (1941).
26. Cosciug, T., *C. A.* **30,** 2733 (1936).
27. Cragoe, E. J., C. M. Robb, and J. M. Sprague, *J. Org. Chem.* **15,** 281 (1950).
28. Crossley, A. W., and N. Renouf, *J. Chem. Soc.* **91,** 81 (1907).
29. Goodman, H., Ph.D. Thesis, The University of Texas, 1951.
30. Huang-Minlon, *J. Am. Chem. Soc.* **68,** 2487 (1946).
31. Knoevenagel, E., *Ann.* **297,** 163 (1897).
32. Kon, G. A. R., *J. Chem. Soc.* **119,** 824 (1921).
33. Lesser, G., *Bull. soc. chim., France* (3) **21,** 549 (1948).
34. Lochte, H. L., and W. G. Meinschein, *Petroleum Eng.*, March 1950, page 725.3.
35. Mukherji, D., *Science and Culture* **13,** 296 (1948).
36. Noller, C. R., and R. Adams, *J. Am. Chem. Soc.* **48,** 1080 (1926).
37. Okazaki, K., *C. A.* **45,** 2884 (1951).
38. Ott, E., *et al.*, *Ber.* **60,** 624, 1125 (1927) and **61,** 2119 (1928).
39. Richter, G., and H. E. Baumgarten, Private communications.
40. Ruzicka, L., *et al.*, *Helv. Chim. Acta.* **25,** 188 (1942).
41. Ruzicka, L., *et al.*, *Ibid.* **30,** 2168 (1947).
42. Ruzicka, L., O. Jeger, and A. Buchi, *Ibid.* **31,** 241 (1948).
43. Sabatier, P., and A. Maihle, *Compt. rend.* **142,** 554 (1906).
44. Sargent, H., *J. Org. Chem.* **7,** 154 (1942).
45. Ungnade, H. E., and A. D. McLaren, *Ibid.* **10,** 29 (1945).
46. Wallach, O., *Ann.* **339,** 111 (1905).
47. Zelinsky, N., *Ber.* **57,** 2057 (1924).

CHAPTER SEVENTEEN

BICYCLIC, POLYCYCLIC, AND AROMATIC ACIDS

While we know that in petroleum acids from the few fields studied, only simple fatty acids are found up to six-carbon acids and a mixture of fatty and naphthenic acids is found in the six to ten carbon range, our actual knowledge about the acids with more than ten carbons is almost nil. Not a single true naphthenic acid has been definitely identified in the regular commercial naphthenic acid range of twelve to twenty carbon atoms. It is assumed that the monocyclic acids related to those with six to ten carbon atoms are found in all regular naphthenic acids and it is known that acids with a higher carbon-to-hydrogen ratio than that of the monocyclic acids occur in some of the commercial types of acids, because this has been shown in a number of cases by elementary analysis and by density and index-of-refraction measurements. In recent years, several workers have converted some of the higher acids to the corresponding hydrocarbons and have then carried out type-analysis determinations which showed that the hydrocarbons contained bicyclic and polycyclic compounds.

In the course of a prolonged research project on Roumanian, North German, California and Galician naphthenic acids, von Braun [3] found that the Galician acids were rich in aliphatic acids, which appeared to be present in all molecular ranges up to twenty carbons and apparently contained no bicyclic acids. The other acids all showed that bicyclic acids were present in the C_{12} fractions and predominated above thirteen carbons. He concluded that fatty acids were absent above the ten-carbon acids, except in the Galician sample. Whether his conclusion that the Galician acids contain no bicyclic acids is correct remains to be seen. Since he depended on analyses and on density and since we know that none of his fractions consisted of a pure acid, his results may have represented average values of aliphatic, monocyclic and bicyclic acids and thus would indicate that bicyclic acids were absent. Von Braun worked not only with acids but also with amines obtained by K. F. Schmidt degradation and with ketones obtained by degradation of his acids and in all of these, he concluded that bicyclic compounds were obtained.

Von Braun was desperately anxious to isolate and identify one of the bicyclic acids or one of the degradation products, but since he depended almost entirely on fractional distillation followed by preparation of solid derivatives which could be purified, it is not surprising that he was unable to achieve his goal. When he could not isolate a solid of any type, he synthesized acids which seemed likely to be present and compared their physical properties with those of the fractions he studied, but he was not able to establish even a probable structure of a bicyclic acid. In one of his letters to J. R. Bailey, he said:

> I continue to have great troubles with the structure of the bicyclic naphthenic acid. I should like to get my hand on enough of this acid, $C_{13}H_{22}O_2$, to get enough pure ketone, $C_{11}H_{18}O$, to prove its structure, but everything that I have done with enormous expenditure of time and effort has led to too little and failure. I have been very much discouraged and depressed and, to recuperate, have from time to time been working on other problems not connected with naphthenic acids.

Muller and Pilat,[10] in an attempt to prepare pure hydrocarbons of high molecular weight to be tested for their lubricating value, decided to proceed by way of complete hydrogenation of pure high-molecular-weight naphthenic acids. As might be expected, they did not obtain any pure naphthenic acid of this type, but they did carry out very careful extractions to remove hydrocarbons, phenols, and other nonacidic compounds. This careful work was followed by careful fractionation of the acids, and then of the methyl esters, all at 1 mm or less pressure. Reduction of the esters by sodium and alcohol, conversion to the iodides, and reduction of these to the hydrocarbons by zinc dust and 20% hydrochloric acid yielded a product still containing a little sulfur and oxygen, and also some aromatic hydrocarbons which were apparently derived from aromatic acids in the original mixture. Ring analysis of the hydrocarbons indicated the presence of bicyclic acids. Meinschein was able to obtain very small amounts of aromatic acids in the fractions with very high index of refraction.[9] His California acids had been very carefully fractionated as acids and as esters and had then been fractionally neutralized and liberated systematically in special equipment. The weakest acids in the ten- to twelve-carbon range had an $n \times d$ value of 1.54 and gave qualitative tests for aromatics, but the amount available was too small to permit definite identification. At best, the concentration was little more than a trace. If aromatic acids, possibly related to tetrahydronaphthalene or indene, were present, they would be expected to lead to wrong conclusions if characterization is confined to analyses, density, and index of refraction, since a small amount of such a compound in a mixture would have an extensive effect on these results.

Since von Braun's contributions, reports on the occurrence of bicyclic acids in samples of naphthenic acids have accumulated. Chichibabin and coworkers reported that they could not find any evidence for bicyclic acid in Baku naphthenic acids. Lapkin [7] studied acids from Grozny petroleum and concluded, in agreement with von Braun, that these acids were monocyclic up to C_{12}, were mixed mono- and bicyclic from C_{12} to C_{14}, and bicyclic from C_{14} to C_{18}. Balada and Wegiel [2] found that acids from Gbley petroleum are aliphatic to C_8, monocyclic from C_8 to C_{10}, and a mixture of mono- and bicyclic acids from C_{10} to C_{14}, after which more complex acids seem to be in the mixture. These investigators appear to have depended on properties of the acids rather than of the hydrocarbons derived from them. Schmitz [12] apparently agreed wholeheartedly with von Braun in his theory that the fundamental ring of the naphthenic acids is the same for all monocyclic acids and probably also for the bicyclic series. His lecture included a discussion of a $C_{13}H_{22}O_2$ acid degraded to a ketone, $C_{11}H_{18}O$, which, in his opinion, had a bicyclic nucleus similar to that of camphor.

Harkness and Brunn [5] carefully removed phenols and nonacidic impurities from high-boiling acids isolated from Texas Gulf Coast crude oil. They fractionated the resulting acids at high vacuum in a molecular still operated so as to obtain six fractions. They combined similar fractions and repeated this fractionation as shown in Figure 9.

Table 28 presents data obtained on the various cuts.

Table 28. **Properties of High Molecular Weight Texas Acids**

Fraction	*D-1*	*D-3*	*I-3*	*E-5*	*J-3*	*H-5*
Saponification Number	230	213	191	175	156	129
Acid Number	221	210	190	176	155	129
n_D^{20}	1.4879	1.4915	1.4950	—	1.4004	1.5020
d_4^{20}	0.944	0.9954	0.9909	—	0.9752	0.9756
$n \times d$	1.479	1.484	1.481	—	1.463	1.465
Molecular Weight (Det.)	219	251	300	320	382	440
Molecular Weight (from Saponification Number)	244	263	294	321	360	435
Carbon, %	74.69	74.46	76.65	77.95	78.86	80.19
Hydrogen, %	10.44	10.81	10.74	11.08	11.07	11.27
Oxygen, %	14.87	12.73	12.61	10.97	10.07	8.54
COOH/Mol	0.90	0.96	1.02	1.00	1.06	1.01
Carbons/Mol	13.6	16.0	19.2	20.8	25.1	29.4
Hydrogen/Mol	22.7	27.0	32.0	35.3	42.0	49.2
Approximate Type Formula	$C_nH_{2n-4}O_2$ and $C_nH_{2n-6}O_2$	$C_nH_{2n-6}O_2$	$C_nH_{2n-4}O_2$ and $C_nH_{2n-6}O_2$	$C_nH_{2n-8}O_2$	$C_nH_{2n-6}O_2$ and $C_nH_{2n-8}O_2$	$C_nH_{2n-8}O_2$ and $C_nH_{2n-10}O_2$

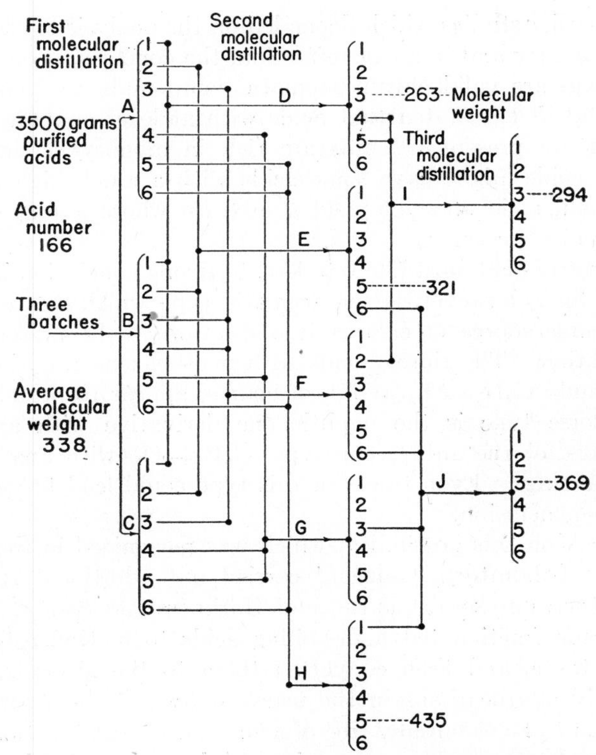

Figure 9. FRACTIONATION BY MOLECULAR DISTILLATION
[Reprinted by permission from *Ind. Eng. Chem.* **32,** 501 (1940).]

A study of the n × d values obtained from the data of Harkness and Bruun and of the value calculated from the data of Adams and coworkers [1] indicates that these values, which had been tested for acids with less than eleven carbons, do not continue to rise with molecular weight, but may actually decrease if long chains are present in the molecule. The values of 1.479, 1.484, and 1.481 would not be unexpected for monocyclic acids containing moderate concentrations of bicyclic or higher acids in the C_{13} to C_{16} range, but values of 1.463 and 1.465 for tri- and tetracyclic acids with twenty-five to twenty-nine carbons would probably indicate that a fairly large portion of the molecules consists of one or two long chains to make up the total number of carbons. An anthracene derivative would have eleven to fifteen carbons available for chains and the single carboxyl group.

As in earlier studies which depended on the analysis of the fraction and the density and index of refraction, the conclusions on the number of rings are valid only if aromatic compounds are known to be absent and if it is stated that acids with at least as many rings as indicated are present. A mixture rich in monocyclic acids or in aliphatic acids might have some acids with a much higher carbon-to-hydrogen ratio and yet yield a mixture whose average analysis indicates bicyclic acids.

The type of acid most likely to lead to wrong conclusion in a study on these lines, however, is the aromatic type which was recognized as a possible source of error and was known to be present in low concentration. The simple acids with one benzene ring, i.e., of the type formula $C_nH_{2n-8}O_2$, would, of course, appear to be tetracyclic; but a worse type is the naphthalene derivative with apparently seven rings, or the anthracene type, $C_nH_{2n-2}O_2$, with analysis indicating ten rings. Even traces of this type could lead to an entirely erroneous conclusion.

To check on this possibility, which was recognized in the Sun Oil Company Laboratory, Goheen [4] carried out additional studies on acids of the same source as those of Harkness and Bruun. He converted some fractionated high-boiling acids to methyl esters, fractionated these, and then converted them to the alcohols, iodides, and finally hydrocarbons in the usual manner. The hydrocarbons were found to be essentially free of acids and of unsaturation toward potassium permanganate or bromine. A low concentration of alcohols was found, but no iodide. Various tests were made on the fractions of crude hydrocarbons, parts of which were treated with cold, concentrated sulfuric acid and again tested. Other portions of the hydrocarbon were subjected to catalytic hydrogenation for 25 hours and then for 50 hours and tested at each stage. Table 28 presents the data obtained in this work.

While these two contributions still give us no information on the structure, or even the type, of bicyclic or polycyclic acids or of the aromatic acids present in low concentration, they together probably give us more information than any other work on bicyclic acids, except possibly that of von Braun who reported degradation of some of his bicyclic acids to bicyclic amines and ketones of unknown type and structure.

Table 29. **Properties of Hydrocarbons from Texas Acids**

	Crude	H_2SO_4 Treated	Hydrogenation 25 Hours	Hydrogenation 50 Hours
d_4^{20}	0.8920	0.8895	0.8835	0.8815
n_D^{20}	1.4849	1.4837	1.4794	1.4781
r_D^{20}	0.3212	0.3215	0.3210	0.3212
$n-d/2$	1.0388	1.0390	1.0375	1.0374
$(n_f-n_c)/d \times 10^4$	107	105	99	98
Carbon, %	86.51	86.54	86.47	86.46
Hydrogen, %	13.08	13.10	13.34	13.41
Average Type Formula	$C_nH_{2n-4.2}O_{0.1}$	$C_nH_{2n-4.1}O_{0.1}$	$C_nH_{2n-3.4}$	$C_nH_{2n-3.2}$

E. Neyman-Pilat and S. Pilat,[11] in a critical study of lubricating oil properties, as related to structure and type of molecule of the hydrocarbons in the oils, extended their previous study of hydrocarbons derived from highly fractionated high-boiling naphthenic acids [10] and concluded, on the basis of ring-type analysis, that there appeared to be a mixture of bicyclic and tricyclic compounds; but in view of the presence of 10 to 15% of aromatic hydrocarbons, they suggested that the original acids were probably bicyclic — possibly of a condensed ring (decalin) type.

Since 1950, Meinschein, and later Jones, have attempted, in the author's laboratory, to determine the nature of bicyclic acids. The first study was on California naphthenic acids, which were found to contain only low concentrations of acids with a high index of refraction. Meinschein fractionated C_{10} to C_{14} acids first as acids and then as methyl esters. To determine whether bicyclic acids actually start at C_{12}, as claimed by von Braun and by others, he studied fractions titrating as C_{10} to C_{12} acids by systematic fractional neutralization and fractional liberation in a bottle machine.[8] Even with solutions containing less than 10% of acids in petroleum ether in the relatively gently agitated bottle machine, there was considerable emulsion trouble and the twenty-stage separation was tedious. A plot of index of refraction against neutralization cut number showed that the weakest (least soluble in alkali) cuts obtained had an $n \times d$ value of 1.54–5, but gave no test for phenols. Nitration, reduction, and diazotization of small fractions led to compounds which could be coupled to yield colored compounds. Oxidation of similar samples led to solid acids in low yield. In addition to the very low yield of what were apparently aromatic acids, there was a 15 to 20% yield of acids with $n \times d$ values of about

1.48 which were still complex mixtures containing, probably, bicyclic acids but, possibly, also aromatic and monocyclic acids with ten to twelve carbons. The yield of any acids with high n × d values from this California acid mixture was so low that large volumes would have to be fractionated to get cuts of sufficient size to permit further fractionation by distribution and by chromatographic methods.

Study was then begun of a Texas Gulf Coast acid which appears to have a much higher concentration of acids with a high index of refraction and high density. No attempt was made to isolate bicyclic acids with ten to twelve carbons and attention was directed instead to the thirteen- and fourteen-carbon range. Because of the higher vacuum required during fractionation and because of greatly increased emulsifying tendency, even in dilute alcohol solutions, Jones [6] found the separation operations here to be even more tedious and time consuming than in work with the California acids. The high n × d fractions finally obtained were then passed through silica-gel columns and yielded a series of fractions differing greatly in properties, with some of them very highly viscous and turning glassy on cooling. It was hoped that some acid could be obtained as a solid which might be purified by recrystallization, but so far no fraction could be obtained in crystalline form.

A study of various possible or plausible types of bicyclic acids previously described or synthesized in the author's laboratory indicates that nearly all possible types of acids are solids, although some are low-melting solids which might not be recrystallized easily. All have properties which could be averaged with other types of acids probably present in naphthenic acid fractions to yield properties observed for such fractions. We must admit that we know little about the concentration of bicyclic acids in naphthenic acid fractions, nothing definite about the nature of such acids, and nothing at all about their structure — unless we accept, in general, the statement by von Braun that they are predominatingly substituted aectic acids.

Bibliography

1. Adams, R., and coworkers, *J. Am. Chem. Soc.* **49,** 2939 (1927); **50,** 1478, 1508, and 2298 (1928).
2. Balada, A., and J. Wegiel, *C. A.* **31,** 2399 (1937).
3. von Braun, J., *Ann.* **490,** 100 (1931).
4. Goheen, G. E., *Ind. Eng. Chem.* **32,** 503 (1940).

5. Harkness, R. W., and J. Bruun, *Ibid.* **32,** 499 (1940).
6. Jones, W. A., Sun Oil Company Fellow, 1952–1954.
7. Lapkin, L. I., *C. A.* **34,** 611 (1940).
8. Lochte, H. L., and W. G. Meinschein, *Petroleum Eng.* March 1950, page 725.3.
9. Meinschein, W. G., Ph.D. Thesis, The University of Texas, 1952.
10. Muller, J., and S. Pilat, *Brennstoff-Chem.* **17,** 461 (1936).
11. Neyman-Pilat, E., and S. Pilat, *Ind. Eng. Chem.* **33,** 1390 (1941).
12. Schmitz, P. M. E., *Bull. assoc. française tech. petrole* **46,** 93–148 (1938).

CHAPTER EIGHTEEN

RECENT STUDIES ON NAPHTHENIC ACIDS

A period of major activity in naphthenic acid research was that of the 20-year span from 1920 to 1940 and thus, this should logically be treated as a unit. Von Braun's work was so important, however, that it has been discussed in separate chapters, even though it occurred in the middle of this period. Important results published between 1920 and 1932 have been covered in other chapters, and the work reported since 1932 will be covered in this chapter on recent studies.

This period saw great advances in separation methods and, mainly for this reason, saw also the first definite identification of a number of individual naphthenic acids.

Chichibabin's Work on Baku Acids

After the appearance of von Braun's main paper [4] on Roumanian acids, Chichibabin [10] published a progress report on work with a number of coworkers on Baku acids. The project was started in 1911, was interrupted by World War I, and was not started again until 1929. He had made two preliminary reports in 1913 and 1929. Although he intended to continue his investigations, there appears to have been no report since 1932.

After giving von Braun credit for results accomplished through the use of the Schmidt hydrazoic acid degradation, the chlorine-number test for hydrogens on the α-carbon atom, and the lactone test for longer side chains, Chichibabin reported on his own study of 416 kg of dark-brown, viscous Baku acids. A batch of it was neutralized, evaporated to dryness in vacuum, and extracted with petroleum ether to remove phenols and neutral impurities. He found about 12% of such impurities.

He topped his main batch of acids from a 30 kg copper still, working at atmospheric pressure to 290°C (554°F). Even below this, he noted some decomposition. These lower acids were purified by steam distillation of their sodium salt solution and a loss of about 24% was found, mainly of hydrocarbons. His hydrocarbons boiled in the same range as the acids and thus could not have been formed by decarboxylation of his acids since such hydrocarbons would boil much lower than the parent acids. He repeated the observation [2] that the naphthenic acids liberate HCl from calcium chloride. Therefore, he dried his fractions with anhydrous sodium sulfate.

He fractionated the acids as such and then very carefully fractionated the methyl esters of these cuts. The esters were fractionated eight times through a 50 cm column and collected as narrow-boiling fractions for which analyses, boiling point, density, index of refraction, and optical activity were noted. Only a few fractions were optically active, but the one boiling at 167° to 170°C (333° to 338°F) showed a rotation of −1.01 in a 100-mm tube. He found that the acids isolated from the ester showed activity in the same direction. Consequently, he concluded that the activity was not due to an impurity (hydrocarbon) in the ester.

He noticed that the density of his lower acids was much lower than would be expected for cyclic acids. Therefore, he calculated an average value for it and for the index of refraction for a number of cyclic acids and a number of fatty acids of the approximate molecular weight range. Using these values and assuming that his petroleum acids were typical acids of either type, he calculated the percentage composition of his fractions. He found that his lower acids were predominantly fatty in nature, but that in the C_{10} range, 90 to 100% of his acids were cyclic. These concentrations were, of course, based on the assumption that hydrocarbons, phenols, and bicyclic acids were absent.

A comparison of his own data with those of Markownikoff [25, 26] and Aschan [2, 3] led him to the conclusion that the other workers had acids with somewhat higher naphthenic acid content than his own.

He first attempted the isolation and identification of individual acids by recrystallization of solid amides. He finally obtained pure amides of 5-methylhexanoic and 3-ethylpentanoic acids and found no depression of melting point when mixing them with the synthetic amides. Later work on other amides showed that sometimes there was no depression of melting point when a different amide was mixed with a known one and he found that this occurred with the amide of 5-methylhexanoic acid which made it necessary for him to confirm its structure by mixed melting points of other derivatives.

Since the amides could rarely be purified by recrystallization, he next tried separation of acid types by difference in solubility of their copper salts in petroleum ether. He concluded that the copper salts of:

1. secondary cyclic acids were slightly soluble in petroleum ether,
2. aliphatic and tertiary cyclic acids were insoluble in petroleum ether, and
3. normal fatty acids and cyclic acids, with the carboxyl group at the end of a long side chain, were soluble only in the presence of free acidity.

Trials on known mixtures showed, however, that the solubility in petroleum ether containing dissolved copper salt was different from that in pure petroleum ether. Therefore, he abandoned this scheme. He and his coworkers apparently did not determine the effect of dilution.

They then tried fractional precipitation of the cadmium salts of acids by slowly adding aliquots of dilute cadmium chloride solutions to sodium salt solutions. They were able to separate *n*-pentanoic acid from cyclohexanecarboxylic acid easily and, by using a second fractionation, also cyclopentanecarboxylic acid from 5-methylhexanoic acid [boiling points 216°C and 217°C to 218°C (about (423°F)].

Cyclohexanecarboxylic Acid

When Chichibabin and his associates worked on acids obtained from highly fractionated methylesters of their petroleum acids, they were able to determine the approximate ratio of fatty acids to cyclic acids readily; but they did not identify any individual acids until they went to the boiling point range including cyclohexanecarboxylic acid [boiling point of ester, 181°C to 184°C (358° to 363°F)]. This range yielded an acid whose amide was purified and whose methylester was dehydrogenated over platinum by Zelinsky's method to yield, finally, benzoic acid, identified by melting point of the acid and odor of the methylester. The final yield of pure compounds was very low, however.

Degradation Experiments

Chichibabin degraded his acids by converting them to the α-bromoacidbromides by the Hell-Volhard-Zelinsky reaction, then changing these to the α-hydroxyacids which were heated to dehydrate them or degrade them to the lower aldehyde or ketone, with formation of formic acid as by-product. These reactions ran smoothly and with good yields to the pyrolysis stage and even here, the yield seems much better than those obtained by von Braun's methods. For instance, 20 g of hydroxyacid yielded 6.8 g of unsaturated acid formed by dehydration, 1.9 g of ketones, and 2 g of crude acids obtained from the aldehydes. Experiments with cyclohexanecarboxylic acid showed that only a trace of cyclohexanone was obtained, while the main product was the cyclohexenecarboxylic acid expected from dehydration of the tertiary alcohol. His high yield of unsaturated acids and low yield of ketones indicated that the Baku acids had largely the carboxyl attached directly to the ring, instead of at least one carbon removed from the ring as in the Roumanian acids of von Braun.

Chichibabin did not identify any of his degradation products.

His work resulted in the identification of 5-methylhexanoic and 3-ethylpentanoic acids among the fatty acids and also cyclohexanecarboxylic acid — the first naphthenic acid, which ironically turned out to be a type considered of minor importance among naphthenic acids. It should be realized that, since it was present in low concentration, if identification through benzoic acid had not been possible, this acid would probably not have been recognized. His work also pointed out the possibility of separating types of acids through fractional precipitation of their cadmium salts and showed that in the case of Baku acids, most of the naphthenic acid molecules have the carboxyl group attached directly to the ring. A few fractions showed optical activity which was proved to be due to the acids and not to hydrocarbon impurities.

Reports of Other Workers

Smirnov and Buks,[44] in 1932, reported on what appears to have been a series of fractional distillations of Grozny acids and their methylesters. They extracted a non-wax-bearing distillate of 0.825 to 0.850 density with alkali, until a total of 12 kg of naphthenic acids had been accumulated. The sodium salts were dissolved in ethanol at a ratio of 100 cc per 150 g of acids and the solution was extracted repeatedly with petroleum ether. The alcohol was recovered and the residual solution evaporated to dryness on a water-bath. The acids were liberated by 50% sulfuric acid and taken up in petroleum ether. The sulfuric acid was removed by washing with sodium sulfate solution and then with water. The acids were dried by anhydrous sodium sulfate, and finally by phosphorus pentoxide, before removing the petroleum ether by distillation. The resulting dry acids were fractionated at 3 mm and cuts collected from 140° to 195°C (284° to 383°F). The total weight of distilled acids was 7,871 g from the 12 kg of crude acids.

The acids were converted to methyl esters in 90% yield and these were repeatedly refractionated at 3 mm pressure to yield, finally, thirty-nine 2° cuts which were saponified; the resulting acids were individually refractionated to yield, in each case, a "heart" cut boiling over a 2° range. These cuts were individually fractionated to yield a total of one hundred fourteen fractions, each boiling over a 2° range and were then combined, on the basis of boiling point and index of refraction, to yield forty-six final fractions of acids.

For the larger fractions that appeared to consist of single sets of isomers, they ran combustion analyses, molecular weight determinations (by neutralization equivalent), and density at three different temperatures, in addition to determining the index of refraction and weight of each sample.

Unfortunately, for some reason, they tabulated by index of refraction instead of by boiling point. When an attempt is made to plot boiling point against index of refraction or any of the other properties, the curve obtained is so irregular that obviously the pressure must have been far from 3 mm at times and the fact that some fractions are much larger than others loses significance, particularly when it is noticed that adjacent cuts may be unusually small.

They tabulated density, index of refraction, and boiling point of acids from $C_{10}H_{18}O_2$ to $C_{15}H_{28}O_2$ and for esters from $C_9H_{17}COOCH_3$ to $C_{14}H_{27}COOCH_3$ obtained in their own work, along with data reported for isomeric Galician acids by Kozicki and Pilat [22] for Roumanian acids by Frankopol,[12] and for Nishiyama and Kurokawa, Japanese acids by Tanaka and Nagai.[45, 46] They concluded that the large differences noted could be due either to the presence, in some cases, of mixtures of isomers or to differences among isomers.

In a short communication, Pilat and Reyman [34] stated that they had decided to prove whether naphthenic acids isolated from crude oil were similar to those isolated from distillates or whether they were "man made" during refining and distilling petroleum, as von Braun had maintained early in his studies.

To do this, they used a special derivative which von Braun [5] had developed and had found to yield nicely crystallizing compounds from fractions of naphthenic acids. The preparation consisted of the following steps:

$$R{-}COOH \rightarrow R{-}COOEt \rightarrow R{-}CH_2OH \rightarrow R{-}CH_2Br$$

$$R{-}CH_2Br + Na{-}CH(COOEt)_2 \rightarrow RCH_2{-}CH(COOEt)_2$$

$$R{-}CH_2{-}CH(COOEt)_2 + \text{Urea} \rightarrow R{-}CH_2{-}CH\begin{matrix} \diagup CO{-}NH \diagdown \\ \\ \diagdown CO{-}NH \diagup \end{matrix}CO$$

The final product was a barbiturate.

Starting with acids from a Galician crude and from a refinery fraction of the same oil, they obtained barbiturates which were obviously very similar in nature, although no individual compound was obtained, or expected, from either source. This demonstration convinced von Braun, who also reported similarity in properties of the higher acids from both crude and distilled Roumanian acids.[4] This aspect of the origin of naphthenic acids has not been extensively questioned since.

Williams and Richter [50] studied acids from a West Texas pressure distillate. As might be expected of a cracking-process product, this sample contained a high concentration of phenols of which phenol, *o*-, *m*-, and *p*-cresol, 1,3,5- and 1,4,2-xylenol were isolated and identified. The acids obtained from the purified and fractionated methyl esters were found to contain 3-methylbutanoic and *n*-heptanoic, *n*-octanoic, and *n*-nonanoic acids, but no cyclic acids could be isolated. The cyclic acids could have been destroyed in the Dubbs process distillation, might have escaped isolation, or might not have been present in this batch of acids. Williams and Richter felt cyclic acids were absent. It would be interesting to study straight-run acids from this field, but it is usually impossible to obtain such acids from refineries.

In later work,[37] Richter and students degraded some Texas C_{10} acids to the lower ketone via the α–β–unsaturated acid and obtained a small yield of ketone whose semicarbazone melted around 162°C

(324°F), but the amount isolated was too small to permit complete characterization. This may have been von Braun's ketone, or it may have been the other ketone he obtained with a semicarbazone melting at 162°C (324°F).

Müller and Pilat [28] were interested in the lubricating properties of high-molecular-weight hydrocarbons of the types found in petroleum. Instead of synthesizing individual compounds presumed to be of the type found in petroleum, as was done by Mikeska,[27] they decided that the best route was through careful fractionation of naphthenic acids and conversion of the fractions to the corresponding hydrocarbons by the familiar series of reactions:

$$R{-}COOH \rightarrow R{-}COOCH_3 \rightarrow R{-}CH_2OH \rightarrow R{-}CH_2I \rightarrow R{-}CH_3$$

Needless to say, they did not expect to obtain individual compounds in this way, but they hoped to be able to arrive at mixtures of isomers and adjacent homologues that, they hoped, would prove as satisfactory as the arbitrary synthesis of individual compounds which might, or might not, be similar in structure to the hydrocarbons found in lubricating oil. In their assumption that the naphthenic acids were similar in structure, but for the functional group, to the hydrocarbons from which they were isolated, they were in agreement with von Braun and many other workers.

They took great care in removing hydrocarbons and phenols by elaborate cross extractions of both phases whenever two occurred and they used a pressure of only 1.0 mm to distill the high-molecular-weight material; but otherwise their separations were essentially the same as those of others. Since they found some decomposition of their esters on distillation, even at 0.1 mm, they finally used a Waterman molecular still to distill the highest-boiling fractions.

Their high-boiling naphthenic acids were from Polish paraffin-free stock and were found to include mainly acids of twenty to twenty-five carbons. Emulsions were very troublesome and even the use of alcoholic instead of aqueous sodium hydroxide gave only partial relief and resulted in partial esterification of the acids by the solvent. Consequently, they had to remove and saponify the ester at a later stage.

The final product, obtained after fractionation of the acids and their methyl esters at 0.1 mm and by the molecular still in the case of the last fractions, was converted fraction by fraction to the hydrocarbons by the reactions mentioned earlier. They studied these in the form of four fractions of hydrocarbons, each of them still contaminated by sulfur and oxygen-containing and aromatic compounds. Treatment with 90% sulfuric acid for an hour left only traces of sulfur in all except fraction 3 — a Waterman still fraction — which also still showed some aromatic content, as indicated by the high index of refraction. They decided to remove sulfur, oxygen, and also aromatics by selective extraction with a 6% solution of water in pyridine. Three such extractions removed all of the sulfur, but seven were required to remove all aromatic hydrocarbons.

They were not able to state whether the aromatic hydrocarbons were formed from aromatic acids or were hydrocarbons which had not been removed in purification and fractionation, although it seems unlikely that such hydrocarbons would not have been removed at some of the many purifications and fractionation steps.

Solid hydrocarbons were obtained on cooling fractions below 0°C (32°F) and recrystallization and analysis of these indicated that they were naphthenic in nature.

In a later report,[33] E. Neyman Pilat and S. Pilat concluded, after further study of the original material, that while the hydrocarbons appeared to be a mixture of bi- and tri-cyclic hydrocarbons, they were probably a mixture of bicyclic and 10 to 15% aromatic hydrocarbons and that the bicyclic hydrocarbons were probably of the alkylated decalin type.

In 1938, Nenitzescu and coworkers[29] reported that so far not a single pure naphthenic acid had been isolated from petroleum. They referred to Chichibabin's identification of hexahydrobenzoic acid through conversion to benzoic acid, but considered that they had never isolated cyclohexanecarboxylic acid. Von Braun also had never isolated his acetic acid. It is not even certain that the acid that he did synthesize from the ketone obtained as degradation product was the isomer which had been a main acid in Roumanian petroleum.

They reasoned that it should be much easier to isolate and identify the lower acids than to start in with acids with ten to twenty carbon atoms, which are probably much more complex mixtures. Presumably, other investigators had thought about this fact, but had probably been content to work with the higher commercial acids, because that was the only type readily available from refineries. Von Braun, alone, specifically stated that he considered it best to work in the C_{10} to C_{12} region.[4]

Nenitzescu's acids had been obtained in the alkali refining of straight-run gasoline from an asphalt-base crude oil from Ploesti. They started with 7.2 kg of crude acids which boiled at 200° to 255°C (392° to 491°F) [80% at 215° to 240°C (419° to 464°F)] and had a density of 0.976 and an acid number of 340.

They converted the acids to their sodium salts with an excess of 10% sodium carbonate and extracted the alkaline solution with petroleum ether boiling below 105°C (221°F) to remove hydrocarbons and some of the phenols. The acids were liberated by dilute sulfuric acid and the lower layer was extracted twice with petroleum ether. This was then combined with the acid layer and shaken repeatedly with dilute sulfuric acid to decompose emulsified or dissolved sodium salts. The total yield of acids and phenols at this stage was 6.5 kg. Esterification yielded 4.8 kg of methyl esters, which were extracted repeatedly with cold dilute sodium hydroxide to remove phenols. On acidification of the sodium hydroxide layer and esterification of the acids liberated, another 200-g batch of esters was obtained. A third similar treatment yielded no additional esters.

The esters were first fractionated through a Widmer column (the first mention of this column in naphthenic acid literature). Distillation was carried out at 20 mm and 5°C fractions were collected. The next fractionation employed a 2-m efficient column operated at a throughput of 0.4 to 0.5 cc a minute, again at 20 mm pressure, and fractions boiling over only 1°C were collected this time. From the boiling-point range of the esters, they calculated (by a method which was not given) that the free acids would show a boiling range of 210° to 240°C (410° to 464°F) at atmospheric pressure. This boiling range includes eleven of the fifteen C_7 fatty acids, all of the seventeen C_8 fatty acids, and all cyclic acids containing six, seven, and eight carbon atoms and a cyclopentane ring. Their ester fractionation had been carried out carefully enough to show definite maxima and minima when the cut number was plotted against the index of refraction; the peaks corresponded, of course, to the cyclic acids with high index of refraction, while minima occurred when one of the aliphatic acids predominated.

They took an aliquot of each ester fraction, added an excess of concentrated ammonium hydroxide solution, and set the stoppered flask aside for as long as $1\frac{1}{2}$ to 2 months — unless a good crop of crystalline solid amide appeared earlier. They observed that when the solutions were saturated with ammonia gas at the end of the amide-formation period, no material increase in yield of amide could be obtained.

Systematic recrystallization from methanol, benzene, and dilute alcohol was used and the process included recombinations of crops of crystals by melting point — evidently an elaborate scheme, since Velrap's thesis was based on this work. The carboxylic acids identified were:

the *amide of 3-methylbutanoic acid* from fractions 45 to 55; melting at 120° to 121°C (248° to 250°F), identified by analysis and mixed melting point with a synthetic sample;

the *amide of cyclopentanecarboxylic acid* from fractions 55 to 64; identified by mixed melting point with synthetic amide;

the *amide of 4-methylpentanoic acid* from fractions 64 to 67; identified by mixed melting point;

from fractions 67 to 82, a low-melting amide of an aliphatic acid and another one melting at 152°C (306°F) which was identified by analysis and mixed melting point as *cyclopentylacetic acid;*

from fractions 82 to 87, as main yield, the amide of another fatty acid; as a minor product, an amide melting at 142°C (288°F); analysis and mixed melting point identified the acid as *3-methylcyclopentylacetic acid.*

Since the reported melting points of the amides concerned are so far apart that mixed melting point determinations are hardly required (except in the last case), their identification is probably correct; but it should be pointed out that it has been found that the amides of naphthenic acids tend to form mixed crystals and a mixed melting point of different amides does not always show a depression.

No other amides could be purified by recrystallization and only a few acids yielded solid amides. The acids identified accounted for only 8% of the total acidity. They converted the acids from the filtrates to their acid chlorides and hoped to be able to identify other acids from them, but they seem to have published no additional results of this study.

The phenolic matter which was worked up was found to contain no phenol, but they did isolate the three cresols and 1,2,3- and 1,3,5-xylenol from this straight-run material.

Since they depended only on fractional distillation and fractional amide formation, their chance of identifying more than a few of the acids present was slight, but they should be given credit for identifying, for the first time, three naphthenic acids.

Harkness and Bruun [15] and Goheen [13] were interested in the number of rings found in the higher naphthenic acids with fourteen to twenty-nine carbon atoms. They studied Gulf Coast acids, which were purified in the usual manner and finally fractionated through an efficient molecular still. Harkness and Bruun studied the acids as such and, on the basis of density, index of refraction, and analyses, decided that they were all monobasic and contained some molecules that had as many as five rings.

Since such an effect could have been produced, in part at least, if an appreciable concentration of aromatic acids had been present, Goheen converted some of the naphthenic acids to the hydrocarbons by the usual series of reactions and then subjected the hydrocarbon mixture to ring-analysis methods.[13] He also concluded that the acids were monobasic, that they had an average of 2.6 rings per molecule, and that the mixture contained about 5% of aromatic hydrocarbons. Between the two reports, they showed that these Texas higher naphthenic acids contained molecules with more than two rings per molecule and that a low concentration of aromatic acids was also present.

Goheen introduced the esterification method of Corson, Adams, and Scott [11] to naphthenic acid work when he passed the vapors of ethanol through the mixture of acids heated at 115° to 120°C (239° to 248°F), thus hastening the reaction and removing the water formed. He obtained a 91% yield of esters in 2 hours.

Work at the University of Texas

During the period 1928 to 1934, the author was a close observer of the work of Professor J. R. Bailey and coworkers on the A.P.I. supervised project on petroleum bases and he was familiar with von Braun's work on petroleum acids through von Braun's publications and frequent correspondence with Professor Bailey. Much of the earlier work on petroleum acids was also reviewed during this period and it seemed that workers who depended almost entirely on fractional distillation to isolate naphthenic acids had a pitifully small chance of success, because no amount of fractional distillation could separate mixtures of isomers and homologues — or even mixtures of different types of compounds — if they boiled at practically the same temperature. It was, of course, known that a change of pressure might result in a small difference in the vapor pressure of different members of such a mixture and that azeotrope formation might sometimes be valuable. The additional separation that can be attained through use of selective precipitation of solid derivatives or salts had been found useful in the separation of bases but, at best, resulted only in the isolation of one or a few individuals out of the complex mixture found in a fraction. It had not found much application in the study of petroleum acids.

What seemed to be needed was not a different method of separation but an independent supplementary one, so that mixtures, which could no longer be simplified by fractionation of the acids, esters, or acid chlorides on the basis of differences in vapor pressure, could now be separated according to some other property, such as acid strength, solubility, or structure. The mixture obtained after two such independent separations should certainly be much simpler than that obtained by fractional distillation alone.

In addition, if typical naphthenic acids were to be isolated, it seemed to the author (as it had, independently, to Nenitzescu) that the proper place to start should be the simplest and lowest-molecular-weight members and not the higher acids with ten or more carbons which had usually been studied. In nearly all cases, this range may not have been the one most desired by the investigator, but simply the one that could be obtained from some refinery. However, von Braun specifically stated that he felt the work should start with the medium range of acids.

Of the two most probably desirable methods of separation that might be found useful after or preceding fractional distillation, chromatography was practically unknown at the time and systematic separation by countercurrent extraction or distribution had but barely been tried. The group studying petroleum bases had employed a distribution method that was so selective, as far as separation of quinolines from pyridines and related compounds was concerned, that no really systematic approach was required. It was the method of separating the hydrochlorides of quinolines and pyridines by extracting the aqueous solution of the hydrochlorides with chloroform. They had also used simple stepwise fractional neutralization or liberation of bases and had obtained fair results in this manner.

This was the situation in 1934, when the author decided to attempt to develop systems of separation using two or more independent methods in the hope that such combinations would make it possible to isolate and identify at least the lower members of the naphthenic acid series. The first problem was that of obtaining a sample of acids from straight-run gasoline or kerosene. This proved a difficult one, since practically all refiners perform alkali washing of gasoline or kerosene only after combining their cracking-process and straight-run fractions.

While the search for a suitable sample of acids was going on, it was found that the Humble Oil and Refining Company would be glad to provide concentrated alkali wash obtained in refining light burner oil, which included matter boiling as low as 100°C (212°F) at 35 mm pressure; a generous sample of these acids was isolated for study until a more desirable batch of acids could be located. The sample used included cracking-process material with a high concentration of phenols. These acids were to serve in the preliminary study of supplementary methods of separation to be used before or after the usual fractional neutralization or liberation.

Work had hardly been begun when it became obvious that some index or method of characterizing fractions obtained had to be found, so that progress in separation could be followed. Neither the boiling point nor the closely related molecular weight or neutralization equivalent could be used, since the method of separation used would not always separate on the basis of molecular weight. Since the compounds with which the work was concerned were the aliphatic acids, naphthenic acids, and phenols, the index of refraction or the density of individual fractions appeared to be the logical choice of property to be used. However, a study of these properties, as reported for a large number of compounds of these classes, showed that neither density nor index of refraction would be as valuable as a combination of the two. After a series of more complex expressions had been tested, it was found that none seemed to serve to indicate what type of acid was in a given fraction better than the simple, purely empirical, product, $n \times d$ (or $n \cdot d$) where both n and d were determined at the same temperature; therefore, this value was used extensively in studying acids of all types with ten or fewer carbon atoms. It was found that pure synthetic fatty acids had $n \times d$ values ranging from 1.28 to 1.35; the naphthenic acids of this range, 1.39 to 1.47, with most acids in the range of 1.41 to 1.44; the phenols gave values above 1.5. Unsaturated acids would fall in the same range as the naphthenic acids; some hydrocarbons would overlap the aliphatic acids; aromatic acids would have the same range of values as the phenols. Fortunately, none of these three interfering types are present in appreciable quantities in purified and fractionated acids.

Using, then, the n $\times$ d values (along with boiling points and neutralization equivalents) as guides, Schutze and Shive [39] tested simple single-stage fractional neutralization of a batch of the Humble acids. The acids were to be neutralized in one stage by adding an aliquot of dilute alkali to a dilute petroleum ether solution of the mixture of acids, stirring for at least half an hour, separating the two phases, cross-extracting the top layer by water to remove sodium soaps dissolved or suspended in petroleum ether, and extracting the bottom layer by petroleum ether to remove free acids drawn into it by the solution of sodium salts.

This scheme was used on crude fractions obtained by simply distilling the acids from a 57 liter steel still having a 275 $\times$ 15 cm unpacked column. The fraction selected was subjected to a six-step fractional neutralization, as described previously, and each of the six cuts was then steam distilled to remove the hydrocarbons and most of the phenols from the sodium-salt solutions. The first three neutralization cuts were then combined and again fractionally neutralized as before. The strongest cuts of these were combined and carefully distilled through an efficient fractionating column to yield eighteen fractions, the last two of which again had n $\times$ d values above 1.5, while the first one had a value of 1.36 and was not studied.

During the fractional distillation, it was observed that a solid separated in the condenser at cuts 9 through 16. This was found to be *dimethylmaleic anhydride*, which was also found a little later in a fraction of California acids.[14] The other fractions yielded valeric and butyric acids, fractionally precipitated as the silver salts, which were decomposed by syrupy phosphoric acid, distilled from it, and identified.

Phenol was found when filtrates from silver salt precipitations were treated with hydrochloric acid and distilled to recover acids with soluble silver salts. The fact that it was isolated from the water layer may indicate why the higher phenols, cresols, and xylenols have been reported frequently, but phenol itself has often been reported to be absent. It may have been lost in water layers obtained in the refinery isolation of the acids or in purifying steps in the laboratory.

It should be noted that these lower-boiling acids were isolated from material which had been fractionated only slightly as acids and had not been converted to esters (and, of course, had not been fractionated as such). The acids had then been fractionally neutralized by a simple single-stage step and had finally been carefully fractionated only as the acids. As might be expected, when an attempt was made to isolate higher-boiling acids, it was found that even careful fractional silver salt formation of neutralization fractions did not lead to pure compounds. Therefore, the acids were converted to the methyl esters and carefully fractionated in a spinning-band Widmer column. *n-Octanoic* and *p-hexahydrotoluic* acids were identified and a number of other solid derivatives were prepared but not identified.

The experimental work had shown that the combination of fractional distillation and fractional neutralization did, indeed, yield much simpler mixtures than had been obtained by fractional distillation alone and the next task was that of development of efficient countercurrent fractional neutralization or liberation methods, whichever might be needed, so that the combination of separation methods would make it possible to isolate not only the lower simple acids but also the lower members, at least, of the naphthenic acid series.

It was just at this time that reports of the brilliant pioneering work in laboratory-scale fractional extraction or distribution of Jantzen and his students at Hamburg became available to American readers through publication of the *Dechema Monograph No. 5* in 1932.[18] The remarkable results obtained in isolation of coal-tar bases with their equipment naturally led immediately to development of this type of apparatus for use in separation of both bases and acids from petroleum at the University of Texas.

In addition to the development of equipment for preliminary tests to determine feasibility of separation by fractional neutralization of particular mixtures of bases,[18] they were using highly efficient laboratory extraction columns of the countercurrent type. In this column, a light phase moves upward and a heavy phase downward through a tall, slender column (6 to 12 mm) in which a glass spinner (4 to 10 mm) is spun at such a speed that the ascending and descending phases pass each other in what appear to be stationary, fine, flat spirals. The appearance of these spirals from top to bottom of the column is used as an indication of the satisfactory operation of the column. Jantzen and Tiedke [47] described only the continuous type of column in which the whole sample is put through the column once to get one cut, which may be any desired fraction of the whole sample. In the author's laboratory a whole series of batch, as well as continuous, types was developed, starting with the ideas of Jantzen and Tiedke.

As applied to petroleum acids, the tendency to emulsification is the worst drawback of the use of these columns. It is not possible, however, to use columns of large diameter because it is difficult to construct these from glass. When metal columns are used, there are not only mechanical difficulties involved in spinning a long tube within such a column, but also, since dimensions appear to be so critical, selection of the correct diameter and the correct speed of rotation to get proper conditions for efficient operation is necessary and this is difficult since it is not possible to see the spirals which indicate the column is operating properly. The long, slender glass columns, while very efficient, have a low throughput which makes separation of large samples difficult because, in the case of petroleum acids, the solutions must be rather dilute to avoid emulsification and, therefore, long runs are required to complete a series of fractionations. These long runs, in turn, involved development of reliable pumping or other equipment that could be depended on to deliver volumes of the order of 100 to 300 cc an hour for long periods of time.

As might be expected, the equipment proved very valuable in work with petroleum bases, which show little tendency to emulsification. It was also applied very successfully to separation of petroleum acids by a number of workers. Specific applications will be described in the following pages. Whenever a combination of acid and ester fractional distillation made it possible to isolate and identify individual acids, this was done to save time, although, if a material balance had been sought, fractional neutralization would have been essential.

Several different batch-process spinning-rod extraction columns were tested on fractions of the Humble Texas acids, but their use did not come until what proved to be an ideal and adequate sample of acids became available for study. The McMillan Petroleum Company, at their Long Beach Refinery, in running Signal Hill crudes, was at that time employing an unique method of refining straight-run gasoline and kerosene. In this method, the fractions were passed in series through two tall columns countercurrently to a sodium carbonate solution.[9] The first tower removed practically all of the carboxylic acids and small amounts of phenols, while the second removed almost entirely the phenols. Attempts to isolate the highly hindered 2,2,6-trimethylcyclohexanecarboxylic acid (which should have behaved as a weak acid and passed through the first column) failed, so that only small amounts of this acid could have been removed in the second column.

Since the material, as received, contained some water and was a low-boiling mixture, it was to be expected that it would contain all petroleum acids from the lowest-boiling to those with perhaps 12 carbons; it was therefore decided to fractionate into a low-boiling cut, a "heart" cut, and a high-boiling one. It was anticipated that the low-boiling acids could be separated and identified with little or no use of fractional neutralization, but it was soon found that acids boiling higher than acetic and present in small amounts were best isolated by the combination of processes.

The original 70.2 liters of main sample was distilled at atmospheric pressure until the top temperature of the column reached 160°C (320°F) and then at 50-mm water-pump vacuum. The fractions collected were:

(a) 3.6 liters at atmospheric pressure
(b) 54.0 liters at 50 mm
(c) 12.4 liters high boiling at 50 mm

From a preliminary sample, 17 liters of high-boiling material was available and was later added to a fraction of the 54-liter cut.

The 3.6-liter fraction of low-boiling matter was selected by Hancock [14] as part of his study. This consisted of a water layer of 760 cc and a top layer of 2,840 cc. The bottom layer, on simple fractionation, yielded a 20 cc acid layer and a new water solution which was neutralized by alkali and evaporated to dryness on a steam cone, yielding 29.1 g of powdered residue. Syrupy phosphoric acid was added to half of the powder and distilled, yielding 4.5 g of dry acids. A few drops of the dry acids reduced silver nitrate and mercuric chloride, showing the presence of formic acid in this fraction.

The remainder of the 4.5 g batch of acids was transesterified by heating with 23 cc of methyl *n*-hexanoate in the presence of a few drops of 18 N sulfuric acid and yielded 3 g of dry methyl ester which was fractionated through a 70 × 5 cm column containing a single long spiral of Chromel wire as packing. This yielded methyl acetate which was identified by its index of refraction and the DuClaux constant of the acid.

The 2,860-cc top layer of the original fraction was titrated and found to contain 14.8 equivalents of acidity. Since it was anticipated that these lower-molecular-weight acids could be fractionated by molecular weight or by fractional neutralization as easily as by fractional distillation and that phenolic and neutral impurities would be removed in the neutralization operations, the top layer was fractionally neutralized by 1.0 N sodium hydroxide solution. Cross extractions to both layers were used to avoid extensive losses due to solubility of the free acids in water, as this would have resulted in extensive distribution of the acids between the petroleum ether (used as organic solvent) boiling at 30° to 60°C (86° to 140°F) and the water.

This procedure yielded five fractions and, for the first time, presented the problem (met frequently in later work) that the first aliquot of base gave a water layer carrying much less than the calculated amount of sodium salt while later ones carried more than the calculated amount, until finally, at some stage in the neutralization, a homogenous solution or emulsion was obtained. This situation is, of course, due to mutual solubility of free acids and sodium salt solutions. It can be avoided by using a sufficiently dilute solution, at the expense of time consumed by working up the large volumes of solutions. Equivalent-weight and index-of-refraction values of the neutralization cuts showed that there had indeed been extensive fractionation (equivalent weight range 104 to 157). The first two fractions of equivalent weight 104 and 116 were again fractionally neutralized, cutting each into three fractions, with the first fraction now yielding neutralization equivalents of 91, 98, and 109 and the second giving 107, 113, and 126.

These six fractions were separately esterified and it was found that a solid crystallized from the portion of the fraction with a neutral equivalent of 90, which was not esterified after one treatment. The solid proved to be dimethylmaleic anhydride, which has been found in Texas acids by Schutze.

The esters were carefully fractionated and combined by boiling point and index of refraction and again fractionated to yield now ten fractions the properties of which are presented in Table 30. The *p*-phenylphenacyl esters of cut 2 and of synthetic propionic acid were prepared and the melting points were found to be 103°, 100.5°, and 102°C (217°, 212°, and 215°F) for the synthetic, the isolated, and the mixture.

Table 30. **Properties of Isolated Fatty Acids**

Cut No.	Boiling Point of Acid °C	°F	n_D^{20} Acid	Melting Point of 2-Alkylbenzimidazole Synthetic °C	Isolated °C	Mixed °C	Acid Present
2	141–2	286	1.3872	175.5	178.5	176	Propionic
4	153–4	308	1.3948	234.0	233.5	234	Isobutyric
6	162–3	324	1.3982	159.0	158.0	158	*n*-Butyric
8	175–6	348	1.4050	190.5	190.0	190	Isovaleric
10	185–6	366	1.4090	154.0	153.0	153	*n*-Valeric

Methylethylacetic and trimethylacetic acids might have been present in these fractions, but could not be found by Hancock or in later work by Quebedeaux. Apparently, even among these lower acids, not all possible isomers are present and we may hope that the petroleum acids do not represent as complex a mixture as some people have assumed.

The 54-liter main cut and the higher-boiling fractions from the low-boiling 3.6-liter cut, as well as 17 liters of similar acids from a preliminary batch of the same acids, were fractionated and recombined by boiling point to get, finally, a series of large fractions. These large fractions were esterified individually to yield batches of esters which, after two fractionations through a 12-foot berl saddle-packed steel column, provided with jacket heat to get adiabatic conditions and controlled high reflux, yielded a series of 720 ester fractions varying in size as shown in Figure 8. Portions of these were refractionated through a 125-double-cone Stedman column, but additional separation was so slight that this was not continued. Meanwhile, small amounts of acids proved unesterifiable during the usual 24-hour treatment with refluxing methanol containing about 2% dry hydrogen chloride.

The first of these hindered acids studied [14] was obtained as a solid melting at 194° to 195°C (381° to 383°F) after recrystallization from dilute alcohol. It remained behind when unesterified material was fractionated. The solid proved to be the highly hindered tertiary 1,2,2-trimethylcyclopentanecarboxylic acid, which had been synthesized in 1933 by Appel.[1] This was the first naphthenic acid isolated and identified from an American petroleum. A total of 62.5 g of the pure acid was isolated at this time; later work on other fractions isolated additional amounts.

In view of the ease of isolation of this hindered acid, there was a frantic search for other hindered acids in these acids and in the 12.4 liters of stored higher-boiling acids from the initial distillation, but only one other acid — in both of its stereoisomeric forms — was isolated and identified.

When Horeczy [40] attempted to esterify all available acids, he obtained a total of 2,830 cc which was not esterified by one treatment. Careful fractionation and cold alkali extraction of the "unesterified" matter resulted in the isolation of an additional 400 cc which had been dissolved in the unesterified portion, 700 cc of liquid acids, and 940 g of acids which solidified in the condenser or the receiver. Systematic fractional liberation of the hindered acids from their sodium salts, at first by treatment with carbon dioxide and later by addition of small amounts of mineral acid, yielded an additional 280 g of camphonanic (1,2,2-trimethylcyclopentanecarboxylic) acid, 500 g of another solid acid melting at 70° to 80°C (158° to 176°F), and 100 g of liquid acids. Purification of the new acid by recrystallization from petroleum ether raised the melting point to 83°C (181°F) ; it was not raised further by recrystallization from acetone or dilute alcohol.

This acid was similar to one isolated from Iranian oil by Kennedy.[20] Since he did not identify it, he furnished a sample which proved to be identical with the new acid. This new acid was also shown to be identical with a $C_{10}H_{18}O_2$ acid obtained by Roberts and Bailey [36] by degradation of their $C_{16}H_{25}N$ base from California petroleum.

While this highly hindered acid (which was not esterified by two 24-hour treatments with refluxing methanol and HCl) had been prepared and described previously by Wallach,[48] the identity was not suspected at first, because it was confidently expected that this would prove to be another tertiary acid. A number of possible tertiary acids were synthesized [41] and found to be different. Since an unusually large amount of the acid was available, its properties and those of its derivatives were studied, and, in particular, tests were run to determine whether the acid was actually tertiary or possibly secondary in nature. The Whitmore test,[49] in which the ratio of carbon monoxide to carbon dioxide liberated on heating an acid with phosphorus pentoxide is determined, indicated that the acid must be tertiary. However, on the basis of von Braun's chlorine-number test,[4, 6] which showed it to have one hydrogen on the α-

carbon atom, it should be a secondary acid. Hydrazoic acid degradation of the acid to the amine with nine carbon atoms and treatment of this amine with nitrous acid yielded an olefin which, on ozonolysis, gave 2-methylcyclopentylmethyl ketone, but this could have been formed directly, or by expected rearrangements, from a number of different acids. Such a secondary acid was 2,2,6-trimethylcyclohexanecarboxylic acid which had been prepared by Wallach.[48] He hydrogenated the cyclohexene over palladium, while Adam's catalyst was used by the Texas group. Wallach obtained an acid melting at 82° to 83°C (179° to 181°F), while hydrogenation over Adam's catalyst yielded one melting at 74° to 75°C (165° to 167°F). A mixture of the acid melting at 74° to 75°C with the petroleum acid melted at 50° to 55°C (122° to 131°F). Hydrogenation of another batch of the synthetic compound over palladium on barium sulfate yielded an acid melting at 40° to 44°C (104° to 111°F) whose melting point was only slightly depressed in a mixed melting-point determination with the petroleum acid. This behavior led to the suspicion that the synthetic acid was probably a mixture of *cis* and *trans* forms melting near the eutectic. Assuming that the acid melting at 74° to 75°C (165° to 167°F) was the lower-melting (probably *cis*) isomer which could perhaps be rearranged to the more stable *trans* form, a sample was heated for 24 hours with an equal volume of glacial acetic acid and a few drops of concentrated hydrochloric acid. The solid acid isolated from this treatment was recrystallized from petroleum ether and found to melt at 82° to 83°C (180° to 181°F) and showed no depression in melting point when mixed with the petroleum acid. The anilides were also shown to be identical. The higher-melting acid was tentatively assigned the *trans* configuration. A highly hindered acid present in high concentration in a naphthenic acid sample would not have been expected to be a *secondary cyclohexane carboxylic* acid, since in the three respects italicized, the acid should not have been a hindered naphthenic acid present in high concentration.

Through a peculiar irony of fate, von Braun [7] mentioned the well-known preparation of 2,6,6-trimethyl-1-cyclohexenecarboxylic acid which, on hydrogenation, would have yielded one of the forms of this C_{10} naphthenic acid, but he was not interested in cyclohexane acids and mentioned it only in connection with the preparation of alkylcyclopentanecarboxylic acids.

Jeger and Buchi [19] later again synthesized this acid in connection with another problem and proved the structure by independent methods.

The liquid hindered acids were reexamined a year later by Rannefeld [32] to determine whether any other acid could be isolated from the 900 cc of remaining material. He refractionated at 5 mm pressure through a 6-foot berl saddle-packed column at high reflux and found that an additional amount of the 1,2,2-trimethylcyclopentanecarboxylic acid (camphonanic acid) crystallized from the first fractions and 2,2,6-trimethylcyclohexanecarboxylic acid from the last fractions, while a smaller amount of a new solid acid appeared in the middle cuts. All solid acids were filtered and the filtrate again fractionated at 5 mm pressure to yield additional amounts of all solids when the fractions were chilled by ice-salt mixtures. Three such treatments yielded a total of 154 g of fresh camphonanic acid, 242 g of the *trans* acid, and only 61 g of the new solid acid which melted at 66° to 70°C (151° to 153°F). Recrystallization from dilute acetic acid and from dilute alcohol yielded colorless rhombic crystals melting at 74° to 75°C (165° to 167°F). Mixed melting points of the acid and its anilide with the synthetic *cis*, 2,2,6-trimethylcyclohexanecarboxylic acid and its anilide showed no depression, but the amide prepared over the acid chloride proved to be identical with the amide of the *trans* acid, thus confirming what had been noticed by the earlier Texas workers — that thionyl chloride rearranged the *cis* to the *trans* acid. In this interesting case, we have both *cis* and *trans* forms of the same C_{10} naphthenic acid isolated from petroleum.

No additional hindered acids were isolated from this or any other boiling range of these acids. The fact that so few hindered acids appear to occur in these acids makes a survey of naphthenic acids from many fields an attractive project, because the presence of either or both acids in most or all petroleum acids would be an interesting lead in the study of the origin of these acids. The relation between carotenoids containing the grouping:

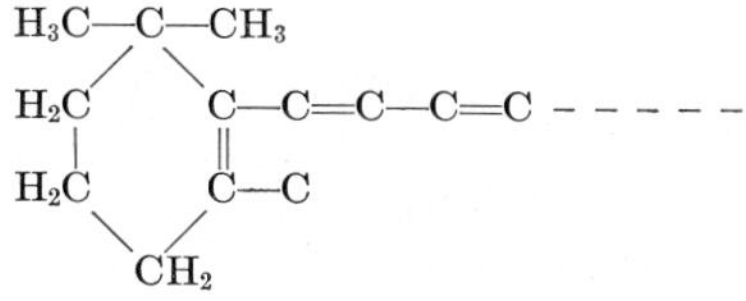

```
H3C—C—CH3
   /   \
H2C     C—C=C—C=C - - - - -
 |      ||
H2C     C—C
   \   /
    CH2
```

and the hindered 2,2,6-trimethylcyclohexanecarboxylic acid:

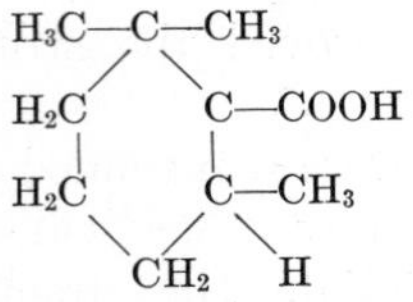

```
H3C—C—CH3
   /   \
H2C     C—COOH
 |      |
H2C     C—CH3
   \   / \
    CH2   H
```

is obvious, although apparently the degradation of a carotenoid to this acid has not been reported.

Isolation and Identification of Acids from the Methyl-Ester Fractions

As indicated in Figure 8,* the volumes of individual esters vary from 35 to 250 cc, depending on the amount distilling per degree rise in temperature. At fraction 500, a 2-liter fraction of methyl-heptanoate was collected.

While the study of the series of esters was carried out simultaneously on various fractions according to agreements made during the tedious cooperative labor of fractionation, the isolation and identification of the acids will be taken up in the order of boiling points — even though in this plan, the naphthenic and the aliphatic acids will not be separated in the discussions.

The sharp drop in index of refraction observed at certain points, such as fractions 220, 380, and 465, does not mean that there is a sudden change in the composition of the fractions but simply that at these points, additional charges of slightly higher-boiling esters were added to the still pot. This was followed by a period during which there was very high reflux and by the time the column reached equilibrium, a liquid of much lower index of refraction had accumulated in the upper part of the column. The boiling point of the fraction does not show a corresponding drop. In a few cases this effect was not noticed.

It was noted very early in the series of fractionations that definite plateaus were forming. The most pronounced plateau was at the boiling point of methyl heptanoate and this region was frequently used to start or stop fractionation series. Peaks in index of refraction were also appearing, pointing to the occurrence of cyclic or naphthenic esters, while in other regions, the index of refraction rose gradually but definitely, indicating that a series of isomeric or homologous esters was distilling. It is, therefore, apparent that, in the case of these highly fractionated esters, the index of refraction

* See page 142.

and boiling point took some of the guesswork out of the isolation of the acids.

Unless otherwise indicated, final identification was by mixed melting points of one or more derivatives prepared from the petroleum acid and from an acid of known structure, usually one synthesized in this work. If only one mixture was examined, it was because other properties of the acids practically established the structure.

Renewed Search for 2,2-Dimethylpropionic and 2-Methylbutanoic Acids

Since Hancock was unable to isolate 2,2-dimethylpropanoic and 2-methylbutanoic acids from fractions in which they should occur if present, it was decided to study anew the lowest-boiling fractions of the long series.[35] Esters boiling at 90° to 130°C (194° to 266°F) (fractions 1 to 16) were refractionated and the location of possible "plateaus" noted. There were only two, one at 115° to 115.5°C (about 239°F) and another at 126.2° to 126.4°C (about 259°F). The methyl ester of 2,2-dimethylpropionic acid boils at 102°C (216°F), which does not correspond to either of the regions found. Methyl-2-methylbutanoate boils at 115°C (239°F); but methyl-3-methylbutanoate boils at 116.7°C 242°F) and the ester boiling at 115° to 116°C (239° to 241°F) had been identified as that of isovaleric acid (3-methylbutanoic acid). It seemed unlikely that both acids were found in the single plateau, and no further work was done on it.

2-Methylpentanoic Acid

This acid from fractions 30 to 55 was indicated by the properties of the free acid and identified by mixed melting points of the *p*-toluides, amides, and *p*-phenylphenacyl esters.

3-Methylpentanoic Acid

Obtained from fractions 80 to 85, this acid was indicated by the properties of the acid and identified by mixed melting point of the *p*-toluidides.

n-*Hexanoic Acid*

n-Hexanoic acid was indicated in fractions 155 to 160 by the large plateau of normal acids and by the properties of the acid. It was identified by mixed melting point of the *p*-toluidides.

Cyclopentanecarboxylic Acid

This acid from fractions 215 and 216 was indicated by the peak in the index-of-refraction curve and by the properties of the acid and its derivatives. It was identified by analysis of the *p*-toluidide and by mixed melting points of the amides and *p*-toluidides. This, the simplest of the naphthenic acids, had just been reported by Nenitzescu, working on Roumanian acids.[29]

2-Methylhexanoic Acid

From fractions 248 and 249, this acid was obtained out of the plateau at 240 to 260 and was indicated by the properties of the acid and its derivatives and identified by analysis of the *p*-toluidide and mixed melting points of the amides and *p*-toluidides.

3-Methylhexanoic Acid

Obtaining 3-methylhexanoic acid from fractions 285 to 290 required refractionation of the acids resulting from saponification of the esters. It was indicated by the properties of the main acid and its derivatives. It was identified by analysis of the *p*-toluidide and by mixed melting points of the amides and *p*-toluidides.

4-Methylhexanoic Acid

From fractions 313 to 314 out of the large plateau between 300 and 340, 4-methylhexanoic acid was obtained and was indicated by the properties of the acid and its amide and *p*-toluidide. It was identified by analysis of the *p*-toluidide and mixed melting points of the amides and *p*-toluidides.

2-Methylcyclopentanecarboxylic Acid

Fractions 377 to 384 yielded 2-methylcyclopentanecarboxylic acid which was indicated by the peak in the index-of-refraction curve at fractions 360 to 385, but it was found mixed with so high a concentration of aliphatic acids, as shown by an $n \times d$ value of only 1.375, that it was necessary to separate these by fractional neutralization in a spinning-rod countercurrent extraction column. The first two out of six fractions had $n \times d$ values above 1.39. Neutralization fraction 1 had an equivalent weight of 129 (the calculated value for a C_7 naphthenic acid is 128). Its amide melted after numerous recrystallizations from petroleum ether-benzene mixtures at 147° to 148°C (297° to 298°F) and its *p*-phenylphenacyl ester at 73.5° to 74°C (163°F), but the *p*-toluidide could not be brought to a sharp melting point. 2-Methylcyclopentanecarboxylic acid was suspected and synthesized by the method of Nenitzescu and Ionescu.[30] The synthetic and the petroleum acid and derivatives were found to be identical. Aschan [2, 3] and Markownikoff,[25, 26] working on samples of different European acids, isolated but did not definitely identify a C_7 naphthenic acid boiling at 215° to 227°C (419° to 440°F) and having a methyl ester boiling at 164° to 167°C (327° to 333°F). The California acid boiled at 220°C (428°F) and the ester boiled at 165° to 167°C (331°F), so that it could well have been identical with the European acid, but Aschan and Markownikoff reported the melting points of their amides at 123.5°C and 121° to 123°C, while the California amide melted at 147° to 148°C. If Aschan's and Markownikoff's amides had been impure, they would hardly have melted so close together.

5-Methylhexanoic Acid

From fractions 392 and 393, 5-methylhexanoic acid was easily isolated, even though so close to the naphthenic acid. The properties of the acid, the amide, and the *p*-toluidide indicated that it was the acid previously reported by Chichibabin [10] and by Nenitzescu.[29] It was identified by analysis of the *p*-toluidide and mixed melting points of the *p*-toluidides and the amides.

3-Methylcyclopentanecarboxylic Acid

Fractions 460 to 470 yielded only one naphthenic acid, although the peak (460 to 490) was a broad one that might have contained more than one acid. Possibly the presence of high concentrations of *n*-heptanoic acid from the region boiling just above this peak made the peak appear broader than it was actually. This was one of the last peaks studied and the most efficient spinning-rod extraction column was used to obtain six neutralization fractions.[32] The n × d values showed that the first three fractions were naphthenic in nature, while the later cuts contained increasing concentrations of aliphatic acids. Neutralization fraction 2 was converted to the amide over the acid chloride and the amide was found to melt at 147° to 148°C (296° to 298°F), the *p*-toluidide at 106° to 107°C (223°

to 225°F), and the *p*-phenylphenacyl ester at 72.5° to 73.5°C (162° to 163°F). The mixed amides of this and of synthetic cyclopentane-acetic acids melted at 146° to 148°C (295° to 298°F), indicating identity, but the mixed *p*-toluidides melted 30°C low; therefore, this was another instance in which the mixed melting point of amides was not depressed. Since hydrazoic acid degradation to the amine gave no clue to the structure, 3-methylcyclopentanecarboxylic acid was synthesized as a likely possibility and was shown by properties of the acids and mixed melting points of amides, *p*-toluidides, and *p*-phenylphenacyl esters to be identical with the petroleum acid.

n-*Heptanoic Acid*

This acid had been tentatively identified early in the fractionation procedures when it was found that large volumes of ester boiled at practically constant temperature. Williams and Richter [50] had identified it in West Texas pressure distillate. The amide was prepared by treating part of the 2-liter batch of fraction 500 with concentrated ammonium hydroxide and leaving it in a stoppered flask for a week, according to the method of Nenitzescu. The amide melted at 94.5° to 95.0°C (201° to 203°F) after four recrystallizations and the acid was identified by its properties and the mixed melting points of the amides and the *o*-phenylenediamine derivatives.

Cyclopentaneacetic and Cyclohexanecarboxylic Acids

These two acids from ester fractions 514 to 520 were isolated from this double peak with considerable difficulty. The esters were saponified and the resulting acids were found to yield no solid derivatives. Six-step neutralization by 7% sodium hydroxide solution yielded a series of cuts, all of which had $n \times d$ values in the lower naphthenic acid range and thus contained mainly naphthenic acids. The neutralization equivalent determined indicated a mixture of C_7 and C_8 acids. The separation of possible aliphatic acids from the naphthenic acids was done in a batch countercurrent spinner extraction column in which 0.5 equivalent of dilute sulfuric acid was added near the bottom of the column to provide 1:1 reflux. This operation was repeated, adding the cuts to the batch column in the order in which they had been obtained the first time, i.e., the strongest or most soluble one first. The aliphatic acids were now found in the last fractions and in the residue in the apparatus, while the first cuts consisted of two series differing in index of refraction. Solid derivatives of both groups of acids could now be obtained but could not be purified without excessive loss during recrystallization. Fractional silver salt formation led to no appreciable separation of the acids boiling at 230° to 231°C (about 447°F). Fractional esterification by isopropyl alcohol was attempted on the assumption that one of the acids was secondary and the other primary and that esterification by a secondary alcohol should proceed slowly with a secondary acid. Neither equilibrium nor rate esterification led to any considerable separation, however.

The acids from all neutralization fractions containing naphthenic acids were then recombined and fractionated very carefully through at 8 × 150 mm spinning-band column at very high reflux in the hope that, although the mixture appeared to boil over a one-degree range, it could be separated nevertheless. The first three and the last two of seventeen fractions obtained now solidified on cooling with an ice-salt mixture. As the mixed melting points of the amides of cyclopentaneacetic and cyclohexanecarboxylic acids with those of the two sets of solidifying fractions were not depressed, it was fairly certain that the first three fractions contained cyclopentaneacetic acid and the last two, cyclohexanecarboxylic acid. However, the *p*-toluidides and *p*-phenylphenacyl esters were still too impure to permit positive identification.

Attempted fractional steam distillation of the sodium salts gradually neutralized by dropwise addition of dilute hydrochloric acid failed to effect a separation. Spinning-band-column fractional distillation of the acid chlorides also failed.

The acids were then degraded to the amines by the hydrazoic acid method and the amines separated via the acid oxalates by the method of von Braun.[4] The corresponding amine oxalates of the two suspected synthetic acids were also prepared and it was found that they were identical with those from the petroleum acids. Analysis of the oxalates agreed with the formulae assigned. Later work showed that cyclohexanecarboxylic acid could be isolated more easily from fraction 535. Previous workers had attempted to isolate this acid on a number of occasions. The nature of the mixture found in California petroleum may explain their failures. Chichibabin [10] was able to identify this acid because dehydrogenation affected only this acid and not the others present.

3-Methylcyclopentaneacetic Acid

From fractions 550 to 570, this acid was isolated after the small peak at fraction 530 could not be resolved. The combined ester fractions 550 to 570 were saponified with an alcoholic solution of sodium hydroxide and adding the esters only as fast as could be done without forming two phases, since it was found on two previous occasions that two-phase mixtures tended to react violently at some critical stage in saponification with loss of material through the condenser. Fractional neutralization by the earlier step neutralization of the batch of acids in 6 volumes of petroleum ether showed that the naphthenic acids concentrated, as usual, in the first three of ten neutralization fractions. The combination of the first three fractions showed an n × d value of 1.405. The amide was prepared and indicated the presence of 3-methylcyclopentaneacetic acid which had been prepared by Nenitzescu.[29] Instead of synthesizing this acid by his rather tedious procedure, it was decided to try to identify it by an indirect method. A fraction of the cyclic acid was carried through the following series of reactions: (where R is H or alkyl)

```
      R                         R                          R
      |                         |                          |
C----CH                   C----CH                    C----CH
:      \                  :      \                   :      \
:       CH—COOH    →      :       CH—NH2      →      :       CH—OH
:      /                  :      /                   :      /
C----CH                   C----CH                    C----CH
      |                         |                          |
      H                         H                          H

      R                         R                          R
      |                         |                          |
C----C                    C----C                     C----C=O
:     \\                  :      \                   :
:       CH        or      :       CH          →      :       COOH    or
:      /                  :      //                  :      /
C----CH                   C----C                     C----CH
      |                         |                          |
      H                         H                          H

      R
      |
C----CH
:      \
:       COOH
:
C----COOH
```

In this case, the only oxidation product that could be isolated was 3-methyladipic acid which was identified by comparison with a synthetic sample. 3-Methyladipic acid could have resulted from the oxidation of 3- or 4-methylcyclohexanol or one of their dehydration products which could have been obtained from 3- or 4-methylcyclohexanecarboxylic acid without rearrangement or from 2- or 3-methylcyclopentaneacetic acid through a Demjanow rearrangement. The properties of the acid and the amide eliminated all but 3-methylcyclopentaneacetic acid, the as yet unknown *cis* or *trans* isomers of 3-methylcyclohexanecarboxylic acid and of 2-methylcyclopentaneacetic acid. Since these isomers were not reported when one form of each was prepared, it was assumed that they were the less stable forms and were thus not likely to be a main form in a petroleum acid mixture and it was assigned the structure of the Nenitzescu acid. The low yield of 3-methyladipic acid isolated after the oxidation and the fact that two unknown stereoisomeric forms have not been completely eliminated make this identification somewhat uncertain, and it is unfortunate that the acid was not resynthesized.

n-*Octanoic Acid*

In 8 liters of petroleum ether 1 liter of acids obtained from fractions 580 to 590 was dissolved and fractional neutralization by the step procedure was attempted. After the first 10% had been added and the sodium salt solution separated, an attempt to obtain a second fraction resulted in such a stable emulsion that the acids were liberated and combined and other methods tried. Only the use of much more dilute solutions or the use of some alcoholic alkali solution would permit fractional neutralization of the higher acids in spinner column or by stepwise neutralization without consuming very much time and effort.

The recombined acids were fractionated as such and those boiling in the range of 236° to 238°C (457° to 460°F) were studied. The n × d value was only 1.317, that is, this fraction was practically free of naphthenic acids. The large volume of narrow-boiling fraction available and these properties indicated that the acid must be the next normal fatty acid-*n*-octanoic. Direct comparison of physical properties of synthetic and petroleum acids and the mixed melting point of the amides were sufficient to identify the acid.

No attempt was made to isolate additional aliphatic acids besides the plentiful *n*-nonanoic, because it was felt that interest in such additional acids did not justify the effort required for their separation and final identification.

2, 3-Dimethylcyclopentaneacetic Acid

The last peak in the index-of-refraction curve of the ester fractions, 630 to 700, was a very large one and probably was due to the presence of more than one naphthenic acid ester. Preliminary separation attempts on various fractions showed that the separation work would be very tedious, even though a total volume of over 10 liters of esters was available in this peak.

Ester fractions 645 to 655 (1,100 cc) were carefully fractionated through the Stedman still. Fraction 3 of this series showed an index of refraction of 1.4483 and a neutralization equivalent of 155.5 — in other words, it was a C_9 naphthenic acid. Six grams of this was converted to the amide which melted at 169°C (336°F).

The first 30% by volume of the Stedman fractions was then tediously fractionated into ten neutralization cuts by the most efficient spinner column in use. To determine the chlorine number of von Braun 5 g of the first of these fractions was used. This number was found to be near 200; therefore, the acid did not have the carboxyl tied directly to the ring.

The Skraup-Schwamberger [43] modification of the Barbier-Wieland degradation, involving the reaction series:

$$R\text{—}CH_2\text{—}COOCH_3 + \text{Phenyl—Mg—Br} \rightarrow R\text{—}CH_2C(\text{Phenyl})_2OMgBr \rightarrow$$

$$R\text{—}CH_2C(\text{Phenyl})_2OH \rightarrow R\text{—}CH{=}C(\text{Phenyl})_2 \xrightarrow{\text{oxidation}} R\text{—}COOH$$

was then carried out, oxidizing the olefin with chromic acid first in acetic and finally in sulfuric acid. This gave a yield of 7 g of an acid boiling at 125°C (257°F) at 12 mm and having an $n_D^{20} = 1.4492$. The amide of this acid melted at 168.5° to 169.5°C (336° to 338°F) after repeated recrystallization from dilute methanol and from benzene-petroleum ether mixtures. The analysis agreed with that of a C_8 naphthenic acid. Since the only amide reported with a melting point as high as 169°C (336°F) is that of 2,3-dimethylcyclopentanecarboxylic acid, it was synthesized by the method of Nenitzescu.[30] As expected, this method starting with 500 cc of methylcyclohexane, 400 g of anhydrous aluminum chloride, and 200 g of acetyl chloride yielded small amounts of by-products due to rearrangements, but careful fractionation gave a main yield of 86 g of 2,3-dimethylcyclopentylmethyl ketone boiling at 103° to 109°C (217° to 228°F) at 54 mm pressure. Oxidation by hypobromide yielded 48 g of acid boiling at 128° to 131°C (262° to 268°F) at 12 mm and having an $n_D^{20} = 1.4520$, $d_4^{20} = 0.9952$, and $n \times d = 1.445$. The corresponding amide, after repeated recrystallization, melted at 169°C (336°F). This melting point was not depressed on mixing with the amide obtained from the petroleum acid. Analysis of the amide agreed with the formula $C_8H_{15}ON$. As additional proof of structure, the C_8 acid was converted to the acetic acid by the reaction series:

$$R{-}COOH \rightarrow R{-}COOCH_3 \rightarrow R{-}CH_2OH \rightarrow R{-}CH_2Br \rightarrow R{-}CH_2Mg{-}Br \rightarrow R{-}CH_2COOH$$

From 44 g of the C_8 acid, a yield of 5 g of the C_9 acid was obtained, the melting point of whose amide was not depressed on mixing with the original petroleum amide; the acid analyzed correctly for $C_9H_{17}O_2N$.

n-*Nonanoic Acid*

From fraction 710, *n*-nonanoic acid, boiling at 213°C (415°F) and having an $n_D^{20} = 1.4257$, was obtained and yielded an amide melting at 99°C (210°F). A mixed melting-point determination with the amide of synthetic *n*-nonanoic acid was not depressed. This, along with the physical properties of the acid (boiling point 254° to 256.5°C (489° to 493°F) and $n_D^{20} = 1.4379$) and the fact that the volume of ester available was large, showed that the product was *n*-nonanoic acid. While there was less of *n*-nonanoic acid than of the normal C_8 and C_7 acids, this does not mean that the higher fatty acids do not occur in California petroleum, because in any distillation fraction like the one from which these acids were isolated at the refinery, there is a tendency for (equal boiling range) fractions near the center of the boiling range to be larger than those near either end of the boiling range — the upper end in this case.

The very tedious and time-consuming fractionations which finally resulted in the seven hundred twenty fractions of methyl esters, yielding all of the acids reported in the last pages, are feasible in academic research only when funds are available to pay technicians for the routine fractionation work or when, as in this case, the unusual situation prevails in which a large group of graduate students cooperates to carry out the work involved in the series of fractionations under the understanding that each student finally is to study a particular range of fractions out of the whole series. In this case, apparatus and methods were developed while the fractionations were in progress and each student became an expert in operation of the equipment. This was particularly important in operating the spinner-column extraction apparatus, because there are so many variables involved that few students master the technique except when working in groups.

The situation which made fractionations by students possible has not existed here since 1940 and the author is convinced, after a number of attempts to carry on this type of work with one or two students, that unless funds of some kind were available to pay for the labor of routine fractionation operations, the isolation and identification of compounds from such complex mixtures as petroleum products would not be practicable in University laboratories.

Henderson, Ayers, and Ridgway,[16] in a survey of the acidity of various Texas and Arkansas crude oils, gasolines, and kerosenes, point to the very large difference in the amount of alkali required to treat 1,000 barrels of crude 7 to 452 lb of sodium hydroxide. They found the acids from crude oils similar to those from distillates, but reported that the total acidity is enormously increased by cracking operations.

Their refinery customarily regenerated the caustic which had been used to remove phenols and mercaptans. Regeneration involved steam distillation to separate the phenols and mercaptans from the stronger carboxylic acids which remained as sodium salts.

They decided to investigate the nature of the organic acids remaining in the spent caustic which had been steam distilled. The addition of an excess of dilute sulfuric acid caused the separation of 4.5% by volume of an "evil smelling" oil from which they removed chloride ions by treatment with silver sulfate and subsequent filtration to discard the insolubles. The filtrate was steam distilled until no additional acids came over; the condensate was then made strongly alkaline and evaporated to dryness.

Table 31. **Lower Fatty Acids Isolated**

Acid Found	*Percentage of Crude Acids*	*Percentage of Spent Caustic Solution*
Formic	trace	trace
Acetic	58	0.74
Isobutyric	16	0.20
Isovaleric	6	0.08
Unidentified	20	0.25

Treatment of the dry sodium salts with syrupy phosphoric acid liberated the acids, which were then distilled and identified by a method similar to that of Hancock.[14] Their data are in Table 30.

They did not identify propionic, *n*-butyric, and *n*-valeric acids which had been isolated from California acids, but they isolated a new compound, *o*-thiocresol, among ordinary phenolic substances.

Gas-Well Acids

The tubing of certain high-pressure gas (condensate) wells, especially in certain fields in Texas and Louisiana, corrodes very rapidly, while that of wells in nearby fields or even other wells of the same field may show no corrosion or only slight damage. In 1945, the Corrosion Committee of the Natural Gasoline Association of America started a careful study of this problem. One of the angles studied was that of the amount and nature of organic acids which, along with carbon dioxide, had been shown to be found in the products of all wells. To the author fell the task of determining what individual acids were present and in what approximate concentration.

This was a welcome assignment, because it afforded an opportunity to study carboxylic acids which were definitely not man made but isolated directly from the well products at or near the well.

The problem of isolating and identifying what proved to be the lower aliphatic acids, when present in total concentrations of 100 to 750 ppm, had not been solved satisfactorily at that time and, although various schemes were tried and found fairly satisfactory, they were not as good as some of the chromatographic methods now used.

The study confirmed — what had been shown repeatedly in the laboratories of cooperating companies — that phenols are produced by practically all of these high-pressure wells. It was further found that when a water sample is alkalized and distilled from a flask equipped with a spray trap, the first 10% or so of distillate gives an alkaline reaction. This alkalinity was shown to be due to very low concentrations of organic bases which boiled above 200°C (392°F), were tertiary in nature, and had the characteristic odor of petroleum bases of the same boiling range; they were presumed to be composed of alkylpyridines.[24]

The acids shown to be present in the water layer produced by all of the wells were formic (trace), acetic (main), propionic, butyric, isobutyric and valeric, and a low concentration of C_6 to C_8 acids.[23]

This result would be expected on the analysis of the water layer obtained from a well, even if higher concentrations of the C_5 to C_8 acids were actually produced, since the higher acids would be expected to be concentrated in the hydrocarbon layer unless the ratio of water to hydrocarbon produced was exceptionally high.

Since high concentrations of higher acids in some wells could possibly serve as polar inhibitors and so protect those wells, while others with lower concentrations or other unfavorable factors would not be protected, the Committee decided to try isolating acids from the hydrocarbon phase as well as from the water phase of the same well or field. The possibility of polar compound inhibition might explain the fact that of two apparently similar wells one was found very corrosive while the other showed no damaged tubing. The logical method of extracting the acids would have been by treatment with caustic, but financial reasons did not permit this operation in the field. Therefore, it was decided to use another method which, as it developed, gave information that caustic extraction would not have done.

The polar compounds, acids, bases, phenols and ketones, as well as aromatic hydrocarbons present in the hydrocarbon phase produced by a Seabreeze field well were adsorbed on the silica gel contained in a length of pipe which held 12.5 lb of silica gel. A total of 75 lb of silica gel was exposed to a total of 56 barrels of hydrocarbon from the well.[42] Unfortunately, the volume of hydrocarbons passed through the 12.5-lb batches of silica gel varied between 1 and 23 barrels, so that the batches of adsorbent could not be considered as identical.

The exposed silica gel was sent to the laboratory and polar compounds and aromatic hydrocarbons were eluted by downward steaming of the silica gel in a similar length of iron pipe which had been connected to a large condenser so that all products could be collected. A total volume of 200 liters of water layer and 2 gal of organic layer were obtained.[23]

The products isolated from this large volume of crude were:

1. About 90 g of phenolic matter which was not separated and identified
2. About 3 g of hydrochlorides of organic bases
3. Eight liters of neutral organic matter including aromatic hydrocarbons
4. A mixture of the common lower fatty acids containing a much

higher concentration of the C_4 to C_6 acids, less of acetic and propionic acids, and only the usual low concentration of acids with seven or more carbons. No naphthenic acid could be isolated from this or other samples studied. This would have been difficult in view of the small amounts available for study.

The experiment showed that in the Seabreeze well the amount of C_4 to C_6 acids produced was comparable to the amount of propionic acid produced. The higher acids were present in low concentration only in both water and hydrocarbon layers, possibly because the vapor pressure of the higher acids was too low to produce them even at a temperature of 80° to 90°C (176° to 194°F) and 2,000 to 3,000 lb/sq in. bottom-hole pressure.

The most interesting development that grew out of the silica-gel adsorbtion experiment was the finding that ketones were being produced in as high or higher concentrations than the acids in this hydrocarbon phase. Work on the 8-liter batch of neutral compounds was carried only through a small series of the lower-boiling fractions obtained on careful fractionation of the mixture, but this was sufficient to show that the methyl ketones from acetone to methyl-*n*-butyl ketone were present and that practically every fraction (except a few that consisted almost entirely of aromatic hydrocarbons) gave precipitates with ketonic reagents, such as 2,4-dinithrophenylhydrazine.[24] A search was made for the cyclic ketones, cyclohexanone and cyclopentanone, but these could not be isolated. The isolation of ketones is hardly surprising to petroleum chemists, but had not been demonstrated as clearly as here, since it is difficult to isolate ketones from ordinary petroleum products.

Bibliography

1. Appel, H., *Z. physiol. Chem.* **218,** 202 (1933).
2. Aschan, O., *Ber.* **24,** 2710 (1891).
3. Aschan, O., *Ibid.* **25,** 3661 (1892).
4. von Braun, J., *Ann.* **490,** 100 (1931).
5. von Braun, J., *Ber.* **66,** 1464 (1933).
6. von Braun, J., *Ibid.* **67,** 269 (1934).
7. von Braun, J., *Oel u. Kohle* **13,** 799 (1937).
8. von Braun J., and H. Ostermayer, *Ber.* **70,** 1004 (1937).
9. Campbell, S. E., *Refiner* **14,** 381 (1935).
10. Chichibabin, A. E., *et al.*, *Chim. et ind.* Special Number March 1932, pages 308–318.

11. Corson, B. B., E. Adams, and R. W. Scott, *Org. Syntheses* **10,** 48 (1930).
12. Frangopol, Thesis, Munich, 1910.
13. Goheen, G. E., *Ind. Eng. Chem.* **32,** 503 (1940).
14. Hancock, K., and H. L. Lochte, *J. Am. Chem. Soc.* **61,** 2448 (1938).
15. Harkness, R. W., and J. Bruun, *Ind. Eng. Chem.* **32,** 499 (1940).
16. Henderson, L. M., G. W. Ayers, and C. M. Ridgway, *Oil and Gas J.* March 28, 1940, pages 114, 118, and 121.
17. Horeczy, J., Ph.D. Thesis, University of Texas, 1941.
18. Jantzen, E., *Dechema Monograph, Vol.* **5,** *Das fractionierte Destillieren und das fractionierte Verteilen als Methoden zur Trennung von Stoffgemischen,* Verlag Chemie, Berlin, 1932.
19. Jeger, O., and G. Buchi, *Helv. Chim. Acta* **31,** 134 (1948).
20. Kennedy, T., *Nature* **144,** 832 (1939).
21. Kozicki, G., and S. Pilat, *Petroleum* **11,** 310 (1915).
22. Lochte, H. L., C. W. Burnam, and H. W. H. Meyer, *Petroleum Eng.* **21,** 225.4 (1949).
23. Lochte, H. L., and H. W. H. Meyer, *Condensate Well Corrosion,* N. G. A. A. Condensate Well Corr. Committee, Natural Gasoline Assn. of America, Tulsa, Okla., 1953, Chapter 15.
24. Lochte, H. L., H. W. H. Meyer, and E. N. Wheeler, *Ibid.* page 167.
25. Markownikoff, W., and W. Oglobin, *Ber.* **16,** 1878 (1883).
26. Markownikoff, W., *Ann.* **307,** 367 (1899).
27. Mikeska, L. A., *Ind. Eng. Chem.* **28,** 97 (1936).
28. Muller, J., and S. Pilat, *Brennstoff-Chem.* **17,** 461 (1936).
29. Nenitzescu, C. D., D. A. Isacescu, and T. A. Volrap, *Ber.* **71,** 2062 (1938).
30. Nenitzescu, C. D., and C. N. Ionescu, *Ann.* **491,** 189, 207 (1931).
31. Ney, W. O., and H. L. Lochte, *Ind. Eng. Chem.* **33,** 825 (1941).
32. Ney, W. O., W. W. Crouch, C. E. Rannefeld, and H. L. Lochte, *J. Am. Chem. Soc.* **65,** 770 (1943).
33. Neyman-Pilat, E., and S. Pilat, *Ind. Eng. Chem.* **33,** 1390(1941).
34. Pilat, S., and J. Reyman, *Ann.* **490,** 76 (1932).
35. Quebedeaux, W. A., G. Wash, W. O. Ney, W. W. Crouch, and H. L. Lochte, *J. Am. Chem. Soc.* **65,** 767 (1943).
36. Roberts, S., and J. R. Bailey, *Ibid.* **60,** 3025 (1938).
37. Richter, G., Private Communication.
38. Schutze, H. G., W. A. Quebedeaux, and H. L. Lochte, *Ind. Eng. Chem. Anal. Ed.* **10,** 675 (1938).
39. Schutze, H. G., W. Shive, and H. L. Lochte, *Ibid.* **12,** 262 (1940).

40. Shive, W., J. Horeczy, G. Wash, and H. L. Lochte, *J. Am. Chem. Soc.* **64,** 385 (1942).
41. Shive, W., W. W. Crouch, and H. L. Lochte, *Ibid.* **63,** 2979 (1941).
42. Shock, D. A., and N. Hackerman, *Ind. Eng. Chem.* **40,** 2169 (1948).
43. Skraup, S., and E. Schwamberger, *Ann.* **262,** 141 (1928).
44. Smirnov, P., and Z. Buks, *Azerbaidzhan Oil Ind.* **1932,** 60; Translation by J. G. Tolpin, *C. A.* **27,** 1492 (1933).
45. Tanaka, Y., and S. Nagai, *J. Am. Chem. Soc.* **45,** 754 (1923).
46. Tanaka, Y., and S. Nagai, *Ibid.* **47,** 2369 (1925).
47. Tiedcke, K., Thesis, Hamburg, 1928.
48. Wallach, O., *Ann.* **418,** 57 (1919).
49. Whitmore, F. C., and H. M. Crooks, *J. Am. Chem. Soc.* **60,** 2078 (1938).
50. Williams, M., and G. Richter, *Ibid.* **57,** 1686 (1935).

CHAPTER NINETEEN

THE ORIGIN OF PETROLEUM ACIDS

The problem of the origin of acids in petroleum, in general, and of the naphthenic acids, in particular, is tied up with that of the origin of petroleum itself. Any theory accepted for the origin of petroleum must be compatible with the theory of the origin of its acids and vice versa. In other words, if we were to accept the theory that petroleum originated solely, without dependence on compounds synthesized by plants or animals, through the formation of acetylene from carbides or from methane of the earth's original atmosphere, little could be gained from a discussion of optically active acids or the formation of acids from long-chain saturated or unsaturated compounds. We may as well admit that at the present state of our knowledge of petroleum compounds, it is not possible to select any one theory as THE correct one when we talk about the origin of petroleum acids. It would be strange indeed if there were only a single mechanism or series of reactions by which petroleum acids are formed.

If we accept the theory that petroleum is formed, almost entirely at least, from animal and vegetable compounds (which is the theory that will be accepted in the following discussion), there are a number of possible relationships of the acids to the hydrocarbons or to other oxygenated compounds in petroleum.

Relationship of Hydrocarbons to Acids

The following three possible relationships of the petroleum acids to other compounds in petroleum will be considered here:

1. The acids might be derived from the hydrocarbons.

2. The hydrocarbons might be derived from the acids.
3. The acids might be derived from other oxygenated compounds.

Considering the first possibility, the acids might have been derived from hydrocarbons underground during the many thousands of years this mixture has been stored in the earth or they might have been formed through oxidation of hydrocarbons during refining operations, i.e., the acids might be man-made in the refinery.

Since there is little or no oxygen in contact with oil underground, and probably has not been much for ages, the acids would have had to be formed from hydrocarbons very long ago, while extensive changes were still going on, or the oil might have come in contact with ferric iron or other oxidizing agents. In any case, it is difficult to see how it could have been the hydrocarbons, rather than other types of compounds, which would have been oxidized to acids.

Again, under the first possibility, the acids might have been produced by oxidation of hydrocarbons during refining operations, especially while air blowing to stir them, sometimes in alkaline solution. That acids may be formed under these conditions cannot be denied and, for many years, this was considered to be the way in which acids formed. At the time von Braun started his work, he was a strong advocate of this theory.

A number of developments have largely discredited this theory, even though some acids may be formed in this way when there is such an opportunity for oxidation.

In the first place, much work has been done in recent years on the development of methods of producing the higher acids from oil. Whenever this is done by methods analogous to those used in refining operations, acids produced have always been mixtures containing hydroxy and unsaturated acids and polymeric compounds formed presumably from unsaturated acids, while the petroleum acids are either completely saturated or contain only low concentrations of unsaturated compounds and there seems to be no evidence for the presence of hydroxy-acids.

In the second place, synthetic acids — ones intentionally prepared in oxidation experiments — do not appear to contain naphthenic acids in concentrations comparable to those found in petroleum acids. No careful work, using modern fractionation methods which might be expected to separate naphthenic from aliphatic acids, appears to have been done, however.

In the third place, there is another observation that is generally familiar and that is not easily explained on the basis of the theory that the acids are man-made. The oil from a certain field always yields essentially the same amount and nature of acids when processed *without cracking operations* by a certain refinery, but oil from a different field yields, perhaps, an entirely different amount and type of acids in the same operations in the same refinery. Furthermore, the relative concentrations of aliphatic and naphthenic acids from a certain fraction of, e.g., Venezuela crude would be the same whether refined there or in Texas, as long as no cracking is involved.

The fact that extensive changes in acids, as well as in the hydrocarbons, take place during cracking (especially thermal cracking) has led the writer to exclude, with few exceptions, data based on studies of cracking process material. Unfortunately, it is now almost impossible to obtain working samples of acids that include only straight-run material, since practically all refineries do any alkali washing only after combining straight-run and cracking-process products.

In the fourth place, in at least two cases, there have been careful studies of acids obtained directly from crude oil and from refinery fractions of the same crude oil. In both cases, the acids isolated were found to be very similar.

Pilat and Reyman [24] not only studied the acids from a crude oil but also prepared the barbiturates of each sample by the method of von Braun and found the two acids and derivatives to be entirely similar. Von Braun [3] was able to compare the acids isolated from Roumanian crude oil with those isolated from a distillate and found that they were evidently the same with the exception of the lower fatty acids, which he considered to have been modified to some extent by formation of additional lower acids through cracking during distillation.

In addition, von Braun and Wittmeyer [4] studied acids isolated from a California distillate boiling at 180° to 260°C (356° to 500°F). The distillate had been treated at the refinery by liquid sulfur dioxide (Edeleanu process) and the acids from the extract and raffinate were isolated and studied separately. It was found that acids from extract and from raffinate boiling in the same range were similar in all respects, as were the amines and other compounds prepared from them Presumably, they were careful to avoid oxidizing conditions in the isolation of the acids. They concluded that these results could be explained only by assuming existence of the acids in the original crude oil. In view of these developments, von Braun abandoned his earlier position and concluded that the acids were in crude oil as produced.[4] In other less carefully studied cases, the same conclusion was reached.[2]

Finally, although the author knows of no cases in which naphthenic acids have been definitely identified from such products, there have been scattered reports of isolation of acids from water produced by oil wells [11, 27], and in the last 10 years, chemists working in the laboratories of many of the natural-gasoline producers of Texas and Louisiana have studied the corrosion of tubing in many of the high-pressure gas wells of this region and have carried out titrations of organic acids produced by the wells. In a number of cases, they have identified acetic, propionic, butyric, and other low molecular-weight fatty acids. In cases studied in the author's laboratory in connection with the corrosion problem, acids up to at least C_7 have been isolated and a number of them identified after isolation from the water phase produced by the wells. In the one case studied, it was found that the same acids were also present in the hydrocarbon layer.[10, 19, 20] Unfortunately, the high-boiling acids were found in too low concentration to permit identification of naphthenic acids, if present.

According to Biske,[2] it is commercially feasible to extract naphthenic acids directly from crude oil that does not have too high a viscosity. In the case of a Peruvian crude oil of 0.30% acidity, he reports that the acidity was reduced to 0.05% by alkali washing, which yielded acids with an acid number of 210 (purified). As would be expected, the acids have a much wider boiling range than the ordinary ones, but otherwise their properties are similar to those of acids from refinery fractions. In view of these facts, it now is generally admitted that petroleum acids are present as such in crude oil, even though small amounts may be formed in refinery operations.[28]

The nature of the compounds involved and of processes by which the acids are formed underground from animal and vegetable products remains to be discussed.

The suggestion that petroleum hydrocarbons and acids are, in some way, derived from long-chain fatty acids — mostly the unsaturated ones — runs like a thread through the history of naphthenic acid studies.

Neuberg [22] concluded that acids are formed underground not only from fats and oils but also, through bacterial action, from proteins.

Engler [14] suggested that naphthenic acids might have been formed by cracking and cyclization of unsaturated fatty acids.

Tanaka and Kuwata [35] were able to identify palmitic, stearic, myristic, and arachidic acids in Japanese petroleum acids in a total concentration of 7.7% of the total acidity. They stated that these acids were probably derived from fish oils, possibly whale and shark oils. Kabayashi [18] had come to the same conclusion. Some of these higher solid acids have been isolated from a number of different petroleum acid samples, in fact these acids have been isolated from more different acid samples than any other acids with more than six carbon atoms.

Petrov and Ivanov [23] carried out experiments in which they heated oleic acid, water, and alumina at 380° to 390°C (716° to 734°F) for 3 to 4 hours at 3,100 to 3,300 lb/sq in. pressure and obtained a whole series of hydrocarbons of the gasoline and kerosene boiling range. This series had an acid content of 10%. From 4 kg of oleic acids, they isolated 130 g of acids boiling at 210° to 250°C (410° to 482°F) and since the acids showed an iodine number of only 11 and a density of 0.922, they were obviously only slightly unsaturated.

A fraction of the acids was converted to the alcohol, the iodide, and the hydrocarbon in the usual way and the properties and analyses of the acid and its derivatives corresponded to those of a C_9 naphthenic acid. The temperature used by these workers was higher than that now usually assumed to have prevailed during the period of formation of petroleum, but the experiment indicates that under proper conditions it is possible to prepare petroleum hydrocarbons and acids from fatty acids.

Petrov and Ivanov also studied the products obtained on air-blowing a gas oil at 100°C (212°F) and obtained acids in this manner, but these were the usual oxyacids which may have contained some naphthenic acids.

During the last 20 years, several other lines of evidence indicating that acids may be formed from one of the most abundant and bacteriologically stable types of natural animal and vegetable products (the fats and oils), through hydrolysis followed by cleavage, polymerization, and isomerization of the unsaturated acids, have been reported.

The main exponent of this theory at this time is B. T. Brooks, who has studied this problem and has defended this hypothesis for years.[5, 6, 7, 8] When this theory is coupled with that of Stevens and Spaulding [33] and others, according to which compounds first formed might be expected to isomerize to new compounds, the common occurrence of cyclopentane and cyclohexane derivatives is plausible. According to this theory, some of the inorganic silicates, aluminates, etc., with which the oil has been in contact for ages or over which it slowly passed during migration from source site to reservoir, act as polymerization and isomerization catalysts — probably through some carbonium ion mechanism — to yield what we now isolate as the petroleum acids and hydrocarbons.

Such a broad and flexible theory is hard to refute and may indeed offer a way of explaining the occurrence of many of the different compounds found in petroleum. If the theory is correct, some petroleum should be found which had not been in contact with these inorganic catalysts under the proper conditions and, therefore, would be a much simpler mixture. For example, such an oil should contain more of the straight-chain hydrocarbons, particularly those with an odd number of carbon atoms, and should contain a higher concentration of the common long-chain fatty acids, like palmitic and stearic acids. There are certain fields (notably in California and Japan) which have been investigated by Japanese workers and which yield acids with a lower index of refraction.[34, 35] The solid fatty acids mentioned previously were isolated from such acids and studies in the author's laboratory [26] gave no reason for assuming that aliphatic acids above [16] the highest isolated (nonanoic) were not present in the Signal Hill, California acids. As far as is known, most of the Eurasian petroleum acids contain practically no aliphatic acids with more than ten carbon atoms, but von Braun found that the Galician acids studied had an abnormally low index of refraction, which he attributed to the absence of bicyclic acids.[3]

If we assume that the properties of certain Japanese and California acids are due to long-chain fatty acids, it should be possible to show that the acids with even number of carbons predominate, as they do in fats and oils. Furthermore, the hydrocarbons with an odd number of carbons should predominate for the same reason, since loss of carbon dioxide should convert an even acid to an odd hydrocarbon. In addition, this type of petroleum should be paraffinic rather than naphthenic. Actually not enough is known about petroleum, especially the acids, to make a definite decision, but what is known does not agree well with these predictions.

While the new theory would explain the occurrence of cyclopentane derivatives among the acids and hydrocarbons, the actual concentration of these, as far as known, appears to be too high to be explained by formation of equilibrium mixtures in the presence of catalysts.[12, 15, 21, 25, 33]

The occasional occurrence of fractions of petroleum acids that are optically active[1, 9, 13] is, of course, very good evidence for the organic origin of petroleum and its compounds, including the acids. The very fact, however, that so few acids are optically active while many, perhaps most, of the petroleum acids have asymmetric molecules and should be optically active if synthesized *as such* by animal or vegetable organisms is one of the best lines of evidence for Brooks' theory. The absence of optical activity in all but a few compounds probably indicates that extensive rearrangements have taken place to form the acids now found from inactive straight-chain compounds or to racemize acids which were once active.

Out of about two dozen petroleum acids of known structure, there are a few in which the relationship to some known compound is so close and simple that it is hard to avoid the conclusion that these acids were obtained by degradation of the known more complex compounds. Whether this conversion was carried out by living organisms, by reaction with other oxygenated compounds, or by contact with inorganic oxidizing agents, like ferric oxide, cannot be decided, but the relationship seems so obvious that the origin of these acids through degradation of other oxygenated compounds seems the only plausible one. Perhaps if we knew the structure of ten dozen instead of two dozen acids, there would be enough obvious relationships to make this one of the recognized ways in which acids originate.

The highly hindered 2,2,6-trimethylcyclohexanecarboxylic acid has been isolated in large amounts from California acids[30, 31] and has also been isolated from Iranian acids by Kennedy.[17] It could perhaps be assumed to have been formed from a long-chain unsaturated acid, but it is obviously related rather simply to the carotenoids found in many plants:

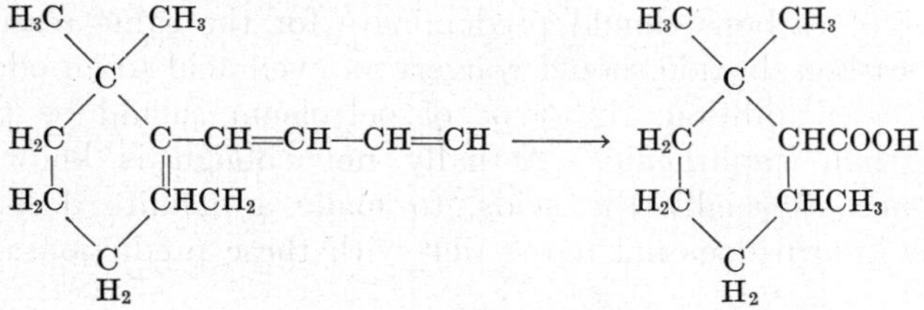

```
H3C     CH3                                  H3C     CH3
   \   /                                        \   /
    C                                             C
   / \                                           / \
H2C   C—CH=CH—CH=CH    ——————→            H2C     CHCOOH
 |    ‖                                     |       |
H2C   CHCH2                               H2C     CHCH3
   \  /                                       \   /
    C                                           C
    H2                                          H2
```

If some future survey should show that this easily isolated acid is found in many different acids and not only in one from California and one from Iran, this compound may become of much greater interest in connection with the origin of petroleum acids.

The same can be said of camphonanic acid (1,2,2-trimethylcyclopentanecarboxylic acid) which is also present in good concentration in the same California acid sample and prepared from camphor to which it is related as shown:

```
              CH3                                CH3
              |                                  |
H2C-----------C-----------C=O      H2C-----------C-COOH
|             |            |        |            |
|        H3C--C--CH3       |   →    |       H3C--C--CH3
|             |            |        |            |
H2C-----------C-----------CH2      H2C----------CH2
              |
              H
```

The fact that the acid is optically inactive, or very slightly levorotatory, may mean that the activity was lost during the conversion to the acid.[16]

The isolation from Texas and also from California acids of dimethylmaleic anhydride,[16, 29] which, at room temperature, is known only in the form of the anhydride, was surprising. Smith and Emerson [32] reported that it can be formed from α-tocopherol, but since it can also be formed from a number of other much simpler compounds, this relationship is probably of no significance.

If hydnocarpic and chaulmoogric acids:

```
HC===CH                              HC===CH
|      \                             |      \
|       CH—(CH2)10—COOH              |       CH—(CH2)12—COOH
|      /                             |      /
H2C---CH2                            H2C---CH2
```

were common compounds in nature, or if we had evidence that they once were, and if we had evidence that the dihydro (saturated) acids occurred in petroleum, these could be considered as parent acids for other naphthenic acids. Unfortunately, no naphthenic acid approaching this molecular weight has been isolated, but according to our present theories on the higher acids, we would not expect to find a side chain as long as these and would expect, instead, to find one or more simple alkyl groups on the ring, in addition to the chain carrying the carboxyl.

In summarizing, it seems fair to say that probably petroleum acids were formed not by a single route or series of reactions but by a number of different ones, each contributing its share of acids, depending on conditions. It is also rather certain that the major portion of the acids was formed underground and is present, as such, in crude oil with only a relatively small portion produced during refining operations. While no one theory is generally accepted, those who hold that the hydrocarbons and the acids were produced underground by degradation, polymerization, and isomerization of long-chain, mainly unsaturated, acids in the presence of inorganic catalysts at moderate temperatures seem to have a fairly simple and flexible theory which is gaining in favor.

Finally, in some special cases, we are almost forced to conclude that the acids are produced by degradation of more complex, known compounds. Since, however, we know very little about individual petroleum acids with more than nine carbon atoms, any theories at this stage are mainly guesses.

Bibliography

1. Albrecht, R., *Chem. Revue der Fett- und Harzindustrie* **18,** 189 (1911); *C. A.* **5,** 3522 (1911).
2. Biske, U., *Refiner* **16,** 72 (1937).
3. von Braun, J., *Ann.* **490,** 100 (1931).
4. von Braun, J., and H. Wittmeyer, *Ber.* **67B,** 1739 (1934).
5. Brooks, B. T., *J. Inst. Petr. Tech.* **20,** 182 (1934).
6. Brooks, B. T., *Am. Assn. Petr. Geol.* **32,** 2269 (1948).
7. Brooks, B. T., *J. Am. Chem. Soc.* **71,** 4143 (1949).
8. Brooks, B. T., *Science* **111,** 650 (1950); *Ibid.* **114,** 240 (1951).
9. Bushong, F. W., and I. W. Humphreys, *8th Int. Cong. Applied Chem.* **VI,** 57 (1912).
10. Burnam, C. W., M.A. Thesis, University of Texas, 1949.
11. Butorin, N. J., and I. P. Buks, *C. A.* **31,** 2399 (1937).
12. Chiurdoglu, G., P. J. Fierens, and C. Henkart, *Bull. soc. chim., Belg.* **59,** 140, 158, 174 (1950).
13. Chichibabin, A. E., F. Tchirikoff, M. Katznelson, S. T. Koriagin, and G. V. Tcholintzeff, *Chim. et ind.*, Special No. March 1932, pages 306–318.
14. Engler, C., *Die Chemie u. Physik des Erdöls*, Volume I, S. Hirzel, Leipzig, 1913.
15. Glasebrook, A. L., and W. G. Lovell, *J. Am. Chem. Soc.* **61,** 1717 (1939).
16. Hancock, K., and H. L. Lochte, *Ibid.* **61,** 2448 (1939).
17. Kennedy, T., *Nature* **144,** 832 (1939).

18. Kabaysashi, K., *J. Chem. Ind. Japan* **24,** 1 (1921); *C. A.* **15,** 2542 (1921).
19. Lochte, H. L., *Condensate Well Corrosion,* N. G. A. A. Condensate Well Corr. Committee, Natural Gasoline Assn. of America, Tulsa, Okla., 1953.
20. Lochte, H. L., C. W. Burnam, and H. W. H. Meyer, *Petroleum Engr.* **21,** 225.4 (1949).
21. Nenitzescu, C. D., and E. Cantuniari, *Ber.* **66,** 1097 (1933).
22. Neuberg, C., *Biochem. Z.* **1,** 368 (1906).
23. Petrov, A. D., and I. Z. Ivanov, *J. Am. Chem. Soc.* **54,** 240 (1932).
24. Pilat, S., and J. Reymann, *Ann.* **499,** 76 (1932).
25. Pines, H., and V. N. Ipatieff, *J. Am. Chem. Soc.* **61,** 1076 (1939).
26. Quebedeaux, W. A., G. Wash, W. O. Ney, W. Crouch, and H. L. Lochte, *Ibid.* **65,** 767 (1943).
27. Reisner, V., *C. A.* **26,** 2308 (1932).
28. Sachanen, A. N., *The Chemical Constituents of Petroleum,* Reinhold, New York, 1945, page 316.
29. Schutze, H., W. Shive, and H. L. Lochte, *Ind. Eng. Chem. Anal. Ed.* **12,** 262 (1940).
30. Shive, W., S. M. Roberts, R. I. Mahan, and J. R. Bailey, *J. Am. Chem. Soc.* **64,** 909 (1942).
31. Shive, W., J. Horeczy, G. Wash, and H. L. Lochte, *J. Am. Chem. Soc.* **64,** 385 (1942).
32. Smith, L. I., and O. H. Emerson, *Ibid.* **62,** 1869 (1940).
33. Stevens, P. G., and S. C. Spalding, *Ibid.* **71,** 1687 (1949).
34. Tanaka, Y., and S. Nagai, *C. A.* **20,** 2744 (1926).
35. Tanaka, Y., and T. Kuwata, *Ibid.* **23,** 4051 (1929).

CHAPTER TWENTY

UTILIZATION OF NAPHTHENIC ACID

The large-scale production of naphthenic acid has resulted in the development of an unusual use pattern. Contrary to the use patterns of most chemicals, this acid has continued to find its major outlet in the single field of driers, since its production on a commercial scale. The reason for this situation appears to lie in the unique properties of the naphthenates of drier metals. Table 32 illustrates the recent use of naphthenic acid as developed from statistics of the U. S. Tariff Commission as well as from private sources.

Table 32. **Consumption of Naphthenic Acid by End Use**

End Use	*Million Pounds* 1947	1948
Driers	18.3	17.4
Flatting Agents	0.5	0.5
Preservatives	2.7	2.0
Rubber Chemicals	1.0	2.0
Detergents and Cleansers	0.6	0.5
Miscellaneous	3.0	3.0
	26.6	25.4

The only major deviation from this end-use distribution occurred during World War II when large quantities of the acid were used in preservatives and incendiary bombs.

The demand for naphthenic acid is also unique in that for some years, it has been in balance with production. It is doubtful that this situation will continue since naphthenate drier demand may be expected to increase in construction and home building associated with the normal increase in population. Additional sources of acid must be developed to satisfy the anticipated increase in demand since the present rate of production appears to have reached a plateau as shown by the annual production figures.

Table 33. **Annual United States Production of Naphthenic Acid**

Year	*Million Pounds*
1943	17.3
1944	28.5
1945	30.0
1946	28.9
1947	32.0
1948	25.4
1949	23.6
1950	24.7
1951	38.5
1952	19.3
1953	23.1
1954	23.7

The preponderant use of naphthenic acid in driers has tended to obscure the many other uses to which this interesting chemical has been put or in which it has been tried. To better review the entire picture of the utilization of naphthenic acid, the subject will be discussed by fields of application.

DRIERS

The position of naphthenic acid in the realm of drier acids is demonstrated by an analysis of the consumption of the more commonly used acids.

Table 34. **Annual Consumption of Acids for Drier Manufacture**

	Million Pounds	
Acid	*1947*	*1948*
Naphthenic	18.3	17.4
Octoic	0.5	1.0
Rosin	1.4	4.9
Tall Oil	0.5	0.5
Linoleic and Others	1.7	1.5

The popularity of naphthenic acid in the drier field is based on the physical and chemical properties of its salts of the drier metals.[19, 32, 35, 37, 49] The salts most commonly used as driers are those of lead, manganese, cobalt, zinc, and iron.[2, 3, 15, 31, 38, 43, 46] While not a drier in the ordinary sense, calcium and magnesium naphthenates are included in this group since they are used in the paint and varnish trade as flatting agents.[4] In general, the naphthenates of the true drier metals are stable to heat, resistant to oxidation, readily and permanently soluble in paint vehicles, and compatible with most resins employed. They contain high percentages of metal and yield low-viscosity solutions in thinners. These properties lead to rapid and uniform drying, reduction in skinning of the paint film, and the formation of the highest quality of film obtainable from the vehicle used.

Naphthenate driers are manufactured by either the fusion or the precipitation method. The choice of method depends on the availability and reactivity of the oxides, hydroxides, or carbonates of the metals concerned and on the metal content or basicity desired in the finished product.[18, 24, 26, 57, 66] As a rule, higher metal content of normal naphthenates can be obtained by the precipitation than by the fusion method. With reactive oxides, such as litharge or zinc oxide, the fusion method may present economies in operation which cannot be neglected. In practice, lead, zinc, copper, calcium, and magnesium naphthenates are prepared from their oxides, hydroxides, or carbonates by fusion, whereas the cobalt, manganese, and iron salts are prepared by precipitation. In principle, the precipitation method is carried out by adding a solution of a water-soluble salt of the drier metal to an aqueous solution of sodium naphthenate and separating the precipitated drier. Since most of the metallic naphthenates are sticky, semisolids, difficult to separate and purify, the reaction is usually carried out in the presence of an immiscible solvent in which the naphthenate is readily soluble. After completion of the reaction, the solvent solution is washed to remove water-soluble salts and dried by distillation of a portion of the solvent from the mixture. After adjustment of the metal content of the residue by the addition of dry solvent or further removal of solvent by distillation, the product is filtered and packaged.

The manufacture of naphthenate driers by fusion consists of adding the oxide, hydroxide, or carbonate of the desired metal to hot naphthenic acid. When the reaction is complete, the product is taken up in a suitable solvent and purified, dehydrated, and its concentration adjusted as in the case of the precipitation method. Elliot [26] has presented a complete discussion of the manufacture of metallic soaps to which the reader is referred for further details.

Naphthenate driers may be manufactured in the form of neutral, basic or "acid" salts, the last being essentially neutral salt solutions containing free naphthenic acid. Several factors govern the selection of the type of salt to be manufactured. Within solubility limitations, the highest metal content is obtained with the basic salts. Their solutions tend to have higher viscosities than the neutral or "acid" type. The basicity of the naphthenate is also related to the absorption of the drier on pigments and is controlled to reduce this effect when necessary.

Basic naphthenates may be formed by either the precipitation or fusion method, although the first is probably easier to control with regard to constancy of composition and uniformity of product. Compounds corresponding to such formulae as Me(OH)(NA) and $Me(OH)(NA) \cdot Me(OH)_2$ can be obtained by maintaining the pH of the reaction at a predetermined value. The formation of basic salts has been observed at a pH of slightly over 7 and is definite at pH 8 to 9. Many of the highly basic salts are insoluble in hydrocarbons.[65] When basic naphthenates are prepared by fusion, an excess of oxide or hydroxide is used. Carbonates are not too suitable for this operation. Care must be exercised in the preparation of basic salts by the fusion method to avoid overheating and subsequent dehydration and decomposition of the product to neutral salt and oxide. The "acid" naphthenates lend themselves readily to the fusion process and usually no complications are encountered.

In common with driers prepared from acids other than naphthenic, the cobalt and manganese salts tend to form viscous solutions in nonpolar solvents. It is not uncommon for such solutions to gel completely.[52] Such a condition apparently does not affect the drier action, but does reduce the ease of incorporation into paint or varnish as well as its removal from the container. The phenomenon of gelling is generally associated with changes in color of the solutions and leads to the assumption that high-molecular-weight coordination compounds may be formed. To avoid the formation of viscous solutions or gels, additives are employed which act as peptizing agents. Such additives as alcohols, phenols, esters of hydroxy acids, naphthenic acid, *p*-toluene sulfonamide, etc., have been found useful.

NAPHTHENATE DRIER PATENTS

Drier	*U. S. Patent No.*	*Year*	*Abstract*
Mn, Pb, Zn, Ca, Ba, etc.	1,694,462 — Alleman	1928	Naphthenates prepared by precipitation method.
Pb	1,875,999 — Gerlach	1932	Lead acetate heated with naphthenic acid and acetic acid driven off.
Co, Mn, Pb, Zn, Al, Ca	1,878,962 — Meidert	1932	Oxides, hydroxides, carbonates and salts of volatile acids heated with naphthenic acid previously dissolved in an immiscible solvent having a boiling point higher than water.
Co, Mn, Pb, Zn, Ca	1,916,805 — Meidert and Schatz	1933	Concentrates obtained by dissolving the naphthenate in a drying-oil fatty acid in the absence of solvents.
Co, Mn, Pb	1,974,507 — Pohl and Isenbeck	1934	Precipitation and washing of undiluted naphthenates carried out in boiling solution. Drying by fusion at 130°C.
Co, Mn, Zn, Pb	1,976,182 — Meidert	1934	Solid, precipitated driers prepared from rosin, drying oil acids, and naphthenic acid are stabilized by the addition of free fatty acid or monocarboxylic aromatic acid.
General	2,025,872 — Krumbhaar	1935	Driers are prepared in drying oil media.
Co, Mn, Pb, Zn, Ca, Mg.	2,049,396 — Meidert	1936	Precipitated and fused driers made with mixtures of wool grease and naphthenic acids.
Pb, Ca	2,071,862 — Fisher	1937	Naphthenates are prepared by fusion in the presence of a small amount of oily, nonalcoholic solvent for the naphthenate to give complete reaction.
Co, Fe, Al, Ca, Mg, etc.	2,075,230 — Schatz	1937	Precipitated or fused driers are prepared in the presence of amines to enhance the metal content of the product.
	2,081,407 — Minich	1937	Antigelling agents for driers — ethyl and butyl citrates, cyclohexyl *p*-toluene sulfonamide.
General	2,019,020 — Shipp	1937	Manufacture of driers from oil-refinery liquors.

NAPHTHENATE DRIER PATENTS (*Continued*)

Drier	*U. S. Patent No.*	*Year*	*Abstract*
Pb (basic) Co, Mn	2,102,633 — Long	1937	Basic lead naphthenate prepared by fusion is used alone or with other driers to obtain solubility in paint film.
——	2,113,496 — Roon and Gotham	1938	Manufacture of driers by precipitation in presence of paint thinner to dissolve the salt as formed.
——	2,116,321 — Minich	1938	Insoluble printing-ink drier of precipitated hydroxy salts of naphthenic acid.
Pb, Mn, Ca, Mg	2,116,884 — Fisher	1938	Precipitated lead naphthenate is used as a flux in the dehydration of unmixed driers by fusion and heating.
Co, Mn, Pb	2,119,753 — Rutherford and Brock	1938	Butyl and amyl alcohols are added to drier solutions to prevent gelling.
Co, Mn	2,138,087 — Burchfield	1938	Salts of organic acids and nitrogen bases are added to drier solutions to prevent gelling.
Hg	2,139,134 — Roon	1938	Mercury naphthenate is prepared by precipitation in the presence of a solvent.
Pb (basic)	2,157,766 — Long	1939	Basic lead naphthenate prepared by fusion in the presence of a small amount of water.
Co, Mn, Zn (basic)	2,157,767 — Long	1939	Basic driers are prepared by fusion from metal oxides or hydroxides in the presence of a high-boiling solvent to remove the water of reaction.
Co, Mn, Zn	2,157,768 — Long	1939	Mixtures of normal and basic driers are used to improve solubility in the vehicle and film.
Ba	2,180,721 — Roon and Minich	1939	Barium naphthenate used as a wetting agent for pigments in oil media.
Co, etc.	2,199,828 — Bogdan	1940	Aromatic amines are used to stabilize driers against discoloration by oxidation and gelling.
Co, etc.	2,199,829 — Bogdan	1940	Aromatic amines and wood-tar fractions boiling above 240°C are used to stabilize driers against discoloration by oxidation and gelling.

NAPHTHENATE DRIER PATENTS (*Continued*)

Drier	*U. S. Patent No.*	*Year*	*Abstract*
Sn, Fe, Zn, Ni, Cr, Mn, Al, Cu	2,205,994 — Towne	1940	Driers are prepared by precipitation from aqueous sodium naphthenate by the addition of dry, water-soluble salts which hydrolyze readily.
Sn, Fe, Zn, etc.	2,206,002 — Deutser and Nelson	1940	An immiscible solvent for the driers is emulsified with the sodium naphthenate solution prior to the addition of dry, water-soluble salts which hydrolyze readily.
Sn	2,311,310 — Bowers	1943	Manufacture of basic and neutral tin naphthenate.
Co	2,360,283 — Rutherford	1944	Polyphenols and alkylated monophenols prevent gelling, discoloration, etc., in cobalt driers.
General	2,368,560 — Minich	1945	Hydroxy naphthenates prepared by a fusion method.
General Cu	2,368,565 — Minich	1945	Hydroxy metal naphthenates are prepared by precipitation in the presence of free alkali.
General Cu, Mg	2,389,873 — Schiller	1945	Basic metal soaps prepared by mixing metal hydroxide and sodium naphthenate and then neutralizing to correct pH. Cu(OH)(NA) is obtained at pH 9.
Co	2,390,830 — Elliot	1945	Sorbitol is used as antigelling agent for cobalt naphthenate.
General	2,411,832 — Linford and Baral	1946	Use of halogenated solvents in drier manufacture.
General	2,466,925 — Brauner	1949	Nitrogen compounds are used as catalysts in drier manufacture.

NAPHTHENATE DRIER AND PROTECTIVE–COATING PATENTS

U. S. Patent No.	*Year*	*Abstract*
1,537,572 — Zernik	1926	Metallic naphthenates in lacquer base.
1,983,006 — Schladebach and Hahle	1934	Copper naphthenate in rustproofing paint.
2,102,633 — Long	1937	Use of basic lead naphthenate as paint drier.
2,157,768 — Long	1939	Mixed neutral and basic naphthenates as driers.
2,180,721 — Roon and Minich	1939	Barium naphthenate as pigment wetting agent.
2,338,892 — Bassford	1944	Plasticizer for shellac.
2,350,520-1 — O'Neal	1944	Mixture of metallic naphthenates and dye pigments.
2,409,774 — Mack and Klebsattel	1946	Metallic naphthenate containing dialkylorthophosphate.
2,482,606 — Adelson and Dannenberg	1949	Pollyallyl naphthenate as ester-gum substitute.
2,518,438 — Beretvas	1950	Moisture-resistant coatings using naphthenate driers.
2,521,675 — Hoogsteen	1950	Naphthenates driers in varnish films.
2,528,429 — Elliott and O'Hara	1950	Stabilized naphthenate driers.
British		
649,694 — Marrian and I. C. I. Ltd.	1951	Cobalt driers for nitrocellulose films.
German		
801,525 — Pallauf	1951	Isotopes in improved naphthenate driers.
2,537,055 — Hiron and Garner	1951	Aqueous emulsion paint containing naphthanate.
2,538,974 — McGlone	1951	Heat resistant coatings.
2,562,062 — Rethwisch et al.	1951	Aluminum paint containing copper naphthenate.

NAPHTHENATE–DRIER LITERATURE

Drier	*Reference*	*Abstract*
Al, Pb, Zn, Mg	*Farben-Ztg.* **37**, 1664 (1932); *C. A.* **26**, 5774 (1932), Kogan and Churdenko	Effect of method of manufacture and metal content on viscosity of solutions and paint pigments.
General	*P. and V. Prod. Mgr.* **9**, No. 11, 28 (1933); *C. A.* **28**, 348 (1934), Klebsattel	Naphthenate driers are superior because of solubility, stability, uniformity, etc.
Pb	*Off. Digest P. and V. Prod. Clubs* No. 136, 158 (1934); *C. A.* **28**, 4612 (1934), Knauss	Lead naphthenate is best for linseed oil because of solubility after precipitation of impurities.
Co, Mn, Pb, Zn	*Off. Digest P. and V. Prod. Clubs* No. 145, 156 (1935); *C. A.* **29**, 4190 (1935), Knauss	Lead and manganese naphthenates cannot be used with some alkyd and phenolic resins; cobalt tends to cause film wrinkling which is minimized by zinc naphthenate.
Co, Mn, Pb, Zn, Ce, Cu, Al	*Oil Color Trade J.* **88**, 1711 (1935); *C. A.* **30**, 1590 (1936), Curwen	Driers are prepared by precipitation and dissolved in a solvent; mixed driers are described.
Co	*Paint, Oil, and Chem. Rev.* **98**, No. 12, 32 (1936); *C. A.* **30**, 8657 (1936), Knauss	Cobalt naphthenate effects faster drying than the linoleate or resinate; refined naphthenic acid gives better results.
General	*Paint, Oil, and Chem. Rev.* **98**, No. 24, 114 (1936); *C. A.* **31**, 4835 (1937), Montreal P. and V. Prod. Club	Naphthenate driers are used in printing inks.
Zn	*Farben-Chem.* **8**, 398 (1937); *C. A.* **32**, 3640 (1938), Müller-Lobeck	Review on the action of zinc driers on pigments and oils.
Co	*Ind. Eng. Chem.* **30**, 114 (1938); *C. A.* **32**, 1951 (1938), Nicholson and Holley	Comparative study of cobalt naphthenate and resinate on linseed oil; practical upper limit of drier concentration is determined.
Co, Mn, Pb	*Paint, Oil, and Chem. Rev.* **99**, No. 23, 72 (1937); *C. A.* **32**, 1951 (1938), Montreal P. and V. Prod. Club	Naphthenates of manganese and lead, the resinate of cobalt, show shortest drying time; mineral spirits accelerate drying with lead and cobalt naphthenate.
Co, Mn, Pb	*Org. Chem. Ind.* (*USSR*) **5**, 421 (1938); *C. A.* **33**, 884 (1939), Zamyslow et al.	Discussion of methods of preparation of naphthenates from distilled acid.
Cu	*Drugs, Oils, and Paints* **54**, 166 (1939); *C. A.* **33**, 5679 (1939), Meckler	Small amounts of copper naphthenate act as antiwrinkling agents in synthetic enamels.
Co	*Nat'l P., V. and L. Assoc. Sci. Sect. Circ.* **568**, 407 (1938); *C. A.* **33**, 8041 (1939), Montreal P. and V. Prod. Club	A detailed study is presented on the storage properties of paints containing cobalt naphthenate.

NAPHTHENATE–DRIER LITERATURE (*Continued*)

Drier	*Reference*	*Abstract*
Co and General	*Off. Digest P. and V. Prod. Clubs* No. 183, 90 (1939); *C. A.* **33,** 9684 (1939), Bryson	Cobalt naphthenate is the best drier for dehydrated castor oil. Review of effect of naphthenate driers on glyptal resins.
Co, Mn, Pb, Ca	*C. A.* **34,** 1499 (1940), Drinberg and Parfenova	The acid salts are preferable as driers because of better solubility. A study of drier efficiency is presented.
Co, Mn, Pb, Zn, etc.	*Paint Varnish Prod. Mgr.* **21,** 166 (1941); *C. A.* **35,** 5331 (1941), Rinse	Cobalt naphthenate was the best accelerator of photopolymerization of linseed oil.
Co	*Paint, Oil, and Chem. Rev.* **104,** No. 23, 94 (1942); *C. A.* **37,** 1281 (1943), N.Y.P. and V. Prod. Club	Cobalt naphthenate is absorbed by TiO_2, but this does not always result in greatly reduced drying power.
Ca, Li, Al	*Off. Digest P. and V. Prod. Clubs* No. 220, 489 (1942); *C. A.* **37,** 1882 (1943), Montreal P. and V. Prod. Club	Lithium and calcium naphthenates act as promoters for cobalt naphthenate.
General	*J. Oil Colour Chem. Assoc.* **25,** 211 (1942); *C. A.* **37,** 1880 (1943), Chubb	The precipitation method is said to yield the best driers. Neutral salts are prepared. Acids of 200–260 acid number are preferable. The advantages of naphthenates are discussed.
Th	*Ann. Repts. Soc. Chem. Ind. Progress Appl. Chem.* **26,** 313 (1941); *C. A.* **38,** 1816 (1944), Howard	Thorium naphthenate in large amounts yields oils bodying like dehydrated castor oil.
General	*Off. Digest P. and V. Prod. Clubs* No. 244, 132 (1945); *C. A.* **39,** 2210 (1945), Tearnley	A review of the preparation of naphthenate driers is presented. The best driers are prepared from acids free from mercaptans, sulfides, and complex phenols.
General	*Chem. Eng. News* **23,** 1164 (1945); *C. A.* **39,** 4217 (1945), Dean	A general review on naphthenates including driers.
Co	*Ind. Eng. Chem.* **39,** 115 (1947); *C. A.* **41,** 7134 (1947), Wheeler	Primary, secondary, and tertiary monoamines do not affect drying time with cobalt driers. Heterocyclic nitrogen compounds with tertiary nitrogen in the ring accelerate drying time.

PRESERVATIVES (Fungicides and Insecticides)

In preservatives, naphthenic acid is used largely in the form of copper naphthenate. Zinc and mercury naphthenates are also used though to a smaller extent because of the lower efficiency of the first and the high toxicity to humans of the second. Copper naphthenate is an efficient fungicide and insecticide and finds its widest application in the preservation of wood and textiles against cellulose-destroying fungi and termites.[11, 12, 56, 63] Conflicting reports on its effectiveness against marine borers make its use against these organisms of doubtful value.[32] In view of the potential importance of copper naphthenate as a factor in the preservation of cellulosic materials and the conservation of our natural resources, it will be discussed in some detail.[28, 42]

Copper naphthenate was probably the first preservative used on a large scale in Europe for the preservation of fishing gear.[13] Before World War II, it was used in this country to a small extent for the preservation of wood though small amounts were also applied to fabrics and cordage. During the War, extensive use was made of this compound for the preservation of sand bags, tents, cordage, and wood.[14, 30, 60, 61, 62, 64] Copper naphthenate achieved its maximum use during the creosote shortage as a wood preservative, but did not maintain its consumption volume mainly because of its high cost of manufacture by the processes then available. New processes have been developed since which may result in making this preservative available in large quantities.[16, 17, 21, 22, 25, 41, 51, 58]

A satisfactory preservative should be efficient, permanent, and nontoxic to human beings at its effective antimicrobiological concentration and under application conditions. It should not adversely affect the end use of the article treated and should be economical in both initial cost and application. Copper naphthenate meets these general requirements and, in addition, yields a clean product when used with wood. It may also be used under paint.[44]

The preservative action of copper naphthenate on cellulosic materials has been the subject of numerous investigations. Typical of the results demonstrating its efficiency against fungi and termites on wood are those reported to the American Wood Preservers Association and by the Forest Products Laboratory, United States Department of Agriculture. The widespread use of copper naphthenate in the immediate postwar period under conditions of inadequate control led to some doubt as to its effectiveness. Since then, however, rapid progress has been made in establishing correct procedures for its use and the minimum or threshold quantities required to secure effective preservation of wood. It is now believed inadvisable, for example, to use copper naphthenate in combination with either creosote or pentachlorophenol. Efforts to determine the threshold value for copper as copper naphthenate have been made through the use of wood blocks by Duncan and Richards at the Forest Products Laboratory. From this work, it was established that adequate protection of the wood for ground-contact service was obtained when the copper content was 0.07 lb per cu ft of wood. Stake tests, reported on by the same laboratory, after 12 years of service, indicate that a lower threshold value may give adequate protection.

In spite of some of the unanswered questions concerning specific points of interest on the use of copper naphthenate as a wood preservative, its value is becoming more widely recognized and several commercial treating plants are now using it. In actual practice, treating is carried to the point where the copper content of the treated wood is 0.07 to 0.10 lb per cu ft for ground-contact service and 0.05 lb for above-ground service. Based on the records available, wood so treated should remain in serviceable condition for many years. Copper naphthenate is not recommended for marine service.

The use of copper naphthenate by noncommercial treaters, such as farmers, is expanding. While such treaters seldom impregnate their wood to the extent obtained under commercial treating practice, the protection obtained extends the life of the wood sufficiently to return a handsome profit on the cost of treatment.

The application of copper naphthenate to textiles probably originated with its use on cordage in the fishing industry where it served as a protection against fungi, slimes, and algae.[13] Its use then spread to fabrics, such as duck, burlap, and cotton bagging. Because of its green color and characteristic naphthenic acid odor, copper naphthenate is not well adapted for use on clothing. The treatment of textiles with copper naphthenate does not contribute to their chemical or physical breakdown except in the presence of sunlight when tenderizing may occur.[6, 7, 8, 59] When, however, a small amount of creosote is added to the copper naphthenate, the tenderizing action is eliminated according to Dean, Strickland, and Berard.[22] Pigmented fabrics are not tenderized by creosote-free copper naphthenate as unpigmented materials are. The resistance of fabrics to microbiological attack after treatment with copper naphthenate is also illustrated by the work of Dean, Strickland, and Berard.[22]

Table 35. **Effective Service in Days of Rotproofed Sandbags** *

Treatment	*Lot No. 1*** *March*	*Lot No. 2*** *June*	*Lot No. 3*** *September*	*Lot No. 4*** *December*	*Average of 20 Bags*
1. Copper Naphthenate	133	78	159	144	128
2. Cuprammonium Carbonate	121	80	162	147	128
3. Copper Ammonium Fluoride	92	53	74	119	85
4. CopperNaphthenate + Creosote	155	217	325	203	225
5. Cuprammonium-Creosote	166	153	300	267	222
6. Copper Oleate	107	73	73	120	93
7. Copper Resinate	78	58	27	69	58
8. Copper Tallate	99	120	85	116	105
9. Controls	—	—	11	—	11*

* Average Soil Temperature during the Test = 72.1 °F; Average Soil Moisture during the Test = 27.6%.

** Figures represent average of 5 bags.

Table 36. **Evaluation of Treatments as Compared with Copper Naphthenate**

Order		*Service Rating*
1	Copper Naphthenate-Creosote (4)	176
2	Cuprammonium-Creosote (5)	173
3	Copper Naphthenate (1)	100
4	Cuprammonium Carbonate (2)	100
5	Copper Tallate (8)	82
6	Copper Oleate (6)	73
7	Copper Ammonium Fluoride (3)	66
8	Copper Resinate (7)	45
	Untreated Controls (September)	

The physical and chemical properties of copper naphthenate make it an exceedingly versatile preservative. Its unlimited solubility in low-cost hydrocarbon solvents permits its application to either wood or textiles by the most economical methods. It may be applied, for example, from a mineral-spirits solution and the solvent recovered, if desired. Such solutions may be used as part of a pigmenting formula to secure a combined treatment. Copper naphthenate may be combined with waxes, resins, or asphalts in the manufacture of waterproof fabrics and, being compatible, will not "bloom" or otherwise separate. For the treatment of wood, copper naphthenate may be applied in a mineral-spirits solution where no solvent retention is desired. When a permanent-type solvent is called for because of economy, increased water resistance, etc., any petroleum oil of satisfactory viscosity may be used, but gas oils or similar distillates are the best. Since most wood-preservative treatments involve heat and pressure to secure the introduction of the required quantity of preservative, it is interesting to note the stability of copper naphthenate under simulated treating conditions. Samples of copper naphthenate solutions in gas oil and containing 0.5% metal, when heated at 300°F (148.9°C) for 24 hours in a steel vessel, showed negligible sludge formation (benzene-insoluble matter) as may be seen from Table 37.

Table 37. **Sludge Formation in Copper Naphthenate Solutions in Gas Oil**

	Colombian	*Heavy Catalytic Cycle Stock*	*Light Catalytic Cycle Stock*
Sludge %	0.06	0.02	0.04
	0.07	0.03	0.04

Copper naphthenate may be applied to fabrics in the form of an emulsion, but whether in solution or emulsion, it is sufficiently stable to heat and air to permit handling at any temperatures usual in textile processing.

Insecticide compositions containing naphthenic acid have been prepared, but not widely used. Copper naphthenate acts both as an insecticide and a fungicide and protects wood against such insects as termites and powder post beetles. Zinc naphthenate serves the same purpose. Neither of these has, however, been used as agricultural insecticides to any great extent, although Rogers and McNeil have presented the following base formula as being useful in the treatment of seeds and plants:

	%
Mahogany Soap	18.5
Paraffin Oil	61.5
Copper Naphthenate	13.5
Water	3.5
Alcohol	3.0

This product forms an emulsion when added to water, which is the final treating agent.

In agricultural insecticides, naphthenic acids have been employed as solvents, derivatives, and emulsifying agents. As an emulsifying agent for agricultural insecticides, sodium naphthenate has some advantages over ordinary soap. Highly concentrated (25 to 50%) aqueous solutions can be prepared without danger of solidification. Since naphthenic acid is somewhat stronger than the higher fatty acids, the equivalent sodium soaps have a lower pH. In aqueous solution, for example, sodium naphthenate has a pH of 8 to 9, whereas that of sodium stearate is 10 to 11. The lower pH of the naphthenate soaps would tend to reduce damage to foliage from excessive alkali. Typical of formulations for agricultural spray oils and emulsifying agents for such oils are the following:

Concentrate	%
Light Lubricating Oil	78–82
Sodium Naphthenate	14–18
Pine Oil	2
Water	2
Emulsifying Agent	
Butyl Alcohol	19.5
Sodium Naphthenate (92%)	80.5

Soap-free emulsifying agents, such as monoglyceryl naphthenate, have also been prepared from naphthenic acid.

Derivatives of naphthenic acid have also been used as agricultural insecticides and fungicides because of their reputed biological activity. Such compounds as phenol naphthenates, glycerol naphthenates, and other esters and salts have been reported as being valuable for this purpose.

Disinfectants, deodorants, and stock dips are related to insecticides and fungicides. Compositions of the following types are easy to prepare:

	%	%
Tar Acid Oil (10%)	70	—
Tar Acid Oil (75%)	—	70
Naphthenic Acid (Acid Number 225)	10	10
Oil-Soluble Petroleum Sulfonate	13	13
Caustic Soda (50%)	3.2	3.2
Water	3.8	3.8

PRESERVATIVE LITERATURE

Naphthenate	*Reference*	*Abstract*
Cu	*Can. Textile J.* **61,** 31, 34, 38 (1944); *C. A.* **38,** 6567 (1944), Tweidie and Bayley	Copper naphthenate gives good protection to cordage of all types, when 0.38–0.66% copper on the fiber is used in the soil burial test.
Cu	*J. Soc. Chem. Ind.* **63,** 220 (1944); *C. A.* **38,** 6570 (1944), Twiselton	Copper naphthenate tenderizes coated carton fabrics in sunlight by catalyzing H_2O_2 and O_2 formation from linseed oil used in the coating.
Cu and Other	*Chem. Age (London)* **51,** 517 (1944); *C. A* **39,** 1051 (1945), Carter	Metallic naphthenates are superior to other metallic organic or inorganic preservatives.
General	*U. S. D. A. Tech. Bull. No. 892* (1945); *C. A.* **39,** 4229 (1945), Marsh et al.	Description of leaching test and comparison between soil burial and inoculation tests for evaluation of preservatives. Soil burial gives most severe conditions.
Cu	*Am. Dyestuff Reporter* **34,** 225 (1945); *C. A.* **39,** 5496 (1945), Bartlett and Goll	Copper naphthenate per se does not oxidize cellulose.
General	*Ibid.* **34,** 195 (1945); *C. A.* **39,** 5497 (1945), Dean, Strickland Berard	Description of soil burial test for fabrics.
Cu	*Ibid.* **34,** 457, 471 (1945); *C. A.* **40,** 1661 (1946), Bayley and Weatherburn	Leaching in water reduces the effectiveness of copper naphthenate on cotton fabrics in the soil-burial test by changing the form of the copper compound and not by loss of copper. The presence of paraffin reduces this effect.
Cu	*Melliand Textilber.* **25,** 314 (1944); *C. A.* **40,** 4223, V. Brandt	Copper naphthenate is used in the preservation of fishing nets.
Cu	*Am. Dyestuff Reporter* **35,** 346 (1946); *C. A.* **40,** 5928 (1946), Dean, Strickland and Berard	Service tests and treated sandbags show copper naphthenate to be the best preservative if 15% creosote is added for sunlight protection.
Cu	*Soc. Dyers Colourists Symp. on Fib. Proteins* **67** (1946); *C. A.* **41,** 1445 (1947), Race	Copper naphthenate is a good bacteriacide when used on protein fibers.
Hg	*Paint, Oil Chem. Rev.* **109,** No. 24, 6, 36 (1946); *C. A.* **41,** 2523 (1947), Minich and Goll	Phenyl mercury naphthenate in 0.05% concentration is an effective fungicide for paint and lacquer.

PRESERVATIVE LITERATURE (*Continued*)

Naphthenate	*Reference*	*Abstract*
Cu	*Can. J. Research* **25F**, 92 (1947); *C. A.* **41**, 2905 (1947), Bayley and Weatherburn	Copper naphthenate used without protective agents on cotton leads to poor weathering but superior soil-burial test results.
Na	*Zt. Hyg. Infectionskrankh.* **125**, 666 (1944); *C. A.* **41**, 2919 (1947), Roelke and Reichel	Sodium naphthenate is reported to have good antibacteriological activity.
K	*Soap* **10**, No. 11, 21 (1934); *C. A.* **29**, 629 (1935), Bachrach	Potassium naphthenate is used as an emulsifying agent for cresylic acid.
Naphthenic Acid	*C. A.* **31**, 4763 (1937), Nemiritskii et al.	Clay impregnated with 10% naphthenic acid was as effective a seed disinfectant as with the copper and iron salts.
Cu	*Soap* (San. Sect.) **14**, No. 11, 86 (1938); *C. A.* **33**, 777 (1939), Smith	Copper naphthenate used as a preservative for wood and fabric.
Cu	*J. Textile Inst.* **30**, 346, 349, 429 (1939); *C. A.* **34**, 2182 (1940), Bryson	Copper naphthenate was superior to the zinc and mercury salts for the preservation of jute.
Cu, Zn	*J. Textile Inst.* **30**, 340 (1939); *C. A.* **34**, 270, 2182 (1940), Elkin and White	Comparative soil-burial tests over a period of 3 weeks showed a superiority of copper over zinc naphthenates and over the oleates.
Cu, Zn	*Ind. Eng. Chem.* **33**, 538 (1941); *C. A.* **35**, 3825 (1941), Furry, Robinson and Humfeld	Copper or zinc naphthenates in Stoddard solvent or in aqueous ammoniacal solution, when applied to fabrics, give full protection against mildew.
Cu	*J. Chem. Met. Mining Soc. So. Africa* **42**, 122 (1941); *C. A.* **36**, 3925 (1942), Bowen	Copper naphthenate is used as rotproofer for fabrics for underground use.
Cu, Hg	*Empire Forestry J.* **20**, 179 (1941); *C. A.* **36**, 4308 (1942), Berry and Carter	2.5–20% copper naphthenate or 0.44–3.5% mercury naphthenate were found to give good protection for at least 20 months under tropical conditions against fungi and termites.
Cu	*Textile Research* **12** Suppl. No 1, 1 (1942); *C. A.* **36**, 5657 (1942), Hock and Harris	Copper naphthenate is the most widely used mildewproofing agent on military fabrics for bags, tents, etc. Based on the fabric, 0.8% copper is used, which concentration is higher than required to counteract leaching.

PRESERVATIVE LITERATURE (*Continued*)

Naphthenate	*Reference*	*Abstract*
Cu	*Ind. Eng. Chem.* **36,** 176 (1944); *C. A.* **38,** 1121 (1944), Marsh et al.	Comparative tests by soil burial show copper naphthenate to be superior to corresponding oleates, tallates, and hydroresinates. Naphthenic acid per se and in combination contributes to the effectiveness of the salt.
Cu	*New Zealand J. Sci. Tech.* **23,** 446 (1946); *C. A.* **41,** 3900 (1947), Kelsey	Copper naphthenate is toxic to certain larvae and mature wood borers and termites and inhibits egg laying. The water-soluble preservatives tested did not inhibit egg laying.
Cu	*U. S. Army Spec. 6–363*, May 1946	Cord and rope: cotton braided.
Cu	*U. S. Army Spec. 100–17*, February 1944	Mildewproofing of fabrics, threads, and cordage, copper processes.
Cu	*U. S. Army Spec. 4–1131*, July 1945	Compound, retreating, water-repellent and mildew resistant (for duck and cotton webbing).
Cu	*U. S. Army Spec. 6–345*, June 1945	Duck, cotton: fire, water, weather, and mildew resistant.
Cu	*Navy Dept. Spec. 52 W 5*, April 1945	Wood preservative, water repellant.
Cu	*British Standard Spec. BS/ARP56*, August 1941	For rotproofing canvas, yarn, and cordage.
Cu	*British Standard Spec. BS/ARP58* August 1941	For rot- and water-proofing jute canvas.

PRESERVATIVE PATENTS

Naphthenate	*U. S. Patent No.*	*Year*	*Abstract*
Cu	1,482,416 — Snelling	1924	Copper naphthenate is used in a nondrying vehicle.
Na	1,502,956 — Jones	1924	Sodium naphthenate as an insecticide and emulsifying agent for use with cresol.
Na	1,577,723 — Hughes	1926	Spray oil base is prepared by dissolving dry sodium naphthenate in an equal weight of oil.
Na	1,582,257 — Frizell	1926	Emulsifiable oil is prepared by dissolving dry sodium naphthenate in cresol and then diluting with oil.
Cu	1,679,919 — Rogers-McNeil	1928	An emulsion of copper naphthenate, mahogany soap, and oil is used as an agricultural spray.
Na	1,694,462 — Alleman	1928	Spray oil is prepared from sodium naphthenate and oil.
Na	1,695,197 — Merrill	1928	An emulsifying agent for mineral oils is prepared from a mixture of sodium naphthenate, mineral oil, and alcohol.
Glycerol	1,949,799 — Knight	1934	Agricultural spray is composed of glycerol mononaphthenate in oil as surface-active agent.
Phenyl	2,015,045 — Teichman	1935	Phenol esters of naphthenic acid are used in oil as an agricultural insecticide.
Na	2,017,391 — Berry	1935	Elemental sulfur is dissolved in naphthenic acid which is then neutralized with soda to give an emulsifiable product.
Naphthenic Acid	2,045,925 — Remy	1935	Mercaptans and disulfides are dissolved in naphthenic acid to form an insecticide and fungicide which is applied as a fog.
Na	2,058,788 — Hendrey	1936	A spray oil is prepared from sodium naphthenate and oil with pine oil as a coupling agent.
Na	2,060,425 — Neukom	1936	A soluble oil is prepared from sodium naphthenate, rosin, oil, and ethylene glycol as coupling agent.

PRESERVATIVE PATENTS (*Continued*)

Naphthenates	*U. S. Patent No.*	*Year*	*Abstract*
Cu	2,188,951 — Kraus	1940	A mixture of naphthenic acid-tar oil is sulfurized, converted to the copper soaps, diluted with oil and emulsified for use as a spray oil.
	2,330,452 — Schiller	1943	Manufacture of mercury naphthenate from the acetate.
Cu and Other	2,335,101 — Belzer and Schiller	1943	The salts are dissolved in a water-soluble alcohol with the aid of ammonia to give a water solution for the treatment of textiles.
Cu and Other	2,364,391 — Schiller	1944	Ammoniacal solutions of heavy metal naphthenates containing an alkyl amine to insure solubility and even distribution are used for textile preservation.
Hg	2,423,044 — Nowals	1947	Bis (naphthenyl mercury) cresol composition as a fungicide.
	2,423,611 — Minich	1947	Basic copper naphthenate fungicide.
	2,466,925 — Brauner	1949	Metallic iminonaphthenates as fungicides.
	2,479,235 — Hampton and Smith	1949	Naphthenate fungicides in asphalt for preservation of cotton.
	2,492,941 — Goll	1950	Fungicidal and bactericidal products.
	2,496,566 — Szwarc	1950	Hot-melt fungicidal coating for paper.
	2,497,579 — Bried	1950	Preservative for polysaccharides in paper sizing.
	2,546,179 — Paine and Hughes	1951	Preservative for raw sugar sirup.
	British 648,400 — Nuodex	1951	Fungicide additive for plastics, paints, waxes, etc.

Lubricants

In the field of lubricants, naphthenic acid is generally used in the form of its salts. To appreciate better the diversity of its applications, a brief summary of the general types and functions of lubricants is presented. For practical purposes, lubricants may be divided into three major groups: greases, lubricating oils, and cutting oils. The differences between the groups reside, basically, in their physical properties and end uses. The function of a lubricant is to provide a low-friction film between moving surfaces, normally metals. Cutting oils, while often classed as lubricants, have a somewhat different function in that they act as coolants and flushing agents, as well as lubricants, in metal cutting and drilling. The choice between a grease and an oil is governed mainly by the method of application to be used as dictated by the design of the machinery to be lubricated. There are also other factors to be considered, such as heat resistance, detergency, etc. Greases are normally used where it is impracticable to maintain a supply of liquid lubricant or where clearances are so large that oil would not be readily retained. Greases are also used where rotating parts operate at such speed that liquid lubricants would be thrown out by centrifugal force or where dripping would be a serious disadvantage, as in a textile mill.

Oils are used as lubricants where machinery is designed for this type of lubrication and where certain specific properties, in addition to lubrication, are required. It is the practice, for example, to include detergents in crank-case oils to maintain the cleanest possible bearing surfaces.[39] Such oils may also contain suspending agents to prevent sludge deposition and agents designed to reduce piston-ring sticking by elimination or solution of engine varnish.[45] It is obvious that liquid lubricants are required for any circulating system.

The cutting oils are, in most cases, polyfunctional emulsifiable oils.[54] They not only lubricate the edge of a cutting tool but also act as a coolant to prevent overheating of the tool and the article being worked. At times, cutting oils are also designed to act as a protective agent to exclude air and reduce oxidation of the part being machined. The addition of water to a cutting oil permits cooling by evaporation and the inclusion of detergents and suspending agents.[54] It is desirable to allow the phases to separate by standing and thus reduce the amount of filtering required in a circulating system.

The greases are normally solid or semisolid and are used in the form of blocks, resting against a moving surface, or as pastes forced on to the surface to be lubricated by external pressure. In order to obtain the desired consistency of a grease, oils are incorporated with a soap. Other substances, such as antioxidants, surface modifiers, etc., may also be added. Hard greases are generally prepared with sodium or alkali soaps while soft or cup greases usually contain calcium or alkaline-earth-metal soaps. The grease soaps may be prepared from many types of acids, including rosin, vegetable and animal fatty acids, and naphthenic acid. Except in special cases, mixed acids are used in grease formulations and mixtures of naphthenic acid and vegetable-oil acids have been reported. A grease usable over a wide range of temperatures can be manufactured by incorporating a mixture of stearic and naphthenic acids according to Kaufman and Puryear. Naphthenic acid, as aluminum naphthenate or zinc naphthenate, is said to improve the adhesiveness of grease. Mixtures of metallic soaps may be used to control the final viscosity of a grease and calcium naphthenate has been added for this purpose to a grease containing aluminum naphthenate. Calcium and aluminum naphthenates have also been used in greases to reduce the tendency of the product to "sweat" oil. The addition of sodium naphthenate to hydrous greases yields a stable product because its concentrated aqueous solution has appreciable solvency for oil.

Metallic salts of naphthenic acid are used, to a great extent, in oils. The most widely used salt is probably lead naphthenate, added to increase the bearing-load capacity of the oil through the formation of a "plate" on the bearing surface.[9, 55] Other compounds, such as oxidation inhibitors, viscosity modifiers, stabilizers, and detergents, are usually employed together with the naphthenate. Zinc and calcium naphthenates, for example, when added to a lubricating oil, yield a detergent-type product. Aluminum naphthenate is added to increase stringiness and to raise the viscosity of the final product.[23] Aluminum naphthenate has been advocated as a crankcase-oil additive to reduce piston-ring sticking as have the tin, zinc, iron, cobalt, and other salts. Several of the naphthenates have also been reported as contributing resistance to oxidation, but this point is not too well established.

In cutting oils, the alkali naphthenates are used mainly as emulsifying agents, but their reported bactericidal action may aid in reducing infections resulting from minor injuries incurred by operators.

LUBRICANT PATENTS

Naphthenate	*U. S. Patent No.*	*Year*	*Abstract*
Naphthenic Acid	1,319,219 — Wells and — Southcombe	1919	Emulsification is prevented by 0.8% naphthenic acid in lubricating oil.
Naphthenic Acid	1,529,658 — McKee and Eckert	1925	Grease is prepared from lubricating oil, vegetable oil, and lubricating oil naphthenic acid soaps.
Naphthenic Acid	1,539,386 — McKee and Eckert	1925	Grease is prepared from a mixture of 85% oil, 9% naphthenic acid (acid number 50), and 15% tallow.
Na	1,552,669 — Becker and Hislop	1925	Grease is prepared from 50% sodium naphthenate, 30% petrolatum, and 20% still wax.
Al	1,582,227 — Rebber	1926	A castor machine oil is prepared from lubricating oil containing aluminum naphthenate to promote stringiness.
Na	1,582,258 — Frizell	1926	The stringiness of castor machine oils is improved by the addition of as little as 0.1% sodium naphthenate.
Cellulose Ester	1,882,816 — Hagedorn	1932	Cellulose naphthenate, when dissolved in oil, yields a product of increased viscosity.
Ca	2,001,108 — Parker	1935	The addition of calcium naphthenate to transformer oils reduced breakdown by light and air.
Na	2,055,795 — Kaufman and Puryear	1936	Sodium naphthenate-stearate mixtures in greases give a product having good lubricating properties over a wide temperature range.
Na	2,058,788 — Hendrey	1936	Cutting oil.
Na	2,060,425 — Neukom	1931	Cutting oil.
Castor Oil Ester	2,068,088 — Steik	1937	The reaction product of castor oil with naphthenic acid is used as a cosolvent for castor oil-lubricating oil mixtures.
Al	2,144,078 — Neely	1939	Aluminum naphthenate, added to crank-case oils, reduces piston-ring sticking.

LUBRICANT PATENTS (*Continued*)

Naphthenate	*U. S. Patent No.*	*Year*	*Abstract*
Al	2,144,855 — Rutherford and Francis	1939	Aluminum dinaphthenate, when added to crank-case oil, is claimed to reduce piston-ring sticking. The excessive viscosity of the product is reduced by blowing with steam.
Aluminum and Other	2,163,622 — Neely	1939	Aluminum, zinc, magnesium, cobalt, cadmium, manganese, and tin naphthenates, in the presence of a small amount of free organic acid, reduce piston-ring sticking.
Glycol Ether Esters	2,173,117 — Johnson	1939	Oiliness of lubricating oils is improved by the addition of naphthenic acid esters of monodiethylene glycol ethers.
Pb	2,176,246 — Kaufman and Philson	1939	E. P. lubricant containing lead naphthenate.
Ni	2,202,826 — Brandes	1940	Bearing corrosion inhibitor for motor oil.
	2,236,120 — Toussaint	1941	Stabilizer for motor oil.
Al, Zn	2,360,631 — Zimmer and Morway	1944	Aluminum or zinc naphthenate, when added to grease, improves its adhesiveness.
Al, Zn	2,374,966 — Zimmer and Morway	1945	As preceding.
Ca	2,379,245 — Morway and Zimmer	1945	Calcium naphthenate is added to greases and oils containing aluminum naphthenate to control the final viscosity.
Al	2,382,694 — Darley	1945	Hydrous aluminum hydroxide is heated with naphthenic acid in the presence of oil to give a dry grease containing less than 0.05% water.
Na	2,380,976 — Korb and Sabina	1945	Stabilized grease containing sodium naphthenate.
Pb	2,388,083 — Reswick	1945	E. P. lubricants are prepared from oils to which 15% lead naphthenate and 25% cresol have been added.
Na	2,388,439 — Cowie and Co.	1945	Soluble cutting oil.
Pb	2,398,429 — Hughes	1946	Cutting oils are prepared from sulfurized petroleum oils to which 0.5% lead naphthenate has been added.

LUBRICANT PATENTS (*Continued*)

Naphthenate	*U. S. Patent No.*	*Year*	*Abstract*
Amine Salts	2,401,993 — Wasson and Duncan	1946	Naphthenates of straight-chain C_8-C_{14} amines are used as corrosion inhibitors in lubricating oils.
Alkaline Earth	2,409,950 — Meyer	1946	Incorporation of alkaline earth naphthenates into lubricants tends to reduce oil sweating and to stabilize the product.
Zn	2,415,353 — Johnston and Wasson	1947	A combination of zinc naphthenate and mahogany soap in turbine oils gives good corrosion resistance. Zinc naphthenate alone is ineffective.
	2,417,088 — Prulton	1947	Extreme-pressure lube.
	2,430,951 — Rouault	1947	Anticorrosive turbine oil.
	2,434,978 — Zisman and Baker	1948	Anticorrosive synthetic lubricants.
	2,466,925 — Brauner	1949	Extreme-pressure lubes.
	2,482,606 — Adelson and Dannenberg	1949	Polyallyl naphthenate as synthetic lube oil.
	2,510,031 — Folda	1950	Stabilizer for sulfurized mineral oils.
	2,560,542 — Bartleson and Hughes	1951	Stabilized motor oil.
	2,554,985 — Heamer and Balvay	1951	Rust-preventative oils.

LUBRICANT LITERATURE

Naphthenate	*Reference*	*Abstract*
Na	*Neft.* **6,** No. 6, 16 (1935); *C. A.* **30,** 3980 (1936), Rubtzov	An emulsifying cutting oil is prepared from sodium naphthenate, machine oil, glycerol, and water.
Pb	*Proc. Royal Soc.* (London) A **177,** 90 (1940); *C. A.* **35,** 1684 (1941), Beeck et al.	Lead naphthenate and other oil-soluble lead compounds are used in lubricants for their ability to form films or to "plate" a bearing surface and improve the load-carrying capacity of the lubricant.
Ca	*Ind. Eng. Chem. Anal. Ed.,* **13** (5), 317 (1941), Lamb et al.	As little as 0.25% calcium naphthenate in a heavy-duty lubricant acts as a detergent or dispersing agent.
Pb	*Ind. Eng. Chem.* **33,** 1352 (1941); *C. A.* **35,** 8272 (1941), Simard et al.	Oils containing lead naphthenate yield lead-containing films on bearing surfaces. In the presence of sulfur, lead-rich lead sulfide and sulfate films are formed.
Al	*C. A.* **39,** 2919 (1945), Derygin et al.	Aluminum naphthenate is used to increase the viscosity of lubricating oils. A differential increase is reported at the friction surface.
Alkaline Earth	*Rec. trav. chem.* **65,** 549 (1946); *C. A.* **41,** 1528 (1947), van der Minne	Alkaline-earth naphthenates are used to peptize suspended particles in a lubricating oil.

Naphthenate Surface-Active Agents

Naphthenic acid has found only limited use as a surface active agent. Its alkali-metal salts, while acting as emulsifying and wetting agents, are not particularly efficient in aqueous media. The characteristic odor has also limited their use as soaps. Despite these unfavorable properties, the acids have been employed as emulsifying agents as noted in the sections on lubricants and insecticides. In addition to these uses, sodium naphthenate has been employed in making asphalt emulsions for road surfacing, as a textile assistant, and for similar purposes.[5, 53]

An unusual result of the surface activity of naphthenates is observed in their use as demulsifiers for oil-well emulsions. Since excessive water in crude oil introduces serious problems in processing, it is desirable to effect dewatering during field storage. This may be done in several ways, one of which involves the breaking of the emulsion normally obtained from the well by the addition of surface-active agents. The emulsions as obtained are of the oil-in-water type and contain relatively small amounts of water or brine. The addition of sodium naphthenate alone or in combination with other compounds, such as sodium sulfonates, causes a separation of the major portion of the water from the crude oil by an inversion of the emulsion, in many cases, to a water-in-oil type. The inversion results in freeing a large part of the water originally present which then settles out since the excess water is insoluble in the oil-invert emulsion solution and since the electrostatic charges on the particles of the original emulsion have been neutralized or reversed.

In the paint industry, metallic naphthenates are used in pigment grinding and dispersion. Lead, zinc, and calcium naphthenates have found extensive application for this purpose and have proved to be quite efficient in reducing grinding time and promoting rapid and even distribution of the pigments in the vehicle.

PATENT REFERENCES TO NAPHTHENATE SURFACE-ACTIVE AGENTS

Naphthenate	*U. S. Patent No.*	*Year*	*Abstract*
Esters	1,596,597 — DeGroote	1926	Naphthenate esters are used to break oil-well emulsions.
Na and Other	1,596,598 — DeGroote	1926	The alkali-metal salt of a sulfonated condensation product of naphthenic acid and an aromatic hydrocarbon is used as an emulsion breaker.
Na	1,671,284 — Lichtenstern	1928	Sodium naphthenate is used as an emulsifying agent for road materials.
Esters and Salts	1,940,391 — DeGroote	1933	Petroleum (oil-in-water) emulsions are broken by the addition of small amounts of naphthenic acid, salts, and esters.
Esters and Salts	1,940,392 — DeGroote	1933	Organic sulfonates are added to the products listed in U. S. Patent 1,940,391 to break emulsions.
Esters and Salts	1,940,395 — DeGroote	1933	Extension of above two patents.
Esters and Salts	1,940,396 — DeGroote	1933	Extension of above three patents.
Na	1,984,023 — Limberg	1934	Sodium naphthenate is used as the emulsifying agent in making asphalt emulsions.
Na	1,984,024 — Limberg	1934	See above patent.
Alcohol	2,000,994 — Schrauth	1935	A wetting agent and detergent was prepared by sulfating the alcohol obtained by reducing naphthenic acid.
Alkaline Earth	2,062,159 — Brizzolara	1936	A readily dispersible carbon black is obtained by adsorption of 1–10% alkaline-earth naphthenates. Better wettability and higher bulk density result.
Pb, Zn, Ba, etc.	2,180,721 — Roon and Minich	1939	Metallic naphthenates are used as wetting agents in pigment grinding.
Pb	2,236,296 — Minich	1941	Lead naphthenate is used as a wetting and dispersing agent in pigment grinding.

Miscellaneous Applications of Naphthenates

There is a great variety of possible outlets for naphthenic acids and their derivatives. With the exception of its use in incendiary munitions, few, if any, of the proposed products have been developed on a commercial scale.[10, 37, 40, 47, 48, 49] We shall review here only a portion of the information available. Bismuth naphthenate has, for example, been proposed as an antisyphilitic while copper naphthenate is reported as being a rust inhibitor in addition to its fungicidal action. Iron, cobalt, and manganese naphthenates are said to reduce soot formation in fuel oil and other salts, when dissolved in alcohol, yield a noncorrosive hydraulic fluid. Esters of naphthenic acid, as well as those prepared from the corresponding alcohol, may be used as plasticizers for lacquers and resins.[29] Several metallic naphthenates have also been used as liquid-phase oxidation and polymerization catalysts.

The use of Napalm during World War II represents an interesting development based on the ability of aluminum dinaphthenate and other aluminum soaps to increase the viscosity of gasoline to the point of gel formation.[27] Masses of gelled gasoline, when dispersed by an explosion as in a bomb, adhere to structures and burn with an intense heat release, thus forming an efficient incendiary munition. The property of stringiness imparted to the gasoline also permits the use of this fuel in flame throwers and increases the range of these weapons.

PATENTS GRANTED ON MISCELLANEOUS NAPHTHENATE USES

Naphthenate	*U. S. Patent No.*	*Year*	*Abstract*
Bi	1,580,592 — Eichhola and Dalmer	1926	Bismuth naphthenate is used as an antisyphilitic.
Naphthenic Acid	1,860,850 — Cassidy	1932	Fluorescent acridine dyes are solubilized for use in coloring gasoline by conversion naphthenate salts.
Cu	1,983,006 — Schladebach and Hahle	1934	Rustproofing paint is prepared with copper naphthenate.
	2,001,108 — Parker	1935	Stabilized transformer oil.
Esters	2,010,727 — Kirstahler and Kaiser	1935	Naphthenyl esters of phenoxy acids are prepared.
Esters	2,016,392 — Schneider	1935	Naphthenic acid esters of ether-alcohols are used as plasticizers.
Si	2,017,000 — Hintermaier	1935	Naphthenic acid is reacted with silicon tetrachloride to yield a silicon carboxylate.
General	2,033,853 — Sherbino	1936	Salts of naphthenic acid in alcohol yield a noncorrosive hydraulic fluid.
Esters	2,047,664 — Barrett and Lazier	1936	Naphthenyl alcohol esters of polycarboxylic acid are used as plasticizers.
	2,052,193 — Rickles	1936	Stabilizer for dyed gasoline.
Cu	2,068,979 — Fisher	1937	Copper naphthenate acts as a corrosion inhibitor in stills.
Cu and Other	2,141,848 — Adams et al.	1938	Copper, calcium, lead, and zinc naphthenates in fuel oil eliminate soot formation by depositing an oxidation catalyst on the soot which then burns off.
Fe and Other	2,230,642 — Fisher and Hulse	1941	Similar to above.
	2,390,609 — Minich	1945	Preparation of stable basic aluminum naphthenate.
	2,390,830 — Elliott	1945	Cobalt naphthenate stabilized with sorbitol.
	2,411,832 — Linford and Baral	1946	Use of halogenated solvent in drier manufacture.
	2,466,925 — Brauner	1949	Nitrogen compound used as catalyst in drier manufacture.
	2,546,421 — Bartholomew and Cross	1951	Gasoline containing naphthenates as wear inhibitors.
	2,562,885 — Barush and Denison	1951	Antiknock motor fuel.

LITERATURE ON MISCELLANEOUS NAPHTHENATE USES

Naphthenate	*Reference*	*Abstract*
Co, Ni and Other	*Kolloid Z.* **77,** 270 (1936); *C. A.* **31,** 5664 (1937), Neyman	The behavior of cobalt and other naphthenates in light petroleum was studied. Naphthenates form stable jellies.
Al	*Ind. Eng. Chem.* **38,** 768 (1946); *C. A.* **40,** 5567 (1946), Fieser et al.	Description of Napalm.
Co	*J. Am. Chem. Soc.* **68,** 2020 (1946); *C. A.* **41,** 707 (1947), Nichols et a.	Cobalt naphthenate is a polymerization catalyst for allyl ethers of carbohydrates.
Alcohol	*Ind. Eng. Chem.* **39,** 55 (1947); *C. A.* **41,** 1203 (1947), Hansley	Methyl naphthenate is reduced by sodium to the alcohol.
Mn	*C. A.* **39,** 2024 (1945), Berezorskaya et al.	Manganese naphthenate is used as a liquid-phase oxidation catalyst.
Ca, Mn	*C. A.* **40,** 2970 (1946), Losev	Calcium naphthenate yields carboxylic acids while manganese naphthenate yields hydroxy acids when used as liquid-phase oxidation catalysts with kerosene.
Al	*Ind. Eng. Chem.* **41,** 1435 (1949), Mysels	Description of Napalm.

Bibliography

1. Anon, *C. A.* **32,** 1951 (1938).
2. Anon, *C. A.* **33,** 8041 (1939).
3. Anon, *Paint, Oil and Chem. Rev.* **104,** No. 23, 94 (1942); *C. A.* **37,** 1281 (1943).
4. Anon, *C. A.* **37,** 1882 (1943).
5. Bachrach, D. J., *Soap* **10,** No. 11, 21 (1934).
6. Bartlett, A. E., and M. Goll, *C. A.* **39,** 5496 (1945).
7. Bayley, C. H., and M. W. Weatherburn, *Ibid.* **40,** 1661 (1946).
8. Bayley, C. H., and M. W. Weatherburn, *Can J. Research* **25F,** 92 (1947).
9. Beeck, O., J. W. Givens, and A. E. Smith, *Proc. Roy. Soc.* (*London*) **A177,** 90 (1940); *C. A.* **35,** 1685 (1941).
10. Berezovskaya, F. I., et al., *C. A.* **39,** 2024 (1945).
11. Berry, A. G. V., and J. C. Cater, *Ibid.* **36,** 4308 (1942).
12. Bowen, J. W., *Ibid.* **36,** 3925 (1942).

13. Brandt, A., *Ibid.* **40,** 4223 (1946).
14. *British Standard Spec.* BS/ARP58, August 1941.
15. Bryson, H. C., *C. A.* **33,** 9684 (1939).
16. Bryson, H. C., *Ibid.* **34,** 2182 (1940).
17. Carter, W. J., *Ibid.* **39,** 1051 (1945).
18. Chubb, L. W., *Ibid.* **37,** 1880 (1943).
19. Curwen, M. D., *Oil Color Trade J.* **88,** 1711 (1935); *C. A.* **30,** 1590 (1936).
20. Dean, J. C., *Chem. Eng. News* **23,** 1164 (1945).
21. Dean, J. D., W. B. Strickland, and E. L. Skau, *C. A.* **39,** 5497 (1945).
22. Dean, J. D., W. B. Strickland, and W. N. Berard, *Ibid.* **40,** 5928 (1946).
23. Deryagin, B., et al., *Ibid.* **39,** 2919 (1945).
24. Drinberg, A., and A. Parfenova, *Ibid.* **34,** 1499 (1940).
25. Elkin, H. A., and W. A. S. White, *Ibid.* **34,** 270 (1940).
26. Elliot, S. B., *The Alkaline Earth and Heavy Metal Soaps,* Reinhold, New York, 1946.
27. Fieser, L. F., et al., *Ind. Eng. Chem.* **38,** 768 (1946).
28. Furry, M. S., H. M. Robinson, and H. Humfeld, *Ibid.* **33,** 538 (1941).
29. Hansley, V. L., *Ibid.* **39,** 55 (1947).
30. Hock, C. W., and M. Harris, *C. A.* **36,** 5657 (1942).
31. Howard, H. L., *Ibid.* **38,** 1816 (1944).
32. Kelsey, J. M., *Ibid.* **41,** 3900 (1947).
33. Klebsattel, C. A., *Ibid.* **28,** 348 (1934).
34. Knauss, C. A., *Ibid.* **28,** 4612 (1934).
35. Knauss, C. A., *Ibid.* **29,** 4190 (1935).
36. Knauss, C. A., *Paint, Oil and Chem. Rev.* **98,** No. 12, 32 (1936); *C. A.* **30,** 8657 (1936).
37. Knauss, C. A., *C. A.* **31,** 4835 (1937).
38. Kogan, A. I., and N. I. Churdenko, *Ibid.* **26,** 5774 (1932).
39. Lamb, G. G., C. M. Loane, and J. W. Gaynor, *Ind. Eng. Chem. Anal. Ed.* **13,** 317 (1941).
40. Losev, I. P., A. I. Danilovich, and A. B. Kon, *C. A.* **40,** 2970 (1946).
41. Marsh, P. B., G. A. Greathouse, K. Bollenbacher, and M. L. Butler, *Ind. Eng. Chem.* **36,** 176 (1944).
42. Marsh, P. B., G. A. Greathouse, M. Butler, and K. Bollenbacher, *U. S. Dept. Agri. Tech. Bull. 892,* (1945).
43. Meckler, J. G., *Drugs, Oils and Paints* **54,** 166 (1939); *C. A.* **33,** 5680 (1939).
44. Minich, A., and M. Goll, *Paint, Oil Chem. Rev.* **109,** No. 24, 6, 36 (1946); *C. A.* **41,** 2523 (1947).

45. von der Minne, J. L., *Rev. trav. chim.* **65,** 549 (1946).
46. Müller-Lobeck, C., *C. A.* **32,** 3640 (1938).
47. Nemiritskii, B. G., et al., *Ibid.* **31,** 4763 (1937).
48. Neyman, E., *Kolloid Z.* **77,** 270 (1936); *C. A.* **31,** 5664 (1937).
49. Nichols, P. L., A. N. Wrigley, Jr., and E. Yanovsky, *J. Am. Chem. Soc.* **68,** 2020 (1946).
50. Nicholson, D. G., and C. E. Holley, *Ind. Eng. Chem.* **30,** 114 (1938).
51. Race, E., *C. A.* **41,** 1445 (1947).
52. Rinse, J., *Ibid.* **35,** 5331 (1941).
53. Roelke, K., and H. P. Reichel, *Ibid.* **41,** 2919 (1947).
54. Rubtzov, G. A., *Ibid.* **31,** 3980 (1936).
55. Simard, G. L., H. W. Russell, and H. R. Nelson, *Ind. Eng. Chem.* **33,** 1352 (1941).
56. Smith, P. I., *Soap* **14,** No. 11, 86 (1938).
57. Tearnley, *C. A.* **39,** 2210 (1945).
58. Tweedie, A. S., and C. H. Bayley, *Ibid.* **38,** 6567 (1944).
59. Twiselton, M. S. J., *J. Soc. Chem. Ind.* **63,** 220 (1944).
60. *U. S. Army Spec. 100–17,* February 1944.
61. *U. S. Army Spec. 6–345,* June 1945.
62. *U. S. Army Spec. 4–1131,* July 1945.
63. *U. S. Navy Dept. Spec. 52W 5,* April 1945.
64. *U. S. Army Spec. 6–363,* May 1946.
65. Wheeler, G. K., *Ind. Eng. Chem.* **39,** 1115 (1947).
66. Zamyslov, R. D., et al., *C. A.* **33,** 884 (1939).

Part II

THE PETROLEUM BASES

CHAPTER TWENTY-ONE

INTRODUCTION

More than 60 years ago, operators observed that ammonia was evolved when certain Galician crude oils were distilled. A few years later, Bandrowsky and Weller independently reported results obtained in their study of basic compounds extracted from Galician and Saxony crude oils.

By 1892, a few quantitative nitrogen determinations had been made and the foundation of the art and science of separating the petroleum bases from other petroleum compounds and from each other was laid. However, both the analytical and the separation methods developed very slowly and little more of permanent value was learned until after 1900. Unlike the naphthenic acids, which early became commercial products routinely isolated in European refineries, the petroleum bases have found no important uses and have had to be extracted specially for each research project. This made the study of petroleum nitrogen compounds difficult and expensive.

Between 1900 and 1920, Mabery and coworkers studied basic compounds isolated from refinery fractions of California petroleum. They utilized then, for the first time, California stocks and isolated their bases from distillates rather than from crude oil. They developed methods of analysis of crude oils for nitrogen and worked with sufficient quantities of petroleum fractions to be able to decide that their bases consisted mainly of alkylated pyridines and quinolines.

When in 1926 the sum of $500,000 became available for fundamental research in petroleum through the generosity of J. D. Rockefeller, the study of petroleum bases was included as project 20 and, for the first time, an adequately financed and supplied study of petroleum bases was made possible. It was found that the extract obtained in the Edeleanu process of refining petroleum fractions with liquid sulfur dioxide could be treated to yield relatively easily and cheaply large amounts of nitrogen compounds which were studied in this project. Separation methods have been developed in the field of alkylated pyridines and it has been found to be easier to isolate pure compounds than to identify them.

Unfortunately even though about three dozen pure basic compounds have been isolated and identified from straight-run products and an equal number from cracking-process gasoline, no large-scale uses for these bases or their mixtures have been found. In the last 5 years, there has been a reawakening of the interest in nitrogen compounds of petroleum. This interest appears to be due very largely to a realization that some nitrogen compounds have a negative value in petroleum, i.e., that they detract from the value of a crude oil because they tend to act as poisons towards cracking catalysts and because they cause increased gum formation in petroleum products. Since some nitrogen compounds may be harmless while others may be valuable, there is a new interest in the nature of the nitrogen compounds — both basic and nonbasic — of petroleum. New analytical methods have been developed which make possible the determination of basic and total nitrogen content. As a result, there has been a rapid increase in data on concentrations of basic and nonbasic compounds found in petroleum and petroleum products. In spite of published analytical data, the petroleum nitrogen compounds are still considered a very complex mixture, from which pyridines, quinolines, and a few other types of basic compounds have been isolated and about which little else is known. They remain an interesting and potentially important class of petroleum by-products which could be produced in large quantities and which may, some day, find a ready market as did the cresylic acids and phenols of petroleum.

CHAPTER TWENTY-TWO

NITROGEN IN PETROLEUM

Analysis

The presence of nitrogen in petroleum was shown as early as 1817, when Saussure [19] observed that ammonia was formed when naphtha was burned. Ammonia was again reported in 1872, when operators noticed that it was evolved during distillation of Galician crude oils.[5] Quantitative determinations of total nitrogen in crude oils or in fractions have been reported from time to time since about 1890, but there is serious doubt whether many of the earlier results can be considered as anything but approximate values. We have a relatively small number of analyses which are correct, another small number which are probably correct because they were obtained by the Kjeldahl method, and a much larger number which may be correct or may be several times as high as they should be.

As in the case of coal, the determination of nitrogen by either the unmodified Kjeldahl or the ordinary Dumas method may lead to erroneous results for reasons to be explained later. The failure of early workers to recognize this fact has resulted in an accumulation of data of unknown value. Where the method used was specified and the results appear to be in line with later results of known accuracy, we may be justified listing such results — a crude oil showing 0.2% nitrogen by Dumas determination could be considered as having not over 0.3% nitrogen.

Since pyridines and quinolines are known to be present in petroleum bases and since it has been known for years that cyclic compounds of this type tend to show low values when nitrogen is determined by the standard Kjeldahl method, we may expect results obtained in the analysis of crude oil by this method to be low but probably not high. When workers observed that in case of analysis of a sample by both the Dumas and the Kjeldahl method, the Kjeldahl results tended to be lower than those obtained by the Dumas method, they assumed that the Kjeldahl method was at fault. A Japanese worker, Takano, claimed in 1900 that nitrogen in petroleum must be determined by the Dumas method.[22] Some of his work was done with Mabery who reported both Dumas and Kjeldahl analyses of a number of California oil samples as shown in Table 38.

Table 38. **Mabery's Analyses by the Kjeldahl and Regular Dumas Methods**

Field	*Kjeldahl N%*	*Dumas N%*
Los Angeles County		
(a)	0.47	1.42
(b)	0.69	1.52
Ventura County		
(a)	0.64	1.91
(b)	0.71	2.39
Summerland	0.88	2.10

Comparison with results which have been obtained since then on similar samples indicates that the Kjeldahl results were probably essentially correct, but Mabery decided that, since nitrogen gas tends to be formed in the Kjeldahl analysis of certain types of compounds, the Dumas values were obviously correct and the others low.

That the Dumas method should be poorly suited for this type of analysis becomes evident when it is realized that we are here attempting to determine nitrogen in a mixture containing less than 0.8% total nitrogen (usually less than 0.3%) mixed with compounds which are known to be difficult to burn completely in a combustion furnace. A sample of crude oil containing 0.1% nitrogen — probably an average value for all but a few fields — would yield only 0.8 ml of nitrogen gas on combustion of a 1 g sample. Even a 5 g sample would then yield only 4 ml of gas. It has been known for many years that hydrocarbons must be heated to a rather high temperature, in contact with copper oxide, to attain complete combustion. The exact nature of the products formed during incomplete combustion is not known, but methane and acetylene have been mentioned. If we assume that methane is formed, it will, of course, not be absorbed by alkali and will thus be measured as nitrogen gas. From a 5-g sample, 1 ml methane formed would represent only about 0.8 mg of unburned material and so would introduce no serious error in an analysis for carbon and hydrogen, but 1 ml of methane measured as nitrogen would make this determination worthless. If the combustion is carried out in a furnace with a long path of copper oxide heated to a bright red heat, a considerable amount of nitrogen will be adsorbed on the large volume of copper oxide and an unknown amount of it will be evolved at red heat. Nitrogen produced during the combustion has to be rinsed out with a large volume of carbon dioxide in view of the large amount of copper oxide. This means that, unless the carbon dioxide is of exceptional purity, an appreciable volume of inert gases introduced with the carbon dioxide will be collected as nitrogen and thus will be the source of another positive error. These positive errors would not be serious if 40 to 50 ml of nitrogen were obtained, but when only 5 ml or less is collected, they may make the observed results worthless. The Dumas method will then lead to acceptable results only if the sample is burned slowly at very high temperature in contact with large volumes of copper oxide, if the combustion tube is rinsed with very pure carbon dioxide, and if errors due to adsorbed nitrogen can be held to a minimum. Thus it is not surprising that many of the Dumas values reported for petroleum nitrogen are obviously three or four times as high as the correct ones.

Table 39. **Mabery's Values Using the Modified Dumas and Kjeldahl Methods**

Crude Oil from	*Kjeldahl N%*	*Dumas N%*
Ohio	0.027	0.022
West Virginia	0.029	0.049
Pennsylvania	0.010	0.010
Texas	0.058	0.068
Oklahoma	0.074	0.082
Kansas	0.035	0.045
Baku, Russia	0.071	0.060

By 1919, Mabery recognized these difficulties in the use of the Dumas method and developed a modified apparatus and procedure that yielded results in close agreement with Kjeldahl values.[14] He used a very hot quartz or steel tube and a special carbon dioxide generator, evacuated the apparatus at the beginning of the run, and in other ways tried to eliminate all of the errors just enumerated. Table 39 shows that the values now obtained in the analysis of a number of crude oils from various fields by his modified Dumas and the Kjeldahl method agree very closely. Most of the Kjeldahl values lie slightly above those of the Dumas method.

Unfortunately, in the same paper, he again referred to old Dumas results, including such obviously incorrect values as 2.55% for a California crude oil.

When the Texas group started work on their project, the first problem was that of selecting a suitable source of bases. A study of published analyses soon showed that values varied over such a wide range even for samples from the same field that they decided to investigate the whole problem of nitrogen determination in crude oils.[16]

Like Mabery, they recognized the weaknesses of the regular Dumas method and modified it mainly through use of two furnaces in series, quartz tubes heated to a bright red heat, high-purity carbon dioxide produced by a different type of generator, and elimination of avoidable rubber connections. Both the Texas and the Mabery modifications gave good results but only with tedious and very careful operation. Apparently, Mabery's more tedious method gave somewhat better results. The Texas modification was used only while a modified Kjeldahl method was being developed and tested, and practically all of the published analyses were carried out with their modified Kjeldahl method.

Gonick and coworkers [7] describe a modified Dumas method which has been used in the Shell laboratories since 1938 and appears to give entirely satisfactory results. In this modification, the nitrogen produced can be recirculated to oxidize any combustible gases formed in the first combustion. Although the apparatus and procedure appear somewhat complicated, they report that six to eight determinations can be made in an 8-hour day, that is, the method is not slower than the ordinary Dumas method.

The Texas modification of the Kjeldahl method used a 5 g sample of crude oil, 150 ml of concentrated sulfuric acid, 50 g of potassium sulfate, and 2.5 g each of mercuric oxide and copper sulfate. Essentially this was the regular method modified to permit use of samples as large as 5 g. The Kjeldahl method, when applied to mixtures in which the ratio of carbon and hydrogen to nitrogen is very high, will probably yield acceptable results even when cyclic compounds, like pyridine, are present. A 5% error would be quite serious in the case of a sample containing 10% nitrogen, but could be ignored in the case of one containing only 0.2% nitrogen. Heterocyclic compounds which cannot be analyzed as pure compounds by the Kjeldahl method are probably completely decomposed in the presence of a large excess of carbon and hydrogen, just as the addition of glucose improves results obtained with nitro compounds.

Lake and coworkers [11] made a critical study of the effect of temperature on the time required for the digestion of petroleum samples and on the accuracy of the results. They found that there had been a tendency to use too little potassium sulfate and thus operate at too low a temperature, and gave detailed directions for a procedure that required much less time than the Poth method developed by the Texas group.

Ball and coworkers [1] modified the Poth method so as to save time through use of a much smaller sample than the 5 g sample recommended by Poth. Their method uses 1 to 2 g of sample, 30 to 40 ml of concentrated sulfuric acid, and 20 g of a mixture of 15 parts potassium sulfate, 3 parts mercuric oxide, and 2 parts copper sulfate. Their digestion time is only 2 hours. Their method is similar to one used by the Universal Oil Products Company [25] and this, or a very similar method, is used by practically all laboratories making large numbers of nitrogen determinations on petroleum.

Finally, there is the ter Meulen method which has been in use in Europe for a number of years and appears to be particularly suited for the accurate determination of nitrogen in samples containing very small amounts of nitrogen — 0.0 to 0.2%, as in the case of many crude oils. In this method, the sample is vaporized or pyrolyzed and carried over a bed of hot nickel catalyst by a stream of hydrogen gas. The nitrogen present is reduced to ammonia which is determined in the same way as in the Kjeldahl method.

Holowchak, Wear, and Baldeschwieler [9] described a modified ter Meulen method with which they obtained excellent results in the analysis of an East Texas gas oil and of synthetic blends prepared to yield nitrogen values similar to those obtained with petroleum products. Their main improvement over earlier ter Meulen procedures appears to be the nickel catalyst precipitated on magnesia and reduced at 340° to 380°C (664° to 716°F). They reported over a hundred satisfactory analyses with the same catalyst mass while formerly it had to be changed after every dozen or so runs. They titrated the ammonia in the usual manner and tested the colorimetric determination by nesslerization or better through the use of the sodium phenate-hypochlorite method which they found superior to the Nessler method. They recommended the colorimetric methods for very small amounts of ammonia, e.g., samples of crude oil containing less than 0.1% nitrogen.

Lake,[12] reporting the results obtained by fourteen laboratories coöperating in testing the Kjeldahl, Dumas, and ter Meulen methods, found that both micro and macro modifications of the Kjeldahl method gave satisfactory results in the analysis of petroleum samples and had the advantage of speed since an analyst with proper equipment could complete about twenty determinations a day.

They found that the macro Dumas method, properly modified for use with petroleum samples, gave more erratic results, but was still satisfactory and had the advantage of being a universal method which was not true of the Kjeldahl method. However, for samples with less than 0.5% nitrogen, the macro Dumas method requires very careful work. The micro Dumas method was found generally unsatisfactory. Both modifications have the additional disadvantage of being relatively expensive, since six to eight determinations an 8-hour day seems to be all a good analyst can do in working with petroleum samples.

The ter Meulen method was tested by only four of the coöperating laboratories, because it was not yet in general use. As recently modified, it gave satisfactory results and was said to permit twelve to fourteen determinations per man a day, if dual equipment is available. It is not considered a general method but is particularly suited for analysis of samples of very low nitrogen content.

Geographical Distribution of Nitrogen in Petroleum

It is customary to talk about the nitrogen content of California, Texas, Pennsylvania, or Wyoming petroleum as if nitrogen in petroleum were distributed geographically rather than geologically or in some other manner. On the basis of what appear to be reliable results, there seems to be much justification for this practice. Mabery,[14] using the modified Dumas method which was found to yield values in close agreement with those obtained by the Kjeldahl method, found for six samples of crude oil from various Ohio fields 0,027,-0.039, 0.016, 0.41, 0.024, and 0.029% nitrogen; for three samples from Pennsylvania fields, he found 0.014, 0.014, and 0.012% nitrogen; and for four samples from Texas fields, he reported 0.058, 0.067, 0.023, and 0.048% nitrogen. In all but one case (0.41), which may well have been a typographical error, he found that the analyses from any one state varied over a narrow range. The Bailey group [16] reported Kjeldahl analyses on thirty-four different California crude oils and found a total range of 0.7 distributed, as shown in Table 40.

Table 40. **Nitrogen Content of Various California Crude Oils**

	Number of Samples
0.100 to 0.199	1
0.200 to 0.299	1
0.300 to 0.399	5
0.400 to 0.449	6
0.450 to 0.500	6
0.500 to 0.549	7
0.550 to 0.599	5
0.600 to 0.690	2
0.802	1

Twenty-nine out of thirty-four samples fall within the range 0.3 to 0.6%, which is a somewhat larger range than that observed for most states, but is to be expected in view of the total range of values and of the geography of California. All reliable results reported seem to indicate that the nitrogen content of petroleum depends largely on the geographical location of the well and, to a lesser extent, on field or geological formation from which the oil is obtained. Thus the common practice of discussing the nitrogen content of *California* or *Baku* petroleum appears to be at least roughly justified.

Concentration of Nitrogen in Various Crude Oils

Apparently, practically all crude oils contain at least traces of nitrogen compounds while none contains more than about 1%. Actually, as far as reported analyses by methods known to be reliable are concerned, no oil has been found to contain more than 0.88%. Table 41 lists analyses of crude oils from various regions with arbitrary omission of values above 1% of nitrogen.

Table 41. **Nitrogen Content of Crude Oils from Various Fields**

Source	*Nitrogen Concentration Pound, % (Range)*	*Reference*
Arkansas	0.051–0.081	16
California	0.101–0.820	16
Colorado	0.000–0.101	16
Indiana	0.026	16
Kansas	0.040–0.096	16
	0.035–0.045	14
Montana	0.042	14
Oklahoma	0.036–0.165	16
	0.074–0.082	14
Ohio	0.000	16
	0.016–0.410	14
Pennsylvania	0.000	14
	0.010	14
Texas	0.014–0.135	14
	0.058–0.068	16
Utah	0.000	16
Wyoming	0.017–0.317	16
Columbia	0.226	5
	0.320	5
Baku	0.050	5
Trans-Caspian	0.140	5
Grozny	0.070	5
German	0.250	5
Italian	0.040	5
Egyptian	0.300–0.600	5
Venezuela	0.231	16
Heavy Mexican	0.358	16
Light Mexican	0.326	16
Mixed Mexican	0.354	16

These results reported by Engler, Mabery, and the Texas group represent older analyses and some of those compiled by Engler may be in error.

Ball and coworkers[1] collected results obtained in a large series of modern analyses mainly by chemists of the United States Bureau of Mines. A summary of these values is presented in Table 42.

Table 42. **Distribution of Nitrogen Content of Crude Petroleums**

Source	*Number of Samples in Per Cent Nitrogen Range* 0.000–0.005	0.051–0.100	0.110–0.300	0.310–0.500	0.510–0.700	0.880
Arkansas	1	4	0	0	0	0
California	2	0	9	6	5	1
Colorado	1	2	0	0	0	0
Illinois	0	3	1	0	0	0
Indiana	0	0	1	0	0	0
Kansas	0	0	2	0	0	0
Kentucky	0	0	2	0	0	0
Louisiana	2	2	0	0	0	0
Michigan	0	0	2	0	0	0
Mississippi	1	1	3	0	0	0
Montana	3	1	0	0	0	0
NewMexico	5	3	1	0	0	0
New York	1	0	0	0	0	0
Oklahoma	3	2	4	1	0	0
Texas	29	5	3	0	0	0
Utah	1	0	0	0	0	0
Wyoming	14	8	14	2	0	0
Iran	0	0	1	0	0	0
Greece	0	0	1	0	0	0
Iraq	0	1	0	0	0	0

All of the early workers had noticed that the nitrogen content of gasoline was negligible, but that it increased with rise in boiling point and that a large fraction of the total nitrogen content was found in the residual tar left on distilling a crude oil. Bailey and coworkers obtained similar results in their survey. Ball [1] collected data on the analyses of fractions and residues from three different Wyoming crude oils. It will be noted that the residues in these cases contain 87.5 to 92.9% of the total nitrogen of the original samples. Table 43 lists boiling-point ranges and analyses of these samples.

Table 43. **Boiling-Range Distribution of Nitrogen in Wyoming Samples of Crude Petroleums**

Boiling Range		*Winkleman Dome*		*Sage Creek*		*Steamboat Butte*	
°*C*	°*F*	% *N*	% *of Total*	% *N*	% *of Total*	% *N*	% *of Total*
Atmospheric pressure							
250–275	482–527	0.005	0.10	—	—	0.003	0.10
40 mm pressure							
175–200	347–392	0.03	0.40	0.01	0.40	0.02	0.60
200–225	392–437	0.04	1.00	0.02	0.40	0.03	1.10
225–250	437–482	0.06	1.40	0.05	1.10	0.06	1.90
250–275	482–527	0.11	2.50	0.09	1.80	0.10	3.10
275–300	527–572	0.14	4.00	0.13	2.90	0.14	6.30
Residue		0.50	91.30	0.53	92.90	0.44	87.50
Crude Oil		0.23		0.28		0.16	

Table 44, from the same paper by Ball and coworkers,[1] shows that the residue from distillation contains 85 to 100% of the nitrogen of a number of crude petroleums from different fields.

Table 44. **Nitrogen Content of Distillation Residues from Various Crude Petroleums**

Source	*Total N* %	*Residue N* %	*Residue N* % *of Total*
Arkansas	0.06	0.16	88.3
Mississippi	0.11	0.24	106.0
Michigan	0.12	0.32	90.8
California	0.58	1.15	84.5
Oklahoma	0.27	0.60	88.9
Texas	0.16	0.34	87.5
Colorado	0.08	0.20	100.0
Wyoming	0.22	0.50	95.0
(Range 4 Samples)	0.28	0.54	96.4
Greece	0.17	0.25	94.1

Isolation

All discussions of nitrogen compounds in this chapter have referred to the total nitrogen present in the sample and so did not distinguish between basic and nonbasic nitrogen compounds. So far as is now known, all of this nitrogen can be removed from petroleum by extraction with concentrated sulfuric acid or by liquid sulfur dioxide. As sulfuric acid enters into reaction with the nitrogen compounds, this method should not be used if the compounds originally present are to be identified. Liquid sulfur dioxide does not appear to produce appreciable chemical changes and its use seems to be the only method now available.

Prior to 1900, practically all studies of nitrogen compounds from petroleum were carried out on bases isolated directly from crude oil with dilute mineral acids. It was assumed that this extraction removed all of the nitrogen compounds. However, all nonbasic nitrogen compounds and possibly some of the slightly basic compounds of high molecular weight were left in the crude oil. The slightly basic high-molecular-weight compounds would probably not be completely removed by simple single-stage extraction, since the salts formed would probably be appreciably soluble in crude oil and would be distributed between the two phases. In all cases, some of the nitrogen may be extracted from crude oil with dilute sulfuric acid, but no worker has been able to account for all of the nitrogen present [16, 19, 20] by simple extraction.

Bandrowsky [2] extracted 22 liters of crude oil from Galician petroleum with dilute sulfuric acid and obtained basic matter from which he prepared 1.29 g of platinic chloride salts. Weller [26] obtained from 500 kg of Saxony crude oil a pyridine-like crude base boiling at 220° to 260°C (428 to 500°F). Schestakow [20] found that Roumanian crude oil yielded 0.006% of bases on extraction with dilute sulfuric acid. These early workers apparently did not know whether they had removed all of the nitrogen compounds or not. It now seems certain that they obtained only a small portion of the total nitrogen compounds since their yield of nitrogen *compounds* was less than the percentage of nitrogen *atoms* in the oil.

Later projects have usually obtained their bases from refinery fractions, such as kerosene or distillate. Chlopin [3] extracted 291 kg of masut* in two equal portions with dilute sulfuric acid. He shook the first portion for 12 days and then shook the other half of the sample with the same sulfuric acid for 22 days to be sure that all of the bases had been removed from the viscous masut. He obtained only 14.59 g of bases which would seem to account for only a small fraction of the nitrogen present in this high-boiling residue. Mabery [13] obtained his bases from Peckham and Salathe who extracted California distillate with dilute sulfuric acid. They found

* Masut or Masout is the residual oil left after distilling a crude oil through the gas-oil range.

48% of their crude bases boiled below 105°C (221°F), 14% at 105° to 125°C (221° to 257°F), and the remainder up to 305°C (580°F). Even in the case of the Edeleanu extract from California kerosene, the Texas group [16] found that only 55% of the total nitrogen content of the kerosene extract could be extracted with 16% sulfuric acid. This failure to extract a greater part of the total nitrogen content with acid was not due to inadequate contact between the phases because quinoline added to the extract was readily recovered by extraction.

The *basic* nitrogen compounds can then be extracted from crude oil or one of its fractions by adequate contact with dilute sulfuric acid. The yield from crude oil is very low, however, and even from California crudes, only about half of the nitrogen compounds appear to be basic. The bases can be liberated from their salts by alkali and the oily layer separated. In practice, partly for economic reasons, the amount of alkali used at the refinery is usually only sufficient to make the solution distinctly basic. In the presence of a very large volume of water compared with the volume of bases it is probable that pyridine and possibly the simple alkylated pyridines will remain largely in the water layer. This may explain the fact that pyridine has not been isolated from petroleum. A similar situation existed for a number of years in regard to phenol in petroleum.

The layer of bases is contaminated with hydrocarbons and probably other neutral compounds which may be removed by:

(1) addition of an excess of dilute sulfuric acid and steam distillation of the volatile impurities

(2) addition of an excess of acid as before, but followed by extraction of the impurities with a low-boiling solvent like ether or chloroform and

(3) use of systematic fractional neutralization to remove the bases from the impurities which remain behind. This is probably the best method either alone or supplementary to method (1), since the higher-boiling impurities do not steam distill readily. The purified bases are then ready for one of the fractionation methods to be discussed in the next chapter.

The Nonbasic Compounds in Petroleum

If instead of isolating only the basic compounds, one wishes to remove all nitrogen compounds from petroleum, this can be done in two different ways. The oil may be extracted with concentrated sulfuric acid which appears to remove all nitrogen compounds from the hydrocarbons, but there is evolution of sulfur dioxide indicating oxidation, heating, and darkening of the solution. Some of the nonbasic compounds like pyrrole are known to be polymerized to gum by this type of treatment and the darkening may be due partly to this reaction. Obviously, the method cannot be used if a study of the nitrogen compounds isolated is to be made. The other method of removing all of the nitrogen compounds consists in extracting with liquid sulfur dioxide (Edeleanu process). This appears to remove all types of nitrogen compounds and so far it has not been found to have produced chemical changes in the compounds extracted. The method is, however, somewhat inconvenient to carry out in the laboratory and the extract contains large amounts of other substances, such as sulfur compounds and aromatic hydrocarbons.

While attempts have undoubtedly been made by various oil-company laboratories to remove nitrogen compounds by adsorption, only one paper by Sauer and coworkers [18] has appeared. They used an alumina column in separating nitrogen compounds from domestic heating oil. This method would presumably also remove most other polar compounds and would yield a highly contaminated mixture of nitrogen compounds.

If the basic compounds are extracted from the Edeleanu extract it is found that 50 to 70% of the nitrogen is left behind as nonbasic material. At various times, the Bailey group found that additional amounts of basic compounds could be isolated from fractions obtained when material which had been extracted with dilute sulfuric acid was distilled. Bailey became firmly convinced that the bases were man-made and formed during refinery or laboratory operations. The theory that complex high-boiling compounds were cracked during distillations seemed to fit in with the fact that most of the nitrogen of the crude oil was usually found in the residue. It ignored the fact, however, that prior to 1900, all bases studied were obtained by direct extraction of crude oil with dilute mineral acids.

The reawakening of interest in nitrogen in petroleum has been centered largely around the nonbasic nitrogen compounds about which very little is known. We have obtained much data on the ratio of basic to total nitrogen in crude oils and in refinery products in the last few years through general use of nonaqueous titration methods.[6, 8, 17, 18] We now know that results obtained by laborious methods by the Texas group can be essentially reproduced in the study of practically any crude oil that contains significant concentrations of nitrogen. Even in the case of shale oil, a rather similar situation seems to prevail.

Richter, Caesar, Meisel, and Offenhauer [17] undertook a careful study of concentrations of basic and nonbasic nitrogen compounds in petroleum and its products. In the course of their study, they suggested that the older Bailey definition of basic nitrogen compounds "as those which can be extracted by dilute sulfuric acid" should be changed to "compounds which can be titrated with perchloric acid when dissolved in a 50:50 solution of glacial acetic acid and benzene." Since basic nitrogen is now easily titrated with perchloric acid, this definition is a logical one. Tests showed that indoles, some pyrroles, and carbazoles behave as nonbasic compounds in this titration while the pyridines, quinolines (and presumably isoquinolines), and some substituted pyrroles can be titrated and thus are considered basic compounds.

They found the ratio of basic to total nitrogen per cent in a number of crude petroleums to be remarkably constant at values of 0.25 to 0.35. Table 45 shows the data obtained in the analysis of fourteen different crude-oil samples.

Table 45. **Ratio of Basic to Total Nitrogen Content of Various Crude Petroleums**

Field	*Basic* N % (N_B)	*Total* N % (N_T)	N_B/N_T
Jackson	0.010	0.04	0.25
Miranda	0.010	0.04	0.25
Scurry County	0.020	0.06	0.33
East Texas	0.020	0.08	0.25
West Texas	0.030	0.11	0.27
Kansas	0.040	0.12	0.33
Mid Continent	0.025	0.10	0.25
Santa Maria Valley	0.190	0.66	0.29
Kettleman Hills	0.140	0.41	0.34
Wilmington	0.140	0.50	0.28
Ventura	0.130	0.42	0.31
Tibur Petrolea	0.033	0.13	0.25
Guico Guaria	0.020	0.08	0.25
Kuwait	0.030	0.12	0.25

They then decided to determine whether this ratio would also vary over such a narrow range within a sample by analyzing a series of fractions and the residue left from distillation of samples of Wilmington and Santa Maria crude oils. Table 46 shows that the range in values of this ratio was slightly broader (0.25 to 0.41), but still so narrow that one wonders what it means in regard to the origin of these bases. Finally they treated the residue with *n*-pentane and found that the *n*-pentane-soluble and insoluble portions of the residue again showed practically the same ratio between basic and total nitrogen content. The Santa Maria crude gave analogous values, the ratios ranging from 0.27 to 0.50 with basic nitrogen contents of 0.01 to 0.35%.

Table 46. **Distribution of Basic Nitrogen in Wilmington Crude Oil**

Boiling Range °C	Boiling Range °F	Volume % of Original	Basic % N (N_B)	Total % N (N_T)	N_B/N_T
To 200	To 392	18.5	0.03	—	—
200–225	392–437	5.0	0.04	0.01	0.25
225–250	437–482	5.0	0.04	0.01	0.25
250–275	482–527	6.0	0.07	0.02	0.28
275–350	527–665*	12.0	0.09	0.04	0.45
350–382	665–720	5.0	0.22	0.09	0.41
382–410	720–770	4.5	0.31	0.12	0.39
410–443	770–830	5.5	0.36	0.14	0.39
Residue		38.5	1.12	0.35	0.31

* Temperatures above 275°C by calculation. Distilled at 10 mm pressure.

A Santa Maria crude oil gave analogous values, with ratios ranging from 0.27 to 0.50 on basic nitrogen contents ranging from 0.01 to 0.35%. In this respect, these results are in good agreement with others which range from about 0.25 to 0.5 for the ratio of basic to total nitrogen content. It appears that even the highest-boiling fractions of petroleum consist of a mixture of basic and nonbasic nitrogen compounds and that the ratio: basic N%/total N% is practically constant from the lowest- to the highest-boiling fractions.

Twenty years ago, when Miss Parker undertook the study of the nonbasic compounds in an Edeleanu extract of California kerosene,[15] nothing was known about the nonbasic nitrogen compounds. She found that extraction with 20% sulfuric acid lowered the nitrogen content of the fraction from 0.045 to 0.025%. Assuming an average molecular weight of 200 and only one nitrogen atom per molecule, this corresponds to 0.4 mol per cent of nitrogen compounds not extractable by dilute sulfuric acid.

Since it would be very difficult to isolate nonreactive molecules from such a dilute solution and since tests for groups or types of molecules would probably fail at this dilution, Miss Parker and Poth, who rendered valuable services for a number of years as fellow, research assistant, and assistant director of the Texas group, attempted to concentrate these compounds so as to be able to learn something about their nature.

The solubility of hydrocarbons in liquid sulfur dioxide is known to decrease rapidly with decrease in temperature. Thus it was hoped that it might be possible to precipitate the hydrocarbons and leave the nitrogen compounds in solution on cooling. Using a laboratory extraction apparatus donated by Edeleanu, they could not obtain a second phase on cooling to − 20°C (− 4°F).

They then attempted to extract the nitrogen compounds with 95% ethanol and were able to remove about 40% of the nitrogen by five simple separatory-funnel extractions, but little more was obtained by further extractions and the extract contained large amounts of other compounds.

Selective adsorption on two grades of natural zeolites and on solid mercuric chloride lead to no separation, but solid phosphotungstic acid removed about 70% of the nitrogen. In view of the expense involved in treating large volumes of extract, this was not followed up. The use of modern chromatographic techniques might have lead to satisfactory results.

They tried 20, 50, 75, and 90% sulfuric acid and found that only the 98% acid removed nitrogen completely, but with heating and sludge formation which ruined the extract as far as further study was concerned.

Table 47. **Nitrogen Content of Fractions of Edeleanu Extract**

Cut Number	*I*	*II*	*III*	*IV*	*V(b)*	*VI(c)*	*VII(d)*
Volume in Liters	20	20	20	12.5	1.5	—	Tar
Median Boiling Point, °C	173	200	218	245.0	—	—	—
Median Boiling Point, °F	343	292	424	473.0	—	—	—
Percent Nitrogen before Extraction (a)	0.012	0.015	0.035	0.090	0.727	0.528	1.50
Percent Nitrogen after Extraction	0.012	0.012	0.016	0.038	0.695	0.340	—
Percent Basic Nitrogen	0.000	0.003	0.019	0.053	0.032	0.188	—

(a) Extraction with 20% Sulfuric Acid
(b) Residue from IV
(c) Pyrolytic Distillation of V
(d) Residue from VI

To learn more about the boiling range and nitrogen content of the Edeleanu extract, Poth [15] fractionated 75 liters of the extract at 60 mm pressure. The data obtained are listed in Table 47.

The ratios of basic to total nitrogen are somewhat higher than those reported by Richter and coworkers,[17] but this might be expected of an extract.

Miss Parker consistently obtained positive pine splinter tests for pyrrole, but other tests for pyrrole and tests for indole and carbazole failed. She found that these tests also failed when she added an equivalent concentration of these compounds to kerosene, i.e., the tests evidently were not sensitive enough to be used for very low concentrations of these compounds.

Since nitriles and amides could be present as nonbasic compounds, she refluxed 10 ml of Edeleanu extract for 10 hours with alcoholic potassium hydroxide. No ammonia could be detected by means of turmeric paper hung into the top of the condenser or by Nessler's solution through which the evolved gases were passed. Some nitriles might not have been hydrolyzed under these conditions, but simple amides ought to have been certainly eliminated in concentrations that would give a positive test from 10 ml of extract. She apparently did not eliminate the possibility that substituted amides, formed by reaction between primary and/or secondary amines and organic acids, might be present. These compounds may be prepared by simple heating of a mixture of high-boiling acids and amines.[10] Primary and secondary amines have not been isolated from petroleum except from cracking-process products. Their absence — except in traces — might indicate that the nonbasic compounds are not substituted amides, or failure to isolate such amines might be due to amide formation with organic acids which are always present. A few preliminary tests have been made in the author's laboratory to determine whether saponification of the nonbasic compounds from petroleum leads to formation of acids and primary or secondary amines, but conclusive results can probably be obtained only after the nonbasic compounds are available in much greater concentration than any yet reported.

Wille [27] reported that phenol and pyridines form 1:1 molecular compounds which are fairly stable and concluded that this type of compound might explain the difficulty met in separating phenols from pyridine bases in coal tar. This type of compound might also explain the observation by a number of the Texas workers that new basic compounds are formed on distillation of a mixture which has been extracted with dilute mineral acids.

Recent interest in the nonbasic compounds has led to experiments which indicate that petroleum bases, like shale oil bases, contain significant concentrations of pyrroles, carbazoles, and indoles. So far, no amides of any type or nitriles appear to have been detected. Since naphthenic acids and bases are normally present together in petroleum, it would not be surprising to find that primary or secondary amines would react with petroleum acids to form high-molecular-weight substituted amides of the type:

$$\mathrm{RCONHR'}, \quad \mathrm{RCONR'R''}, \quad \text{or} \quad \mathrm{RCO{-}N}\begin{matrix}\diagup \mathrm{CH_2{-}CH_2} \diagdown \\ \\ \diagdown \mathrm{CH_2{-}CH_2} \diagup\end{matrix}\mathrm{CH_2}$$

Thompson, Chenicek, Druge, and Symon [23] were able to get positive tests for pyrroles in fuel oils and Sauer, Melpolder, and Brown [18] used the mass spectrometer to detect and to get a semiquantative estimate of the concentrations of various nitrogen compounds. Their results are listed in Table 48.

Table 48. **Nitrogen Compounds in Domestic Heating-Oil Distillate**

	Concentration mg/l	
Compound	*Catalytically Cracked*	*VirginKuwait*
Carbazoles	350 ± 40	530 ± 50
Indoles	40 ± 8	170 ± 40
Pyrroles	30 ± 15	160 ± 80
Pyridines	50 ± 10	790 ± 160
Quinolines	30 ± 15	150 ± 75

While these results are obviously only approximate, they indicate that in crude Kuwait petroleum the nonbasic carbazoles are present second in concentration to the basic pyridines, while in the cracked material from the same crude, the carbazoles are found in five times the concentration of the next type — the pyridines. The ratio of basic to total nitrogen which has been determined by other methods for a large number of samples is 0.52 which is somewhat higher than that usually found, but is in satisfactory agreement in view of the approximate nature of the data.

Bibliography

1. Ball, J. S., M. L. Whisman, and W. J. Wenger, *Ind. Eng. Chem.* **43,** 2577 (1951).
2. Bandrowsky, F. X., *Monatsh.* **8,** 224 (1887).
3. Chlopin, G. W., *Ber.* **33,** 2837 (1900).
4. Dinneen, G. W., and W. D. Bickell, *Ind. Eng. Chem.* **43,** 1604 (1951).
5. Engler-Hofer, *Die Chemie und Physik des Erdöls,* Leipzig, 1913.
6. Fritz, J. S., *Anal. Chem.* **22,** 1028 (1950).
7. Gonick, H., D. Tunnicliff, E. Peters, L. Lykken, and V. Zahn, *Ind. Eng. Chem. Anal. Ed.* **17,** 677 (1945).
8. Hall, N. F., *J. Am. Chem. Soc.* **52,** 5115 (1930).
9. Holowchak, J., G. E. Wear, and E. L. Baldeswieler, *Anal. Chem.* **24,** 1754 (1952).
10. Hunter, B. A., W. I. Harper, and H. Gilman, *Proc. Iowa Acad. Sci.* **47,** 263 (1940); *C. A.* **35,** 7372 (1941).
11. Lake, G. R., P. McCutchan, R. van Meter, and J. C. Neel, *Anal. Chem.* **23,** 1634 (1951).
12. Lake, G. R., *Ibid.* **24,** 1806 (1952).
13. Mabery, C. F., *J. Soc. Chem. Ind.* **19,** 505 (1900).
14. Mabery, C. F., *J. Am. Chem. Soc.* **41,** 1690 (1919).
15. Parker, I., M.A. Thesis, The University of Texas, 1931.
16. Poth, E. J., W. D. Armstrong, C. C. Cogburn, and J. R. Bailey, *Ind. Eng. Chem.* **20,** 83 (1928).
17. Richter, F. P., P. D. Caesar, S. L. Meisel, and R. D.Offenhauer, *Ibid.* **44,** 2601 (1952).
18. Sauer, R. W., F. W. Melpolder, and R. A. Brown, *Ibid.* **44,** 2606 (1952) 19.
19. Saussure, T., *Ann. chim. phys.* **4,** 314 (1917); from Engler-Hofer, page 479 (see reference 5).
20. Schestakow, P. J., *Chem. Ztg.* **23,** 41 (1899).
21. Smith, H. M., *Ind. Eng. Chem.* **44,** 2577 (1952).
22. Takano, S., *Petroleum Congress,* Paris, 1900, page 58; from Engler-Hofer, page 479 (see reference 5).
23. Thompson, R. B., J. A. Chenicek, L. W. Druge, and T. Symon, *Ind. Eng. Chem.* **43,** 935 (1951).
24. Thompson, R. B., T. Symon, and C. Wankat, *Anal. Chem.* **24,** 1465 (1953).
25. *Universal Oil Products Company, Laboratory Test Methods for Petroleum and Its Products,* Chicago, 1947, Method A120–40.
26. Weller, A., *Ber.* **20,** 2098 (1887).
27. Wille, H., *Brennstoff-Chem.* **23,** 271 (1942).
28. Wilson, H. N., *J. Soc. Chem. Ind.* (*London*) **67,** 237 (1948).

CHAPTER TWENTY-THREE

SEPARATION OF BASIC COMPOUNDS

The crude bases isolated by the procedure described in Chapter 22 always contain considerable amounts of hydrocarbons and other neutral petroleum-derived compounds. Pure bases can be isolated from this mixture only after extensive fractionation. The steps involved in the isolation of pure bases include one or more of the following procedures, not necessarily in the order listed:

(1) Separation of the basic from neutral compounds by neutralization of the bases.

(2) Fractional distillation to separate the mixture on the basis of difference in vapor pressure which, in the case of homologous series, approximates separation by difference in molecular weight.

(3) Some system of separation based on differences in a property other than vapor pressure or solubility of a salt.

(4) Separation on the basis of differences in the solubility of salts or other solid derivatives — fractional precipitation and recrystallization.

Usually removal of neutral impurities precedes fractional distillation and this, in turn, precedes extraction or fractional neutralization, but, as will become apparent later, this order may sometimes be changed with profit.

It has not been possible to isolate crystalline salts or addition compounds from bases which have merely been converted to salts and separated from the neutral compounds. Even when the bases were neutralized in steps the mixture of bases obtained from each step was far too complex to permit purification by recrystallization of a salt.

Modern fractional distillation, through efficient controlled-reflux columns, leads to remarkably close separation of compounds boiling only a few degrees apart, but even these fractions rarely contain one compound in sufficiently high concentration to convert it to a crystalline salt or other derivative that can be purified by recrystallization without losing practically all of the compound in the process or becoming involved in a very tedious series of crystallizations. Prior to the work of the Bailey group, fractional distillation, supplemented sometimes by simple fractional neutralization, was the only separation method used and not a single pure compound was isolated. Bailey and coworkers were able to isolate a few pure compounds directly from distillation fractions, but these compounds were present in unusually high concentration and had salts which showed much lower solubility than those of other bases in the mixture. Such compounds were the $C_{16}H_{25}N$ base of Thompson and Bailey [47] and a few substituted quinolines.[3, 22, 37]

Since azeotrope formation between bases is absent or very rare,[40] this probably plays a minor role in the separation of mixtures containing only bases, but azeotropes with other types of compounds do form and limited use has been made of this fact.

In the separation of coal-tar bases, J. Idris Jones [20, 21] made use of the azeotropes formed by pyridines and the lower aliphatic acids. He used azeotrope formation between acetic acid and propionic acid and some of the methyl pyridines. Such azeotropes were reported as early as 1890, but apparently Jones was the first to make systematic use of the method. Since its usefulness depends on the selective azeotrope formation with a few of the bases in a mixture, it is not a general method of isolating bases from complex mixtures.

A method depending on a combination of azeotrope formation and carrier liquid distillation was developed by Bratton, Felsing, and Bailey.[7] In this method, carefully refined hydrocarbon is fractionated into 5 or 10° fractions which are then compounded to obtain about 20 volumes of hydrocarbon boiling uniformly from about 20° below to about 5° above the boiling range of the bases to be separated. One volume of the mixture of bases is then added to the hydrocarbon and fractionated carefully. Azeotrope formation may or may not help in separation of individual compounds of the mixture, but the main advantage of the method is that as each lower-boiling base distills, it is rinsed out of the column by a higher-boiling hydrocarbon mixture instead of by a higher-boiling base as in fractionation without the added hydrocarbon. This eliminates the usual mixing of compounds in the intermediate cuts provided the bases to be separated boil a few degrees apart and the hydrocarbon mixture boils uniformly. It was expected, and found, that bases boiling only 1° apart could not be separated unless one of them formed an azeotrope. This so-called "amplified" distillation was found very effective in the separation of the simple alkylated pyridines isolated from cracking-process gasoline [6] and from pyrolysis of cotton-seed meal.[35] One of the most important advantages of the method is that a few grams of such mixed bases can be fractionated through an ordinary column instead of requiring the use of semimicro stills.

If we consider the fact that the neutral compounds present in crude bases may perform, to some extent, the function of the hydrocarbon mixture added in "amplified" distillation, it is obvious that there may sometimes be a gain in efficiency of separation of the lower-boiling bases if they are distilled before removal of the neutral impurities. Usually any such advantage is more than balanced by loss due to tar formation from neutral nonbasic nitrogen compounds which tend to form gums and turn dark.

The methods in which separation is based on differences in vapor pressure must be usually supplemented or preceded by separation based on some property other than vapor pressure. Several such methods have been developed recently in connection with the study of the separation hydrocarbon mixtures and biological mixtures. Methods which have been found applicable to mixtures of bases include fractional neutralization of the bases or their fractional liberation from their salts, fractional precipitation of salts or other derivatives, distribution of the bases or their salts between immiscible solvents, and the various chromatographic separation methods developed mainly by biochemists, but also used extensively in separation of petroleum hydrocarbons. Since fractional distillation is usually the most rapid and convenient method, it normally precedes other methods of separation, but the order may be reversed under special circumstances. This is the case, for example, met when it is desired to study only the quinolines or other compounds with high index of refraction of a certain fraction. In most cases these compounds are present to the extent of only 10 to 30% of the total volume of bases. It may be better in this case to convert the bases to their hydrochlorides and extract the dilute solution with chloroform which removes the pyridines and other bases with low index of refraction and leaves only the desired bases in the aqueous layer.

The relatively small volume of these bases can then be fractionally distilled more easily and possibly more efficiently since the mixture should be much less complex than the original.

Methods depending on difference in stability or rate of formation of addition compounds were sometimes useful in concentrating bases to the point where recrystallizable solid salts or derivatives could be obtained. A method commonly employed in the separation and purification of coal-tar bases depends on the difference in stability of zinc chloride addition compounds.[16, 24] Mahan[29, 30] made a special study of this method as applied to petroleum bases and found it valuable in a few cases, but it was not employed extensively.[23]

Methiodide formation, when methyl iodide is added to a mixture of bases, takes place at rates that vary over a wide range. Usually the method as used consisted simply in addition of an excess of methyl iodide to a mixture of bases and removal of the methiodide formed from time to time. In some cases, it was necessary to heat the mixture in a sealed tube to obtain the desired fractions. This method was used by Roberts[38] in the 200° to 210°C (392° to 410°F) range to isolate a $C_{10}H_{15}N$ base and was tried as an intermediate fractionation method by several others.

A much more useful and industrially promising method depending on difference in stability of an addition compound was that of fractional sulfite formation or fractional decomposition of sulfites. This method should perhaps be considered as a special case of fractional neutralization or liberation of bases. It has been used in the separation of coal-tar bases and is mentioned frequently in the patent literature. Unlike such expensive procedures as fractional methiodide formation, this one may become industrially important if and when petroleum bases are isolated from petroleum commercially, because the sulfur dioxide required is cheap and can be recovered readily.

In this process, sulfur dioxide was passed into a solution of the bases and the sulfites formed removed from time to time, or the solution was saturated with sulfur dioxide which was then permitted to escape slowly with gradual liberation of bases which were removed as fractions. This method was referred to as fractional sulfiting or degassing, depending on which plan was used. Biggs[4] added sulfur dioxide to a dry ether solution of the bases and then permitted ether and sulfur dioxide to escape with gradual formation of oily layers of sulfites which were removed from time to time. Later workers used various modifications, including aqueous or brine solutions of sulfurous acid in which the bases gradually dissolved.[2, 3, 10]

A remarkably simple but efficient method of separating bases by distribution between immiscible solvents was developed by Perrin and Bailey.[36] The process very effectively separates bases with a high index of refraction (their "aromatic" bases) from bases with a lower index of refraction (their "nonaromatic" bases). Apparently, in practice, this amounts usually to separating quinolines and related compounds from pyridines.

In the method as used routinely by practically all members of the Texas group after 1933, 1 volume of 1:1 hydrochloric acid, 1 volume of a mixture of bases, and 1 or 2 volumes of chloroform were mixed and separated in a separatory funnel or equivalent apparatus. The bases isolated from the water layer were found to have a higher index of refraction than those isolated from the chloroform layer.

The method as used obviously involves the distribution of a mixture of bases and their hydrochlorides between water and chloroform, since 1 volume of 1:1 hydrochloric acid does not contain enough acid to neutralize 1 volume of bases with an equivalent weight of less than 200 and the method has been used most extensively on such mixtures. Bratton,[6] in 1936, modified the method slightly by using only sufficient 1:1 hydrochloric acid to obtain a homogeneous solution which also does not usually involve as much as one equivalent of acid. However, since the method has been used effectively with mixtures of bases with a molecular weight above 200 and also in experiments in which an excess of 1:1 hydrochloric acid was employed, the equilibria involved when free bases are present do not seem to be essential for the success of the method. Unfortunately, a systematic study of the equilibria involved with simple mixtures of known bases has not been made.

For a fairly complete separation of the two types of bases, an extraction according to the standard procedure, followed by several cross extractions of the water layers with chloroform and of the chloroform layers with water, is required, but even a single simple separatory-funnel extraction leads to extensive separation as shown by the data of Figure 10 obtained by Perrin.[36] He placed 10 ml of distillation fractions in a small separatory funnel, added 10 ml of 1:1 hydrochloric acid and 20 ml of chloroform, shook, and separated. The bases isolated showed the differences in index of refraction indicated for each fraction.

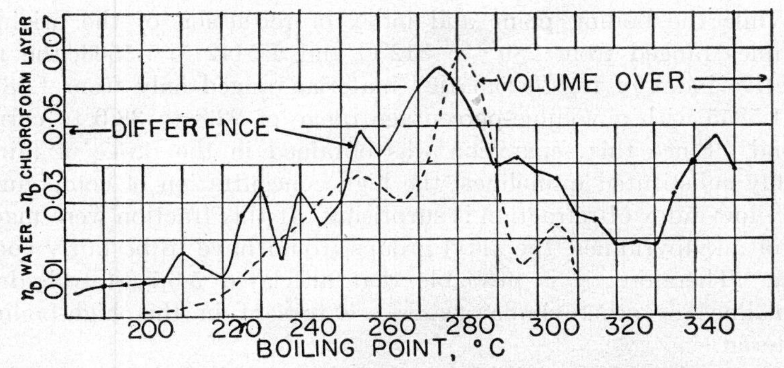

Figure 10. Distribution of Base Hydrochlorides from Bases Boiling at the Temperatures Shown between Water and Chloroform (Difference in index of refraction of bases isolated from water and chloroform layers)

At each maximum in the difference curve, we may expect to find that quinolines or other bicyclic compounds are present or have accumulated. It is an interesting fact that the difference and distribution curves are approximately the same in shape.

An even more striking illustration of the effectiveness of the method is shown by the data of Table 49 obtained by Axe [1] in work with material that had been highly fractionated by distillation and, in some cases, also by other methods not including chloroform extraction.

Table 49. **Chloroform Separation of Bases from Highly Fractionated Material**

Cut Number	n_D^{25}	*Boiling Range* °C	°F	*Original Volume ml*	*Bases from Water Layer* Volume ml	n_D^{25}	Volume %
144	1.5442–1.5457	290–295	554–563	3490	850	1.5850	24.4
146	1.5480–1.5486	290–297	554–566	2920	980	1.5842	38.6
147	1.5485–1.5492	292–297	557–566	2180	835	1.5845	29.3
148	1.5493–1.5495	296	565	2060	520	1.5830	25.2
149	1.5503–1.5518	289–295	552–563	2450	560	1.5848	22.8
152	1.5514–1.5526	301–305	573–581	3030	900	1.5832	29.7
153	1.5531–1.5538	298–306	566–583	3590	850	1.5832	23.7
154	1.5542–1.5547	307	584	2510	760	1.5832	30.0
155	1.5540–1.5549	298–302	566–575	1980	650	1.5850	32.8
156	1.5552–1.5557	302–306	575–583	2900	780	1.5855	27.0
157	1.5553–1.5566	300–305	572–581	3500	1100	1.5848	31.5
158	1.5555–1.5563	307–312	584–594	3060	800	1.5847	26.0

While the boiling point and index of refraction of the original samples ranged from 289° to 312°C and 1.5442 to 1.5563, the refractive index of the "aromatic" material ranged only from 1.5830 to 1.5855 with a volume-percentage range of 22.8 to 38.6 to correspond. Since this separation was obtained in the range of fairly highly substituted quinolines, the high concentration of compounds with low index of refraction is surprising. If this fraction were made up of alkylpyridines, the alkyl groups would have to be fairly complex. Therefore, it is probable that alkylated 5,6,7,8-tetrahydroquinolines or other similar bases are present in this high-boiling material.

If only the "aromatic" bases are to be studied, it is obvious that only about one third of the total volume of bases need be carefully fractionated if the roughly distilled material is converted to hydrochlorides and extracted by chloroform before being subjected to elaborate fractional distillation which may also be more effective now, because many of the bases have been removed by the extraction process.

At the time the names "aromatic" and "nonaromatic" were proposed for the fractions obtained in the chloroform extraction, the only compounds which had been identified were the aromatic quinolines from the water layer. The chloroform layer contained such unidentified compounds as the $C_{16}H_{25}N$ base which were thought to belong to a new type of hydroaromatic compounds. Although it is possible that a number of as yet unidentified types of bases occur in petroleum, all that can be said on the basis of data now available is that the water layer contains compounds with higher index of refraction, like the quinolines, while the chloroform layer contains compounds with lower index of refraction, like the alkylated pyridines. A more rigorously correct statement appears to be that compounds with a high carbon-to-hydrogen ratio are found in the water layer while those with a lower ratio go into the chloroform layer. The complex pyridine, 2,4-dimethyl-6-(2,2,6-trimethylcyclohexyl)-pyridine, was found in the chloroform layer after separation from the substituted quinolines boiling at 276° to 280°C (529° to 536°F).

Recent work in the author's laboratory by Wheeler [49] has shown that synthetic 3- and 4-cyclopentylpyridines, prepared in large amounts and carefully purified, are not identical with the two bases identified by Truitt [26] as 3- and 4-cyclopentylpyridines. The structure of the two bases isolated from the water layer in a chloroform extraction of a dilute aqueous solution of hydrochlorides of bases has not been determined, but their properties agree with those of methylated 5,6,7,8-tetrahydroquinolines.

Through the use of fractional distillation, followed or preceded by chloroform-water distribution, it was possible to obtain a number of new crystalline salts as well as get better yields of bases previously isolated. It was soon found, however, that even with refractionation of the bases obtained after chloroform extraction of narrow-boiling distillation cuts, many of the new fractions were still too complex to permit isolation of pure picrates or other salts from them and it became apparent that other efficient separation methods would have to be developed before much additional progress could be made.

Another group working in the Texas laboratories had meanwhile run into similar difficulties in the separation of petroleum acids. They developed the methods first used by Jantzen at Hamburg in 1928–1932 [18] in his separation of coal-tar bases and soon found that systematic countercurrent neutralization of acids dissolved in petroleum ether achieved extensive separations based on what appeared to be a combination of difference in molecular weight or solubility in the two phases and difference in ionization constants.

Prior to the development of the countercurrent spinner columns (to some extent for preliminary fractionation in later work), various modifications of multistage fractional neutralization or liberation schemes were used.[23, 42] Large quantities can be processed if carboys with bottom removed are used in place of separatory funnels, but the method is time consuming if many stages and is not very effective if only a few are to be used. Several bases were isolated by these methods, but whenever more than five or six stages of neutralization were needed, various modifications of the Jantzen [18] and the Tiedcke [46] spinner columns were used. These columns are described and discussed in some detail in the corresponding chapters (7 and 18) on petroleum acids and the original literature should be consulted for further details.[13, 25, 34, 41]

Recently, a U-tube apparatus, described very briefly by Jantzen [18] and employed by him for some of his work with coal-tar bases, has been used in the separation of gas-well acids and shale-oil bases.[27] It obtains twenty stages of fractional neutralization or liberation and yields results analogous to those obtained in the Craig apparatus.[9] A larger machine, in which the individual mixing and settling vessels are quart bottles, has been used for larger quantities of acids or bases.[28] A still later modification now in use employs tall liter separatory funnels instead of quart bottles in the same machine. This permits much faster operation and can be used as a single twenty-two-stage machine or as a number of lower-stage machines in parallel. It can operate with center feed for preliminary fractionation of large volumes of material and has the advantage of being easily operated and used by a laboratory helper.

Relatively few compounds were isolated by recrystallization of picrates or other salts obtained from fractions separated by the simple fractional neutralization schemes,[23] but a series of new quinolines and pyridines were isolated from fractions obtained by use of the very efficient spinner columns.[2, 13, 25, 41] They proved to be the most powerful tool tried for separation of bases boiling at practically the same temperature. Hackman and Wibaut used fractional neutralization and chloroform extraction in combination with good effect in separation of cracking process bases.[15] No spinner or other countercurrent equipment seems to have been used by them. In their excellent work, Hackman and Wibaut used fractional neutralization only when it was found that very precise fractional distillation was not sufficient.

Hackman and Wibaut studied,[15] but apparently did not actually use, in the separation of cracking process bases, a modified chromatographic analysis scheme in which they passed the vapors of a mixture of bases through a tube filled with activated carbon heated to 189°C (372°F). The first vapors passing through the column and condensing were found to be an essentially pure individual base. Repetition of the process improved the purity. The method is obviously closely related to the frontal analysis scheme of Claessen and Tiselius. There have been several very recent publications dealing with this method of chromatographic analysis[14, 17] which appears to be a powerful tool in the separation of compounds like most of the bases which can be vaporized.

Another new and very powerful method of separation, use of the thermal diffusion column, should be considered when very difficult separation problems are encountered.[19, 45] In this method, the mixture in the narrow space between two concentric pipes or tubes, one of them heated and the other cooled, separates as molecules rise along the hot surface and move downward along the cold one. So far a number of such units have to be used when liters of mixture must be separated and, as expected, equilibrium is reached after many hours of operation, but the equipment requires very little attention while in operation and a large number of such tubes could be run simultaneously by the same operator. The method has apparently not yet been used in the separation of mixtures of nitrogen compounds.

Purification and Identification

After the individual base has been concentrated by a combination of the methods mentioned to the point where a solid salt or other derivative is obtained, the problem becomes one of purifying the precipitate by recrystallization, fractional precipitation, or liberation and conversion to another solid form. In the case of the petroleum bases, the picrates have been most commonly employed as initial salts because they appeared to be least likely to precipitate as a smear. Wibaut [15] found that the picrolonates were the most convenient derivatives as far as crystalline nature and ease of recrystallization were concerned, but admitted that a serious drawback of their use was the fact that they have rarely been used previously and so it was not often possible to predict the structure of a base from the melting point of its picrolonate. Another disadvantage of this derivative is the high cost of the reagent. In addition to these general reagents, such special ones as concentrated sulfuric acid, hydrochloric acid, nitric acid, zinc chloride, methyl iodide, and mercuric chloride have sometimes been found to yield solids which could be purified by recrystallization. In some cases, mixed crystals were formed and it was not possible to use a crystalline salt that at first appeared very promising. At best there is always a considerable loss during recrystallizations, since no method of concentration yields pure compounds and recrystallization of mixtures is almost always accompanied by serious losses. It is then difficult to obtain a close estimate of the amount of any particular compound present originally.

From the point at which a pure base is obtained, problems involved in the identification of bases are not peculiar to petroleum bases and the structure of a pure known or previously unknown base has to be determined.

The compound is, of course, analyzed and the usual physical properties, such as boiling or melting point, density, index of refraction, and molecular weight determined. In a few simple cases, the exact structure could be predicted at this stage, but usually tests to determine the presence or absence of active methyl groups and identification of acids formed on oxidation of the base had to be carried out. Since all bases so far isolated proved to be tertiary amines, tests for type of amine have not been helpful. Even if the exact structure was predicted at this stage, the compound was usually synthesized to confirm the prediction. Sometimes probable structures could be narrowed down to two or three and these were then synthesized one by one until the correct one was found.

In the case of the quinolines, methyl iodide usually does not add if the 2 position is occupied and the 8 position carries a substituent larger than methyl. Phthalones form only with 2-methylquinolines. In the case of the quinolines, substituents have so far been found only at positions 2, 3, 4, and 8 and position 2 was always substituted and the substituent was always a methyl group. Stated differently, the alkylquinolines were found to be 2-methylquinolines with other substituents, if any, at positions 3, 4, or 8. In view of the isolation of 7-methylquinoline from cracking-process gasoline by Hackman and Wibaut,[15] it would be surprising if quinolines with a different orientation of substituents were not found to occur in petroleum, but the fact remains that this rule has been found very convenient in the past.

In the case of the simple alkylated pyridines, very extensive use was made of Eguchi's rule by which the structures of various methylpyridines have been predicted from the boiling points.[11] Eguchi's formula reads as follows:

$T = 115.3 + 14m + 28m' + C$ where

T is the calculated boiling point at atmospheric pressure in degrees Centigrade,

m = number of α-methyl groups,

m' = number of β- and γ-methyl groups, and

C is a constant depending on the other substituents and has a value of 4 for 2,3-, 8 for 3,4-, and 2 for 2,4-substituents.

A modified formula applies to the ethylmethylpyridines. Professor Bailey used this formula extensively in predicting the structures of various pyridines isolated.

In most cases, the information gained from analyses and various tests did not permit prediction of the exact structure but usually all but two or three structures were eliminated or made unlikely. At this stage, if the suspected bases were easy to synthesize, this was the next step, but if, as in the case of many of the alkylpyridines, this was difficult, degradation reactions had to be used. Most of the quinolines were isolated in quantities of 100 g or more and degradation reactions were indicated and were used in practically all cases, followed by synthesis of the petroleum base.

In the case of the alkylated pyridines, isolation of individual bases was found to be much more difficult and was rarely accomplished until after the spinner columns came into use. Rather dilute solutions had to be used and the throughput of the columns was 300 to 400 ml per hour. Thus, the amount of pure base isolated usually amounted to 1 to 5 g and was sometimes even less. Degradation reactions were used only as a last resort and the results were sometimes misinterpreted due to failure to obtain completely purified products from the small batch. Synthesis always had to be employed for final identification.

Degradation reactions employed were:

1. Vigorous oxidation to convert side chains to acids. This method was used with both quinolines and pyridines and sometimes made it possible to predict the structure since the acids formed are known. In some cases, one or more of the carboxyl groups were lost during oxidation and led to wrong conclusions.

2. Ozonolysis of the substituted pyridines leads to the amide of the acid formed from the alpha carbon atom and any attached side chain. Any α-isopropylpyridine yields isobutyramide or the acid, depending on the conditions under which the isolation was carried out.[43, 44] In some cases, other cleavage products, such as diketones or ketoacids, may be isolated and may help determine the exact structure.

In the quinoline series, ozonolysis has been used only a few times. Schenck [41] used it in determining the probable nature of the substituent in position 8 of a substituted quinoline.

3. Hydrogenation to the completely hydrogenated pyridine or quinoline followed by one of the ring-opening methods can be used. If the structure of the compound formed after elimination of the nitrogen atom in this way is readily determined, it may be a very valuable method and has been used frequently in alkaloid chemistry. It was used in several cases by Hackman and Wibaut [15] in determination of the structure of bases from cracking-process gasoline and by Shive, Roberts, Mahan, and Bailey [43] in the determination of the structure of the $C_{16}H_{25}N$ base.

Synthesis of suspected compounds proved relatively easy for

quinolines and was almost universally used by the Bailey group; but in the pyridine field, the isolation of pure compounds was more easily accomplished than the determination of the structure of the pure compound isolated.[25, 26] Methods developed or tested by Wheeler[49] have made 3- and 4-alkylpyridines more readily accessible than formerly, but the synthesis of polyalkylated pyridines is still rather difficult and will continue to make the identification of these compounds rather tedious and uncertain.

Bibliography

1. Axe, W. N., Ph.D. Thesis, The University of Texas, 1938.
2. Axe, N., and J. R. Bailey, *J. Am. Chem. Soc.* **60,** 3028 (1938).
3. Axe, W. N., and J. R. Bailey, *Ibid.* **61,** 2609 (1939).
4. Biggs, B., and J. R. Bailey, *Ibid.* **55,** 4141 (1933).
5. Begeman, C. R., and P. L. Cramer, *Ind. Eng. Chem.* **47,** 202 (1955).
6. Bratton, A. C., Ph.D. Thesis, The University of Texas, 1936.
7. Bratton, A. C., W. A. Felsing, and J. R. Bailey, *Ind. Eng. Chem.* **28,** 424 (1936).
8. Bratton, A. C., and J. R. Bailey, *J. Am. Chem. Soc.* **59,** 175 (1937).
9. Craig, L. C., and O. Post, *Anal. Chem.* **21,** 500 (1949).
10. Edens, C. O., Jr., M.A. Thesis, The University of Texas, 1939.
11. Eguchi, T., *Bull. Chem. Soc. Japan* **2,** 176 (1927); **3,** 227 (1928).
12. Elving, P. J., *Anal. Chem.* **23,** 1206 (1951).
13. Glenn, R. A., and J. R. Bailey, *J. Am. Chem. Soc.* **63,** 637 (1941).
14. Griffiths, J., D. James, and C. Phillips, *Analyst* **77,** 897 (1952).
15. Hackman, J. T., and J. P. Wibaut, *Recueil* **62,** 229 (1943).
16. Heap, J. G., W. J. Jones, and T. B. Speakman, *J. Am. Chem. Soc.* **43,** 1936 (1921).
17. James, D. H., and C. G. S. Phillips, *J. Chem. Soc.* **1954,** 1066.
18. Jantzen, E., *Das fraktionierte Destillieren und das fraktionierte Verteilen als Methoden zur Trennung von Stoffgemischen*, Verlag Chemie, Berlin, 1932.
19. Jones, A. L., *Ind. Eng. Chem.* **47,** 212 (1955).
20. Jones, J. Idris, *J. Soc. Chem. Ind.* (*London*) **65,** 169 (1946).
21. Jones, J. Idris, *Ibid.* **69,** 99 (1950).
22. King, W. A., and J. R. Bailey, *J. Am. Chem. Soc.* **52,** 1245 (1930).
23. Lake, G., and J. R. Bailey, *Ibid.* **55,** 4143 (1933).
24. Lang, W., *Ber.* **21,** 1578 (1888).
25. Lochte, H. L., W. Crouch, and D. Thomas, *J. Am. Chem. Soc.* **64,** 2753 (1942).

26. Lochte, H. L., D. Thomas, and P. Truitt, *Ibid.* **66,** 550 (1944).
27. Lochte, H. L., and H. W. H. Meyer, *Anal. Chem.* **22,** 1064 (1950).
28. Lochte, H. L., and W. G. Meinschein, *Petroleum Engr.* March 1950, C–41.
29. Mahan, R. I., and J. R. Bailey, *J. Am. Chem. Soc.* **59,** 2449 (1937).
30. Mahan, R. I., Ph.D. Thesis, The University of Texas, 1938.
31. Maier-Bode, H., and J. Altpeter, *Das Pyridin und seine Derivaten in Wissenschaft und Technik,* Wilhelm Knopp, Halle, 1934.
32. Morton, A. A., *Laboratory Technique in Organic Chemistry,* McGraw-Hill, New York, 1938, page 200.
33. Ney, W. O., and H. L. Lochte, *Ind. Eng. Chem.* **33,** 825 (1941).
34. Ney, W. O., W. Crouch, C. E. Rannefeld, and H. L. Lochte, *J. Am. Chem. Soc.* **65,** 770 (1943).
35. Parker, I., C. L. Gutzeit, A. C. Bratton, and J. R. Bailey, *Ibid.* **58,** 1097 (1936).
36. Perrin, T. S., and J. R. Bailey, *Ibid.* **55,** 4137 (1933).
37. Poth, E. J., et al., and J. R. Bailey, *Ibid.* **52,** 1243 (1930).
37A. Ray, N. H., *J. Applied Chem.* **4,** 21 (1954).
38. Roberts, S. M., and J. R. Bailey, *Ibid.* **60,** 3025 (1938).
39. Roberts, S. M., Ph.D. Thesis, The University of Texas, 1939.
40. Scheibel, E. G., *Chem. Engr. Progress,* **44,** 927 (1948).
41. Schenck, L. M., and J. R. Bailey, *J. Am. Chem. Soc.* **63,** 1364 (1941).
42. Schutze, H. G., W. A. Quebedeaux, and H. L. Lochte, *Ind. Eng. Chem. Anal. Ed.* **10,** 676 (1938).
43. Shive, W., S. M. Roberts, R. I. Mahan, and J. R. Bailey, *J. Am. Chem. Soc.* **64,** 909 (1942).
44. Shive, W., E. G. Ballweber, and W. Ackermann, *Ibid.* **68,** 2144 (1946).
45. Sullivan, L. J., T. C. Ruppal, and C. B. Willingham, *Ind. Eng. Chem.* **47,** 208 (1955).
46. Tiedcke, K., Ph.D. Thesis, Hamburg, 1928.
47. Thompson, W. C., and J. R. Bailey, *J. Am. Chem. Soc.* **53,** 1002 (1931).
48. Walker, C. A., M.A. Thesis, The University of Texas, 1940.
49. Wheeler, E., Ph.D. Thesis, The University of Texas, 1953.
50. Wyler, M., *Ber.* **60B,** 398 (1927).

CHAPTER TWENTY-FOUR

EARLY INVESTIGATIONS

Bandrowsky[1] mentioned that operators in refining Galician crudes had often noticed that ammonia was evolved during distillation of oil at atmospheric pressure. Since the original oil was neutral, they concluded that the ammonia was formed by pyrolysis of nitrogenous compounds. We find thus here, as in the case of the petroleum acids, very early development of the theory that these compounds were not present in crude oil but were formed in some way during distillation.

Bandrowsky extracted 22 liters of crude oil with 10% sulfuric acid, then extracted the aqueous layer with ether to remove hydrocarbons dissolved or still emulsified in the salt layer after several weeks of settling. He finally neutralized the excess mineral acid with sodium carbonate and then liberated the bases with sodium hydroxide. The bases were then extracted with ether which, on evaporation, left a dark-red oil from which no crystals could be obtained on cooling to −20°C. Chloroplatinic acid added to the bases from 12 liters of crude oil yielded 1.29 g of a heavy gum which could not be induced to crystallize. Alkaloidal tests were positive and Bandrowsky concluded that the bases were alkaloidal in nature — an idea that was later emphasized by Bailey.

Bandrowsky's publication prompted Weller[13] to report on work he had begun previously but then abandoned. He studied Saxony crude oil of which he extracted 500 kg with dilute sulfuric acid using essentially the same procedure as employed by Bandrowsky. Since he extracted a much larger volume of oil, Weller obtained a sample large enough to permit distillation which yielded a series of colorless fractions with a sharp unpleasant odor and densities ranging from 0.98 to 0.99. The oily fractions would not crystallize at −11° and were found to be free of oxygen and sulfur. Heating with concentrated potassium hydroxide did not result in the evolution of ammonia thus indicating that the ammonia observed by other workers must have been obtained from different or higher-boiling compounds. Weller's bases were found to react with bromine and with methyl iodide. Attempts to prepare pure crystalline salts were, however, fruitless, although oxalates could be obtained in solid form.

Zaloziecki [14] reported that petroleum had been found to contain as much as 1% total nitrogen although the usual concentration was about 0.1% of crude oil. He worked on acid sludge obtained in refining Boryslow petroleum. He was then apparently the first to use bases obtained from refinery products rather than crude oil. He was also the first to avoid emulsification troubles by diluting his bases with low-boiling hydrocarbons.

Zaloziecki liberated his bases with an excess of calcium hydroxide and steam distilled the material which he said had a pyridinelike odor. He observed that this process yielded a product essentially free of color and tar. This would be expected since the tar would not distill and the petroleum bases, as distilled, are usually colorless. The higher petroleum bases are not volatile with steam, however, so that Zaloziecki probably failed to recover most of his higher-boiling bases.

Various salts, such as mercuric chloride, gave him solid precipitates, but none of them could be purified by recrystallization. He analyzed various chloroplatinic acid precipitates and obtained results that agreed with such formulae as $(C_{10}H_{15}NCl)_2PtCl_4$, $(C_{10}H_{17}NCl)_2PtCl_2$, and $(C_{10}H_{19}NCl)_2PtCl_2$ and concluded from this that he had a mixture of platinum salts of quinoline and possibly hydroquinoline instead of the expected pyridines. While not much stress can be laid on analyses of compounds that were obviously impure or at least were mixtures of bases, it should be noted that $C_{10}H_{15}N$ corresponds to an alkylpyridine — a type that appears to be very common among petroleum bases.

Schestakow [11] found 0.006% of basic compounds in a sample of Roumanian crude oil. The mixture of bases could be distilled without decomposition at 260° to 370°C (500° to 698°F) and was optically inactive. The mixture was said to be hygroscopic and distinctly basic. He analyzed a fraction and found it to contain 85.7%

carbon, 8.08% hydrogen, and 6.6% nitrogen, which corresponds to an average empirical formula of $C_{15.2}H_{17}N$ or approximately C_nH_{2n-13} which would indicate that the sample was composed largely of alkylated quinolines. He could not obtain crystalline salts with mineral acids, but picric and chromic acids and mercuric and gold chlorides yielded gums. He obtained semicrystalline derivatives with platinic chloride, lead tetrachloride and with prussic acid, but could not recrystallize them.

Chlopin [2] said he and others had frequently found bases in low concentration in Caucasian crudes, but had never obtained enough material to permit study. He now divided 291 kg of crude oil into two portions and shook one for 12 days and the other for 22 days with the same 20-liter batch of 15% sulfuric acid. He then filtered, neutralized the strong acid, and liberated the bases with concentrated sodium hydroxide. He extracted the liberated bases with ether and obtained 14.58 g of dry bases — a yield of 0.005% based on crude oil. The mixture of crude bases was a thick, dark-brown oil with green fluorescence and a sharp pyridinelike odor. Since his method of isolation would tend to include considerable amounts of hydrocarbons, the fluorescence is probably of no significance. He was able to obtain gummy precipitates with platinum, palladium, cadmium, iron, and mercuric salts, but none could be obtained in crystalline form. Analyses of such gums indicated a molecular weight of 215 to 250, corresponding to bases with fifteen to eighteen carbon atoms. To indicate the complexity of the mixture, fractional precipitation of the platinic chloride derivatives yielded molecular weights of 280, 247, 227, and 218 for four fractions.

Chlopin found the bases rather poisonous for fish, but harmless to cats and bacteria. He concluded that his bases belonged to the series $C_nH_{2n-15}N$, with $C_{22}H_{29}N$ as a typical base. If his results are significant, they might indicate the presence of quinoline with a cyclic substituent or a tricyclic compound, like acrydine.

The first American publication dealing with research on petroleum bases appeared in 1900 and was a preliminary report on the work of Mabery.[5] This was followed, in 1920, by a final report on this project.[7] This was one of a long series of studies on compounds in petroleum. One of the first problems encountered was that of quantitative detérmination of nitrogen in crude oils. Mabery recognized the problems involved and developed a modified Dumas method that gave results in agreement with Kjeldahl results.[6] He found that the nitrogen of petroleum is present mainly in the higher-boiling fractions in which there is a considerable concentration of nitrogen compounds on the basis of Kjeldahl results and fantastic concentrations on the basis of some of the obviously incorrect Dumas results which were quoted, for some reason, after the results obtained in 1919.

The petroleum bases studied by Mabery and Wesson had been isolated for them by Peckham and Salathe from California distillate with dilute acid. Mabery found that the bases began to boil at 225°C (437°F) at atmospheric pressure and less than half of the material had distilled at 300°C (572°F). Therefore, he decided to fractionate in vacuum. After several distillations at 50 to 90 mm pressure he found that bases accumulated at 130° to 140°, 197° to 199°, 209° to 211°, 215° to 217°, 223° to 225°, 243° to 245°, and 270° to 275°C. Analyses of these fractions led to the following empirical formulae:

Fraction Boiling °C, at 50 to 90 mm	*Empirical Formula*
130–140	$C_{12}H_{17}N$
197–199	$C_{13}H_{18}N$
215–217	$C_{14}H_{19}N$
233–235	$C_{15}H_{19}N$
243–245	$C_{16}H_{19}N$
270–275	$C_{17}H_{21}N$

The number of hydrogen atoms was admittedly uncertain and from what is now known about petroleum bases, it is highly probable that these bases contained considerable concentrations of hydrocarbons.

They observed a pungent nicotinelike odor and at other times described the odor as that of an "empty cigar box." This characteristic odor is well known to workers in this field and adheres tenaciously to clothing, skin, and hair of laboratory workers. He found his bases were completely soluble in dilute mineral acid in the form of salts, but could not obtain any crystalline salt or other derivative. In his preliminary report, he thought the bases were probably present as such in crude oil, but were difficult to extract because of emulsification. In their final paper,[7] they referred to the great difficulty of isolating petroleum bases present in very low concentration from large volumes of hydrocarbons present. Since they used essentially only fractional distillation this difficulty would be a serious one.

In their study of the chemical reactions of various fractions, they found that potassium permanganate in aqueous solutions, produced acids whose silver salts were analyzed with the results shown in Table 50.

Table 50. **Mabery's Tetrasilver Salts of Pyridinepentacarboxylic Acid**

Boiling Point °C, at 50–90 mm	C %	*H* %	*N* %	*Ag* %	*O (Difference)* %
130–132	16.37	0.25	2.08	60.66	20.64
176–178	16.34	0.31	2.38	60.53	20.48
200–202	15.52	0.23	1.58	59.73	22.94
209–211	15.99	0.24	1.84	60.48	21.45
Average	16.06	0.26	1.97	60.35	21.38
Calculated for $C_{10}HNAg_4O_{10}$	16.51	0.14	1.93	59.41	22.01

The silver analyses, in particular, show that the salts were not identical, but they pointed out that the silver content of a salt of this type would not be expected to agree well from preparation to preparation since the analyses show that the compounds are *tetra*-silver salts of pyridine*penta*carboxylic acid so that the amount of silver might vary. To confirm their identification, they converted the silver to the barium salt and this to pyridine which was identified through the picrate which could not, however, be obtained in pure form, but gave no depression in a mixed melting point determination with known pyridine picrate. Less vigorous oxidation led to a methylpyridinetetracarboxylic acid which was easily converted to its silver salt. They could not obtain either of the acids as crystalline acids, even though they started with the pure silver salt and even though the crystalline acids are known.

In the case of one fraction, they found that careful oxidation by chromic oxide and dilute sulfuric acid resulted in the formation of an acid which, on decarboxylation by heating the barium salt, yielded 3-methylquinoline. This base was converted to the picrate which melted at the reported melting point of the picrate of 3-methylquinoline.

In qualitative tests, they obtained gums but no crystalline precipitates on treating the bases with platinum, palladium, mercury, cadmium, and ferric-iron salts, potassium dichromate, ferro- and ferricyanide, and with picric and oxalic acids. Liquid bromine reacted vigorously with evolution of hydrobromic acid while bromine in carbon tetrachloride solution yielded no hydrobromic acid, but a dark oil which was soluble in hot alcohol, but separated on cooling. Analysis indicated the presence of four bromine atoms per mole of base. Phthalic anhydride reacted to yield a phthalone in a reaction typical of 2-methylquinolines.

Reduction with sodium amalgam and alcohol, zinc and acetic acid, tin and hydrochloric acid, or hydroiodic acid converted these tertiary amines to secondary amines. These could be benzoylated and acetylated, but they were not able to open the ring with PCl_5, or at least they were not able to isolate any products.

While Mabery and Wesson were not able to identify any individual compounds among these bases, they made more progress in this field than had been made up to that time. Some of their conclusions were undoubtedly too broad but, in general, what we now know about California bases fits fairly well into the picture they outlined. They made the following conclusions:

1. No significant yields of aliphatic acids were obtained on oxidation of these bases and those obtained in traces included none higher than propionic. From this, they concluded that none of the side chains on the nucleus had more than four carbons.
2. Chromic acid oxidation of one fraction yielded an acid which was decarboxylated to 3-methylquinoline.
3. Reduction led to secondary amines.
4. Analysis of oxidation products indicated the presence of alkylated quinolines and isoquinolines.
5. Oxidation of a number of different fractions led to an acid whose silver salt was analyzed and found to be derived from pyridinepentacarboxylic acid. If formed from a quinoline, this must have been one which was completely alkylated on the pyridine ring. Since only the lower acids could be isolated from oxidation products, they decided that the alkyl groups on the pyridine ring must all be lower ones like methyl or ethyl.
6. They suggested that the general behavior of the bases was not contrary to what would be expected of highly alkylated quinolines or isoquinolines.

Richardson [10] examined oil from the Olinda field of California and decided that it differed from others in that all but the lowest fractions obtained on distillation yielded phenols and large amounts of bases which caused low color stability. He claimed pyridines were present and that concentrated sulfuric acid removed quinoline and aromatic hydrocarbons, but he apparently made no effort to identify any of the bases.

Griffiths and Bluman [3] extracted gasoline from Roumanian petroleum with 25% sulfuric acid and isolated a base boiling at 117°C (243°F) which could be reduced by metallic sodium to what they considered to be piperidine. They admitted that their base was only slightly soluble in water while pyridine is miscible, but this may have been due to admixture of large amounts of hydrocarbons. Since they did not definitely identify their compound as pyridine, they should probably not be credited with having isolated it from petroleum.

In 1922, Pyhala [9] reported briefly on some experiments performed in 1912 in the isolation of bases from the acid sludge obtained in the refining of kerosene from Baku crudes. He separated hydrocarbons, acids, and bases by simple dilution of the acid sludge with water followed by steam distillation of the bases. The distillate, as might be expected, contained hydrocarbons and some acids as well as bases. The bases were then extracted from the distillate with "not too concentrated" sulfuric acid and liberated in a closed system in which all gases and vapors were condensed or absorbed. The flask in which the bases had been liberated as finally heated to steam distill the higher-boiling bases. The vapors and all fractions obtained on distillation fumed on contact with concentrated hydrochloric acid. Fractionation of the liquid bases resulted in the following fractions:

I, boiling at 75° to 120°C (167° to 248°F) (mainly at 75° to 80°C)
II, boiling at 127° to 147°C (261° to 296°F)
III, boiling at 147° to 187°C (296° to 369°F)
IV, boiling above 187°C (369°F)

All fractions were partially soluble in dilute sodium hydroxide. Platinum and palladium chlorides yielded gums which were soluble in chloroform giving various colored solutions.

Pyhala concluded that he had obtained a mixture of amines which were not present as such in the crude oil, but were probably formed during refinery or laboratory distillations since they were lower boiling than the kerosene fraction from which they were isolated. According to his theory, their source might have been amides which had been formed underground from ammonia and organic acids. Obviously, the amines which he obtained could not have originated from simple amides of this type but might have been formed from substituted amides or perhaps ammonium salts obtained from primary and secondary amines and organic acids. It is doubtful whether either the separation by simple dilution of the acid sludge or steam distillation removed all of the bases in his mixtures.

Bibliography

1. Bandrowsky, F. X., *Monatsh.* **8,** 224 (1887).
2. Chlopin, G. W., *Ber.* **33,** 2837 (1900).
3. Griffiths, A. B., and M. N. J. Bluman, *Bull. soc. chim.* (**3**) **25,** 725 (1901).
4. King, W. A., and J. R. Bailey, *J. Am. Chem. Soc.* **52,** 1245 (1930).
5. Mabery, C. F., *J. Soc. Chem. Ind.* **19,** 505 (1900).
6. Mabery, C. F., *J. Am. Chem. Soc.* **41,** 1690 (1919).
7. Mabery, C. F., and L. G. Wesson, *Ibid.* **42,** 1014 (1920).
8. Perrin, T. S., and J. R. Bailey, *Ibid.* **55,** 4136 (1933).
9. Pyhala, E., *Chem. Ztg.* **46,** 953 (1922).
10. Richardson, C., *J. Soc. Chem. Ind.* **19,** 123 (1900).
11. Schestakow, P. J., *Chem. Ztg.* **23,** 41 (1899).
12. Thompson, W. C., and J. R. Bailey, *J. Am. Chem. Soc.* **53,** 1002 (1931).
13. Weller, A., *Ber.* **20,** 2098 (1887).
14. Zaloziecki, R., *Monatsh.* **13,** 498 (1892).

CHAPTER TWENTY-FIVE

INVESTIGATIONS AT THE UNIVERSITY OF TEXAS

All of the research work done prior to 1900 was financed and carried out by individual laboratories with samples furnished by some oil company. The supply of bases was quite small and the tedious fractionation had to be carried out with inadequate staff and facilities.

In 1926, John D. Rockefeller and the Universal Oil Products Company made available the sum of $500,000 to be used in furthering fundamental research in the chemistry of petroleum. The fund was administered by the American Petroleum Institute in coöperation with the National Research Council. Among the projects financed by this fund was Project 20 under the direction of the late Professor J. R. Bailey of the University of Texas. This project was to undertake the isolation and identification of petroleum nitrogen compounds. When Project 20, as such, was discontinued, the work was carried out for some years in coöperation with the Union Oil Company of California who had isolated and supplied the bases and who, on Professor Bailey's death, made the large existing supply available for further research at the University of Texas. Since the work of Bailey and coworkers, from 1928 to 1940, was and has remained the main contribution to our knowledge of petroleum bases, it will be the basis of most of the discussion in this chapter and the group will often be designated as the Texas group.

The $C_{16}H_{25}N$ base which has been shown to be 2,4-dimethyl-6-(2,2,6-trimethylcyclohexyl)pyridine is of such importance, occurs in such abundance, and occupied so much of the attention of the Texas group, that it will be considered in a separate section.

Since the nitrogen content of most crudes was known to be quite low, Bailey's group first undertook a survey of possible supplies of bases to be studied. This involved accurate determination of the nitrogen content of crude-oil samples and, after a critical study of the Dumas and Kjeldahl methods, they modified the Kjeldahl method mainly to make possible the use of samples weighing 1 to 5 g. Results obtained by this method indicated that some California crudes should prove most satisfactory. They also found that only very small amounts of basic compounds could be extracted from crude oil by dilute mineral acids and that even refinery fractions, like kerosene, contained only about half of its total nitrogen in the form of basic compounds that could be extracted with 16% sulfuric acid.[27, 28]

The Union Oil Company furnished numerous small preliminary samples, a 50-liter kerosene-base sample, the main 3-barrel batch of similar bases and many later special samples, such as crude Edeleanu extracts from various refinery products. The crude oil from which most of the samples were isolated contained about 0.55% nitrogen. The 3-barrel batch was made from 4,175* barrels of Edeleanu extract which, in turn, was obtained in refining kerosene derived from 240,000 barrels of crude oil.

According to the detailed description of the procedure used by the Union Oil Company at their Oleum refinery as given by Lake,[17] the isolation of the main sample from Edeleanu extract was accomplished as follows:

Three hundred barrels of sea water and 19,145 lb of concentrated sulfuric acid were carefully mixed in a light-oil agitator. Edeleanu extract obtained in refining kerosene was pumped through the water layer through a lead-pipe distributor with 256 holes each 3/32 in. in diameter. The distributor rested near the conical bottom of the tank. The supernatant extract layer was removed from a point about 6 in. above the level of the acid layer. The cooled extract was pumped through the acid at a rate of about 350 barrels a day. At the end of 12 days, 4,175 barrels of extract had passed through the acid which had decreased in strength from 16.04 to 15.15%. The floating layer of spent extract was removed as completely as possible before addition of 7 tons of dry caustic at such a rate as to

* A number of theses and papers state that it was obtained from 3,000 barrels of extract, but Lake's thesis [17] quoting original reports shows that 4,175 barrels were involved.

avoid overheating. After several days of settling, the aqueous layer was drawn off and found to contain about half a barrel of bases which were not studied. Pyridine and the lower alkylated pyridines which may have been present were probably discarded at this stage, but their isolation would have been rather expensive.

The bases were then roughly distilled and combined according to boiling point to yield 3 barrels of 50 to 53 gal each before shipment to the University of Texas. Table 51 lists data pertaining to this large batch of bases.

Table 51. **Data on the Supply of Crude Bases**

Barrel	*Mean Atmospheric Boiling Point* °*C*	°*F*	*Density*	*Index of Refraction*	*Sulfur* %*	*Nitrogen* %**
I	252	486	0.951	1.5208	0.21	7.74
II	281	538	0.971	1.5323	0.33	7.70
III	290	554	0.995	1.5531	0.36	7.83

* Analyses at the refinery.
** By modified Kjeldahl method at Austin.

From the crude bases, 2 to 10% of neutral matter was easily removed by a three-step partial neutralization with 20% sulfuric acid. All of the crude base, one barrel at a time, was treated by stirring in a 200-gal crock with three batches of 19 gal of 20% sulfuric acid which was just sufficient to remove all of the basic matter.

The group then undertook the long and tedious series of batch fractionations through variable reflux columns packed with "jack chain" and operated at an undetermined efficiency — probably not over fifteen plates per distillation. Since some of the material was finally fractionated as much as six times, the total amount of fractionation employed was probably as much as can be used profitably. Any slight additional fractionation would have been obtained at too great expense of time and at increased thermal hazard to the compounds.

As usually happens when any petroleum product is carried through fractional distillations, the cuts obtained from each barrel showed a definite peak in the curve when the volume over per degree was plotted against temperature. Figure 11 shows this effect. As fractionation was continued, additional peaks — minor ones on the sides of the main peak — appeared. These corresponded to the plateaus obtained when boiling point was plotted against constant-volume

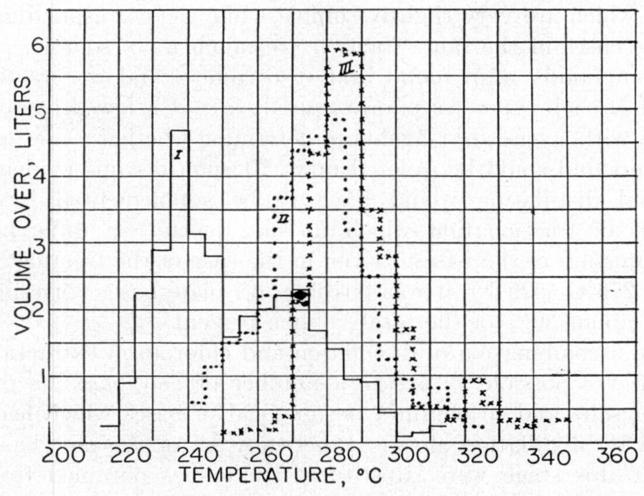

Figure 11. Plot of Volume Distilled per 5° Boiling Range for Three Batches of Petroleum Bases

cut number, for instance. Inspection showed that barrels II and III included far more material per degree of boiling range than the first one. The 270° to 280°C (518° to 536°F) fractions were very large and pure bases were most easily isolated from them.

The first crystalline salts were obtained from the bases boiling just below 280°C (536°F). Thompson [38] found that when dilute sulfuric acid was added to a sample of bases boiling at 276°C (529°F), a white crystalline precipitate formed. After recrystallization from alcohol, it was found to be the acid sulfate of 2,3,8-trimethylquinoline which was isolated in nearly quantitative yield in this form. The sulfates of the other bases boiling in this range are much more soluble and it was possible to isolate 2,3,8-trimethylquinoline in approximately 20% yield from this fraction. If there were a large demand for this base, it could be isolated rather easily from California bases in this way.

Fractional precipitation of the picrates of the bases from another sample boiling in the same range permitted isolation of the same quinoline followed by a new base, $C_{16}H_{25}N$, and finally a third base, 2,4,8-trimethylquinoline,[15] which was definitely identified by Perrin and Bailey.[25] These successes started a frantic search for other salts or other derivatives with peculiar solubility behavior, i.e., compounds which are very slightly soluble while the corresponding salts of other bases in the same fraction are much more soluble. A few such compounds were found and it became standard practice, at the end of each type of separation used, to test a few drops of each fraction with a series of reagents to determine whether a new crystalline compound could be isolated now. It soon became evident that fractional distillation would have to be supplemented by other methods of fractionation which do not depend on differences in vapor pressure of the bases. Even in the case of the fraction boiling around 276°C (529°F), it was possible to isolate bases corresponding to only about 30% of the total volume present.

When a combination of distillation and chloroform extraction was used, it was possible to isolate a number of new bases as picrates or other salts and to obtain a better yield of bases which had been isolated by distillation alone. Practically all of the new bases isolated at this stage were substituted quinolines obtained from the water layer of the chloroform extraction and for a number of years, the only bases isolated from the chloroform layer were $C_{16}H_{25}N$ and a few of the lower pyridines. In view of the mysterious nature of the so-called "nonaromatic" bases, there was a tendency to assign to each new worker some fraction of the chloroform-soluble bases in the hope that somebody would find a suitable method of separation or a salt that would permit identification of some of these bases.

Perrin and Bailey [25] used chloroform extraction on 1 liter of bases boiling at 273°C (524°F) and isolated from the water layer:

44 g 2,3,8-trimethylquinoline hydrochloride
37 g 2,3,8-trimethylquinoline picrate
30.5 g 2,4,8-trimethylquinoline picrate
80 g mixed quinoline picrates
30 g of a picrate melting at 174°C (345°F)

From the chloroform layer, they obtained 144.5 g of $C_{16}H_{25}N$ base as hydrochloride and they found that from both layers a total of 736 g of bases yielded no solid salts. This separation represents the closest approach to a complete separation of any of the fractions boiling above the boiling point of quinoline.

Biggs and Bailey [5] had isolated 2,3- and 2,4-dimethylquinoline from bases boiling at 265° to 269°C (509° to 516°F) through fractional sulfite formation. When Perrin treated the residual bases from this work by the chloroform-extraction method, he was able to isolate additional large amounts of 2,3-dimethylquinoline, but could not obtain any 2,4-dimethylquinoline. Lake and Bailey [18] isolated 2,8-dimethylquinoline from bases boiling around 253°C (488°F), even though only 5% of quinoline-type (water-soluble) bases were found to be present in this range and it would have been almost impossible to separate 5% of this type of base from 95% of other bases by any other known method. Axe [3] isolated 2,4-dimethyl-8-ethylquinoline from a fraction boiling at 290° to 291°C (534° to 556°F) by adding concentrated sulfuric acid to an alcohol-acetone solution of the bases from the water layer of chloroform extraction. He obtained an 18.5% yield of purified acid sulfate of 2,4-dimethyl-8-ethylquinoline.

It soon became evident that even in the bases isolated from the water layer, there often were so many isomeric and homologous quinolines and similar compounds that no crystalline salts or other derivatives could be obtained, while for the bases from the chloroform layer, this was true in practically all cases.

Other types of fractionation were tried which, in some cases, gave valuable separations on material that had been carefully fractionated by distillation and extracted by chloroform. Fractional neutralization by reaction with sulfurous acid or fractional liberation of the bases from their sulfites was used by a number of workers in the Texas group. The method was used with coal-tar bases and was developed by Biggs [5] for petroleum bases. He isolated 2,3- and 2,4-dimethylquinoline from a fraction boiling at 265°C (509°F) by dissolving the sample in ether, saturating the solution with sulfur dioxide, transferring to a flask, and evacuating the flask to remove sulfur dioxide. The solution turned cloudy and the stable acid sulfites separated as an oily layer. This layer was washed with petroleum ether and then dissolved in the minimum amount of 95% alcohol. A hot saturated solution of picric acid in alcohol was then added and the solution boiled to decompose the sulfites and precipitate the insoluble picrates. After 24 hours, 2% of his original bases had been converted to picrates from which 2,3- and 2,4-dimethylquinolines were isolated.

Edens [8] used a similar method in the reisolation of 2,3,8- and 2,4,8-trimethylquinolines from bases boiling in the 271° to 273°C (519° to 523°F) range. His yield, like that of Biggs was rather low, however. Axe [2, 4] and Schenck [32, 33] used a modification of the Biggs method on bases boiling at 300° to 320°C (572° to 608°F) which had been extracted with chloroform. They used later modifications, such as two-phase systems with water or brine as one phase and the mixture of bases alone or in an organic solvent as the other. Sulfur dioxide was bubbled into this system until all of the bases had been neutralized. On removal of the sulfur dioxide, the bases were fractionally liberated. They isolated 2,3-dimethyl-8-ethylquinoline, 2,3-dimethyl-8-n-propylquinoline, and 2,3-dimethyl-4-ethyl-8-n-propylquinoline in this way. Axe and Bailey [2] used the method on the 295°C (563°F) range to isolate 2,3,4,8-tetramethylquinoline and 2,3-dimethyl-8-n-propylquinoline, while Meadows [23] used it, again in combination with other methods in the isolation of 2,3- and 3,5-dimethylpyridines, 2,3,5- and 2,3,4-trimethylpyridines, and 3-methyl-5-ethylpyridines from the 170° to 200°C (338° to 392°F) range of bases. The method may prove to be of industrial importance in this field whenever the time comes when bases will be fractionated commercially because the sulfur dioxide can be recovered and recycled.

While each new method of separation appeared to lead to some new bases there was a need for a general method to be used after separation by chloroform extraction and careful fractionation by distillation. Some system of solvent extraction or, probably more conveniently in this case, some system of extraction by systematic fractional neutralization appeared more promising than chromatographic separations which had not been used extensively at that time. Since Jantzen [13] and Tiedcke [39] had been able to use such methods with notable success with coal-tar bases and another Texas group had tested suitable apparatus for countercurrent fractional neutralization, it was used after fractional distillation and chloroform extraction and it was found possible to isolate a large number of new bases and to reisolate a number of others that had been obtained by earlier methods.

This method in one form or another was used by Garland [9] in isolating $C_{11}H_{15}N$ and $C_9H_{11}N$ bases and a 5,6,7,8-tetrahydroquinoline from bases which boiled at 220° to 230°C (428° to 446°F) and were found in the chloroform layer after extraction; by Walker [40] in isolating quinoline and isoquinoline from bases boiling at 230° to 240°C, and 2,8-dimethylquinoline from bases boiling at 240° to 243°C (464° to 470°F); by Glenn [12] in isolating 2-methyl-8-ethylquinoline from bases boiling at 258° to 264°C (496° to 507°F); by Axe,[3] Glenn [11] and Schenck [31, 32, 33] in isolating about a dozen highly alkylated quinolines from the complex fractions boiling at

300° to 365°C (572° to 688°F). The method was also used in the isolation of *dl*-2-sec-butyl-4,5-dimethylpyridine by Crouch and Thomas [19] and by Barton and Roberts [21] in the isolation of 2,3-dimethyl-6-isopropylpyridine. Crystalline salts had been obtained in only a few cases after use of earlier methods of separation of alkylated pyridines.

The two bases isolated by Crouch and Thomas [19] and erroneously identified as cyclopentylpyridines by Truitt [20] had an index of refraction too low for quinolines and too high for simple alkylpyridines. They were found in the water when bases which had been in the chloroform layer were re-extracted and analyzed correctly for $C_{10}H_{13}N$, but since they were isolated in low yield, proper degradation was difficult and they were identified as cyclopentylpyridines. When Wheeler [22, 41] recrystallized the isolated bases and their derivatives and compared them with the cyclopentylpyridines which had been prepared in ample quantity, he found that they were not identical. The physical properties of the bases agree with those of methyl 5,6,7,8-tetrahydropyridines, but the correct one has not been synthesized and thus the structure of these two bases is unknown. The lowest member of this series — 5,6,7,8-tetrahydroquinoline — was first isolated by Garland [9] and its presence was later confirmed by ultraviolet absorption spectra by Pickard.[27]

The $C_{16}H_{25}N$ Base

One of the first bases isolated and apparently one of the most abundant of the petroleum bases from kerosene was one originally isolated by Thompson [38] and known for years as Thompson's base. Because of its relatively high concentration in the 275° to 280°C (526° to 536°F) range, because of the ease with which it was isolated and purified, and because of its remarkable lack of reactivity toward many reagents this base was studied intensively for a number of years.[16, 23, 38]

The results obtained from year to year by the investigators led to a number of tentative formulae by Bailey in which he attempted to reconcile reactions and properties with structure. Many of these early structures proved to be far from correct. The peculiar formulae were proposed to account for the remarkable stability and inert nature of the compounds and were due only to a minor extent to faulty interpretation of laboratory data.

Since several gallons of the base were available at one time or another, all of the characterization and degradation reactions employed in alkaloid chemistry for the determination of structure were tried. The structure was finally established through a series of degradations and shown to be 2,4-dimethyl-6-(2,2,6-trimethylcyclohexyl)pyridine :[38]

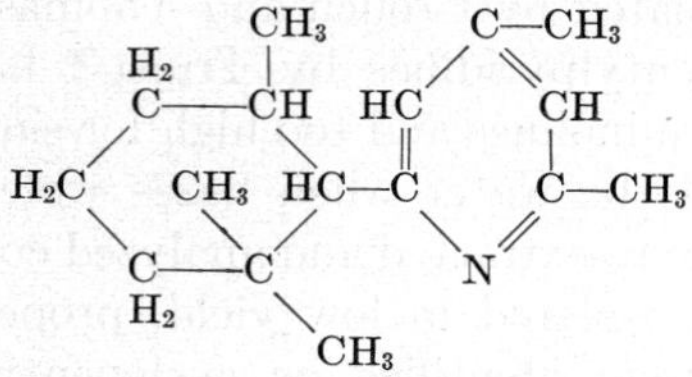

```
              CH3        C—CH3
       H2      |        /  \\
       C——————CH    HC      CH
      /         \    ||      |
  H2C      CH3  HC——C       C—CH3
      \         /       \  //
       C——————C          N
       H2      \
                CH3
```

A few years later, the structure was confirmed by synthesis at Zurich by Prelog and Geyer [30] and at present the structure seems to be well established.

Thompson, in his original work in which the empirical formula was suggested to be $C_{16}H_{25}N$, showed that the base, like all other petroleum bases so far isolated, was a tertiary amine and the index of refraction indicated a rather high C:H ratio which, in the absence of active unsaturation, could indicate the presence of a pyridine or bz-tetrahydroquinoline or similar nucleus. Quinolines have a much higher index of refraction and so were excluded from the start. One, and then two, active methyl groups were indicated by early results. Failure to react with ammonium iodide and hydroiodic acid eliminated an N-alkyl structure.

The problem proved to be a difficult one because, while its physical properties did not exclude some substituted pyridine structure, it was extremely stable to oxidizing agents under conditions under which ordinary pyridines are oxidized and reduction by sodium in ethyl or amyl alcohol did not take place while pyridines are usually reduced to piperidines. Even after vigorous oxidation led to a pyridinetricarboxylic acid which indicated that a pyridine ring was present either in a substituted pyridine or in a hydroquinoline or similar compound, unusual types of compounds were postulated to explain the behavior of the base. Alpha- and gamma-methyl groups were indicated early in the work, but these could have been substituents on a hydroquinoline as well as on a pyridine.

Lackey [16] condensed the base with two molecules of formaldehyde and oxidized the resulting product to a dicarboxylic acid — obviously

6-(2,2,6-trimethylcyclohexyl)-2,4-pyridinedicarboxylic acid—but this compound could have been formed from some of the postulated incorrect structures.

Between 1931 and 1941, when the structure was finally established, only a few days before the unexpected death of Professor Bailey who had directed research on this base almost continuously for 10 years, a number of important positive contributions had been made to the knowledge of its structure.

1. High-pressure hydrogenation,[38] first performed in the laboratory of the late Professor Homer Adkins, showed that three molecules of hydrogen were absorbed and the resulting base was dehydrogenated to yield the original base. The presence of a pyridine ring was proved almost conclusively.

2. 2,4,6-Pyridinetricarboxylic acid was shown to be formed on vigorous oxidation of the original base, indicating alkylation or fusion with another ring at these positions.

3. Formation of a phthalone established the presence of a methyl group at position 2 or 6 and confirmed previous data indicating active methyl groups at these or 2 and 4 positions.

4. Condensation with formaldehyde, followed by oxidation of the product obtained, led to a dicarboxylic acid which was decarboxylated first to a monobasic acid and finally to a new base, $C_{14}H_{21}N$, again confirming presence of active methyl groups at 2 and 4 or 6. This conclusion was confirmed by formation of a dibenzal derivative.

The complete structure was finally elucidated by degradation of the base obtained on hydrogenation of the original compound and by ozonolysis of the original base.[38] The hydrogenated base was subjected to a von Braun [6] ring-cleavage reaction, involving preparation of the benzoyl derivative of the secondary amine, bromination with phosphorus pentabromide and distillation in vacuum to yield benzonitrile and $C_{16}H_{30}Br_2$. Hydrogen bromide was eliminated from the dibromide to yield an olefin, $C_{16}H_{28}$, which was shown to be $C_9H_{17}CH{=}CHCH(CH_3)CH{=}CHCH_3$ since ozonolysis yielded acetaldehyde and an acid, $C_9H_{17}COOH$. Ozonolysis of the original base yielded the amide $C_9H_{17}CONH_2$, derived from the acid. This amide was also prepared from the acid $C_9H_{17}COOH$. Ozonolysis of a C_9H_{17} substituted pyridine could not have yielded this amide, unless the radical C_9H_{17} had been attached to a carbon next to the nitrogen atom of the pyridine ring of Thompson's base. This established the structure of this base as:

$C{-}CH_3$

$HC \quad CH$

$C_9H_{17}{-}C \quad C{-}CH_3$

N

This left only the structure of the C_9H_{17} radical to be determined. This might have been a difficult problem had not, as a good illustration of the advantages of coöperation between groups working on related subjects, the $C_9H_{17}COOH$ acid and its amide been isolated from another California petroleum. It was definitely identified as 2,2,6-trimethylcylohexanecarboxylic acid.[36] It was now merely necessary to show the identity of the two acids to prove the structure of the petroleum base as 2,4-dimethyl-6-(2,2,6-trimethylcyclohexyl)-pyridine.

In view of the very highly hindered nature of the $C_9H_{17}COOH$ acid, its amide, and its esters and the fact that this interesting pyridine has a methyl group at position 2 and the hindered group at position 6, it is not surprising to find that Thompson's base is a remarkably stable compound that fails to show many of the reactions expected of a substituted pyridine.

The molecule formed when any group is tied to the C_9H_{17} radical at position 1 should be able to exist in the following two isomeric forms:

$H \quad CH_3$ / $C{-}C$ / $H \quad H \quad R$ / $HCH \quad C$ / $H \quad CH_3 \quad H$ / $C{-}C$ / $H \quad CH_3$

cis

$H \quad CH_3$ / $C{-}C$ / $H \quad H \quad H$ / $HCH \quad C$ / $H \quad CH_3 \quad R$ / $C{-}C$ / $H \quad CH_3$

trans

In the case of the petroleum acid, both forms have been synthesized[37] and both forms have been isolated from petroleum.[25] The more stable form was present in greater concentration in petroleum and was tentatively assigned the *trans* configuration. When R is a 4,6-dimethyl-2-pyridyl group, this formula represents Thompson's base which should then also exist in two forms. The form which was easily isolated from petroleum and was plentiful was again designated as the *trans* form. The other isomer was probably partially isolated by Mahan[23] who did not know the true structure of the base and thus was unable to predict that an isomeric base might exist in petroleum. He isolated 20 g of a base hydrochloride from 28 liters of the plentiful base through repeated distribution of the hydrochloride between chloroform and dilute hydrochloric acid and fractional liberation of the base from its hydrochloride. The physical properties indicated that his base was far from pure, however.

This fact was brought out when Prelog and Geyer,[30] in 1947, synthesized and studied both forms of the base and found that the plentiful form, as submitted to them by the author, was essentially pure and showed properties agreeing satisfactorily with those found for the synthetic base. The only important difference between the properties reported by Bailey and coworkers and Prelog and Geyer was Prelog's observation that the petroleum base showed optical activity $[\alpha]_D = q - 0.47 \pm 0.02$, which had not been reported by Bailey's group. Mahan reported a rotation of −1.29 for his impure isomer. The reported properties of Thompson's base, Mahan's isomer, and of the *cis* and *trans* forms synthesized by Prelog and Geyer are listed in Table 52.

Table 52. **Properties of the $C_{16}H_{25}N$ Base and of Its Salts**

			Prelog and Geyer	
	Thompson	*Mahan*	*Trans*	*Cis*
Melting Point	23–24°C (L) 24.5 (R)	—	—	—
Boiling Point	279–280°C (R)	277°C	70–72°C at 0.05 mm	65–70°C at 0.003 mm
d_4^{20}	0.9391 (T)	0.9306	0.9382	0.9382
n_D^{20}	1.5106 (R) 1.5100 (P)	1.5114	1.5101	1.5085
M_D	73.8 (P)	—	73.9	73.9
Rotation	−0.47 (P)	−1.29	—	—
Maximum absorption	2660 Å(P)		2660 Å	2660 Å
	Melting Point, °C			
Picrate	164 (T) 163.5–164.5 (P)	162	165.0–165.5	174–176
Picrolonate	219–222		219–221	203–205 (Decomposition)
Hydrochloride	251 (T)	210 (Sublimation)		
Chloroplatinate	240 (T)	230 (Decomposition)		
Methiodide	250 (T)	265 (Sublimation)		
Base. $HgCl_2$	157.5 (T)	156		
Base. $ZnCl_2$	171 (T)	216		
Phthalone	220 (A)	205		
$KMnO_4$	Stable at 100°C for 48 hours	Easily reduced at 50°C		

(A), Armendt; (L), Lackey; (P), Prelog; (R), Roberts; (T), Thompson.

Table 53 lists the boiling-point range, final method of separation, salt used in final purification, and other information on all of the bases isolated from the Edeleanu extract of straight-run California kerosene and distillate. The boiling-point range from which the bases were isolated was 5° to 15° lower than the boiling point of the pure base. Even though the bases had been highly fractionated, it was sometimes possible to isolate the same base from fractions boiling 10° to 15°C apart. Relatively little work was done on fractions boiling at 200° to 260°C (392° to 500°F) and none at 323° to 355°C (617° to 671°F). Modern methods of separation would probably permit isolation of a number of additional bases even from the intensively studied 270° to 280°C (518° to 536°F) range.

Table 53. **Compounds Isolated frcm California Straight-Run Extracts by the Texas Group**

Boiling-Point Range, °C	Approximate % Aromatics	Part of Range Used °C	Part of Range Used °F	Final Fractionation Method	Base Isolated as	Base Identified	Reference
170–200	40	170–200	338–392	Amplified distillation	Picrate	3,5-Dimethylpyridine	24
		170–200	338–392	Amplified distillation	Picrate	2,3-Dimethylpyridine	24
		170–200	338–392	Amplified distillation	Picrate	3-Methylpyridine	24
		170–200	338–392	Amplified distillation	Picrate	2,3,5-Trimethylpyridine	24
		170–200	338–392	Amplified distillation	Picrate	3-Methyl-5-ethylpyridine	24
200–210	29	198–205	388–401	Fractional methiode formation	Methiodide	2,3-Dimethyl-6-isopropylpyridine	21
210–220	33	210–213		Countercurrent fractional neutralization			
		210–215	410–419	Fractional neutralization	Picrate	dl-2-(1-Methylpropyl)-4,5-dimethylpyridine	19
		216–221	421–430	Amplified distillation	Picrate	2,3,4-Trimethylpyridine	31
		216–221	421–430	Amplified distillation	Picrate	2,3,5-Trimethylpyridine	31
220–230	32	220–230	428–446	Fractional neutralization	Picrate	A 5,6,7,8-Tetrahydroquinoline	9
		220–230	428–446	Countercurrent fractional neutralization	Picrate	$C_{11}H_{15}N$ and $C_9H_{11}N$ bases	9
230–240	14	230–232	446–450	Fractional neutralization	Picrate	A 5,6,7,8-Tetrahydroquinoline	27
		232–234	450 453	Fractional neutralization	Picrate	Quinoline	7
		230–240	446–471	Fractional sulfiting	Picrate	Quinoline	41
		230–240	446–471	Countercurrent fractional neutralization	Sulfate	2-Methylquinoline	41
240–250	11	243–244	469–471	Fractional neutralization	Picrate	2,8-Dimethylquinoline	41
250–260	14	253	487	Base. $ZnCl_2$ salt	$ZnCl_2$ salt	2,8-Dimethylquinoline	18
		258–264	496–507	Countercurrent fractional neutralization	Picrate	2-Methyl-8-ethylquinoline	12

Table 53 (Continued)

Boiling-Point Range, °C	*Approximate % Aromatics*	*Part of Range Used* °C	°F	*Final Fractionation Method*	*Base Isolated as*	*Base Identified*	*Reference*
260–270	11	265	509	Fractional sulfiting	Picrate	2,3- and 2,4-Dimethylquinolines	5
		265–269	509–516	Chloroform-water extraction	Picrate	2,3-Dimethylquinoline	26
270–280	16	271–273	520–524	Fractional sulfiting	Nitrate	2,4,8-Trimethylquinoline	8
		271–273	520–524	Fractional sulfiting	Hydrochloride	2,3,8-Trimethylquinoline	8
		273	524	Chloroform-water extraction	Hydrochloride	2,3,8-Trimethylquinoline	26
		275	527	Distillation alone	Sulfate	2,3,8-Trimethylquinoline	26
		275	527	Distillation alone	Picrate	2,3,8-Trimethylquinoline	15
		275	527	Chloroform-water extraction	Picrate	2,4,8-Trimethylquinoline	26
		275	527	Chloroform-water extraction	Picrate	2,4-Dimethyl-6-(2,2,6-trimethylcyclohexyl) pyridine	23, 26, 38, 39
280–290	30	285	445	Amplified distillation	Picrate	2,3-Dimethyl-8-ethylquinoline	14
290–300	30	290–291	554–556	Chloroform-water extraction	$ZnCl_2$ salt	2,4-Dimethyl-8-ethylquinoline	3
		292–293	557–559	Countercurrent fractional neutralization	Picrate	2,4-Dimethyl-8-propylquinoline	4
		295–300	563–572	Distillation alone	Nitrate	2,3-Dimethyl-8-propylquinoline	2
		295–300	563–572	Distillation alone	Solid base	2,3,4,8-tetramethylquinoline	2
300–310	30	300		Fractional sulfiting			
		300		Fractional sulfiting			
		308–315	586–599	Countercurrent fractional neutralization	Picrate	2,3,4-Trimethyl-8-ethylquinoline	10
		308–315	586–599	Countercurrent fractional neutralization	Picrate	2,3,4,8-Tetramethylquinoline	10
		308–315	586–599	Countercurrent fractional neutralization	Picrate	2,3,8-Trimethyl-4-ethylquinoline	11
310–320	27	317	602	Countercurrent fractional neutralization	Nitrate	2,3,4-Trimethyl-8-propylquinoline	32

Table 53 (Continued)

Boiling-Point Range, °C	*Approximate % Aromatics*	*Part of Range Used*		*Final Fractionation Method*	*Base Isolated as*	*Base Identified*	*Reference*
		°C	*°F*				
		319	606	Fractional sulfiting	Nitrate	2,3-Dimethyl-4-ethyl-8-propylquinoline	35
		319	606	Countercurrent fractional neutralization	Picrate	2,4-Dimethyl-8-*sec*-butylquinoline	33
		319	606	Countercurrent fractional neutralization	Nitrate	2,3-Dimethyl-4,8-diethylquinoline	35
		319	606	Countercurrent fractional neutralization	Solid base	2,3,4-Trimethyl-8-isopropylquinoline	34
320–330	18	323	613	Countercurrent fractional neutralization	Picrate	2,3,4-Trimethyl-8-ethylquinoline	32
330–340	12			Not studied			
340–350	20			Not studied			
Above 350	20	355–365	671–688	Countercurrent fractional neutralization	Nitrate	2,3-Dimethyl-benzo-h-quinoline	36
		355–365	671–688	Countercurrent fractional neutralization	Sulfate	2,4-Dimethyl-benzo-h-quinoline	36

Bibliography

1. Armendt, B. F., and J. R. Bailey, *J. Am. Chem. Soc.* **55,** 4145 (1933).
2. Axe, N., and J. R. Bailey, *Ibid.* **60,** 3028 (1938).
3. Axe, N., *Ibid.* **61,** 1017 (1939).
4. Axe, N., and J. R. Bailey, *Ibid.* **61,** 2609 (1939).
5. Biggs, B., and J. R. Bailey, *Ibid.* **55,** 4141 (1933).
6. von Braun, J., *Ber.* **37,** 3210 (1904); **38,** 2339 (1905).
7. Crouch, W. W., Ph.D. Thesis, The University of Texas, 1943.
8. Edens, C. O., M.A. Thesis, The University of Texas, 1939.
9. Garland, F. M., Ph.D. Thesis, The University of Texas, 1939.
10. Glenn, R. A., and J. R. Bailey, *J. Am. Chem. Soc.* **61,** 2612 (1939).
11. Glenn, R. A., and J. R. Bailey, *Ibid.* **63,** 637 (1941).
12. Glenn, R. A., and J. R. Bailey, *Ibid.* **63,** 639 (1941).
13. Jantzen, E., *Das fraktionierte Destillieren und das fraktionierte Verteilen als Methoden zur Trennung von Stoffgemischen,* Dechema Monograph No. 5, Verlag Chemie, Berlin, 1932.
14. Key, C., and J. R. Bailey, *J. Am. Chem. Soc.* **60,** 763 (1938).
15. King, W. A., and J. R. Bailey, *Ibid.* **52,** 1245 (1934).
16. Lackey, R. W., and J. R. Bailey, *Ibid.* **56,** 2741 (1934).
17. Lake, G., M.A. Thesis, The University of Texas, 1933.
18. Lake, G., and J. R. Bailey, *J. Am. Chem. Soc.* **55,** 4143 (1933).
19. Lochte, H. L., W. W. Crouch, and D. Thomas, *Ibid.* **64,** 2753 (1942).
20. Lochte, H. L., D. Thomas, and P. Truitt, *Ibid.* **66,** 550 (1944).
21. Lochte, H. L., A. D. Barton, S. M. Roberts, and J. R. Bailey, *Ibid.* **72,** 3007 (1950).
22. Lochte, H. L., P. F. Kruse, and E. N. Wheeler, *Ibid.* **75,** 4477 (1953).
23. Mahan, R. I., Ph.D. Thesis, The University of Texas, 1938.
24. Meadows, J. L., Ph.D. Thesis, The University of Texas, 1937.
25. Ney, W. O., W. W. Crouch, C. E. Rannefeld, and H. L. Lochte, *J. Am. Chem. Soc.* **65,** 770 (1943).
26. Perrin, T. S., and J. R. Bailey, *Ibid.* **55,** 4136 (1933).
27. Pickard, P. L., and H. L. Lochte, *Ibid.* **69,** 16 (1947).
28. Poth, E. J., W. D. Armstrong, C. C. Cogburn, and J. R. Bailey, *Ind. Eng. Chem.* **20,** 83 (1928).
29. Poth, E. J., W. A. King, W. C. Thompson, W. M. Slagle, W. W. Floyd, and J. R. Bailey, *J. Am. Chem. Soc.* **52,** 1243 (1930).
30. Prelog, V., and U. Geyer, *Helv. Chim. Acta.* **29,** 1587 (1946).
31. Rettig, W. A., M.A. Thesis, The University of Texas, 1939.

32. Schenck, L. M., and J. R. Bailey, *J. Am. Chem. Soc.* **61,** 2613 (1939).
33. Schenck, L. M., and J. R. Bailey, *Ibid.* **62,** 1967 (1940).
34. Schenck, L. M., and J. B. Bailey, *Ibid.* **63,** 1364 (1941).
35. Schenck, L. M., and J. R. Bailey, *Ibid.* **63,** 1365 (1941).
36. Schenck, L. M., and J. R. Bailey, *Ibid.* **63,** 2331 (1941).
37. Shive, W., J. Horeczy, G. Wash, and H. L. Lochte, *Ibid.* **64,** 385 (1942).
38. Shive, W., S. M. Roberts, R. I. Mahan, and J. R. Bailey, *Ibid.* **64,** 909 (1942).
39. Thompson, W. C., and J. R. Bailey, *Ibid.* **53,** 1002 (1931).
40. Tiedcke, K., Ph.D. Thesis, Hamburg, 1928.
41. Walker, C. A., M.A. Thesis, The University of Texas, 1940.
42. Wheeler, E. N., Ph.D. Thesis, The University of Texas, 1952.

CHAPTER TWENTY-SIX

CRACKING-PROCESS BASES

All of the results reported in previous chapters were obtained in the study of bases isolated directly from crude oil or from refinery products made by simple distillation. The study of bases isolated from products obtained during thermal cracking-process operations should provide data needed in the development of theories on the origin of petroleum bases. Additional studies, including cracking-process material boiling in the kerosene range, may also indicate that if petroleum pyridines should become commercially important, they will be isolated from cracking-process rather than straight-run material. No effort appears to have been made to study bases which may be formed during various catalytic cracking processes.

Bratton and Bailey [2] undertook a study of bases extracted from gasoline produced by cracking gas oil remaining after the removal of gasoline and kerosene from California crude oil. Since the crude oil was similar to that from which the kerosene bases studied by the Texas group had been obtained, a comparison of the bases isolated from this gasoline and from straight-run kerosene is particularly interesting.

The gas oil was cracked at 455°C (842°F) under a pressure of 800 to 850 pounds per square inch to yield a gasoline with a boiling range of 38° to 176°C (100° to 349°F). Extraction of various such gasoline fractions showed 0.016 to 0.079% by volume of basic nitrogen compounds of which 154 ml were isolated from 4 barrels of gasoline and sent to Professor Bailey by the Union Oil Company.

Bratton found that the bases, as received, showed an $n_D^{25} = 1.4881$ and a boiling range of 93° to 204°C (199° to 399°F). Simple distillation of the sample showed that 14.5 ml boiled below 115°C,

61 ml at 115° to 185°C (239° to 365°F), and 79 ml at 185° to 235°C (365° to 455°F). He tested all fractions for primary, secondary, and tertiary amine nitrogen, but could detect only tertiary amines as in previous work with straight-run products. He then fractionated the whole batch carefully after adding 20 volumes of special refined hydrocarbon mixture boiling uniformly at 90° to 240°C (194° to 464°F), collecting constant volume cuts. Each fraction was then converted to the hydrochloride and extracted with chloroform in the usual manner. Each of the resulting fractions was converted to the picrate and purified by recrystallization if possible. In cases in which the picrate could not be purified in this manner, the fractions were converted to the mercuric chloride double salt and purified as such. Initial crystallization of the picrates was usually carried out by controlled stepwise cooling of the solution to obtain a series of crops of crystals which were treated as separate fractions. This fractional cooling was carried to the point at which previous tests had shown that picric acid begins to crystallize. This procedure, without such refinements as avoidance of contamination by picric acid, had been used previously by Ganguli and Guha [4] in working with coal-tar bases.

Bratton made a special effort to isolate all individual bases present, but obtained smeary precipitates in some cases and may have lost some compounds during recrystallization. A material balance is very difficult to obtain in cases like this, since pure bases are never isolated as such until after solid salts have been obtained and purified by crystallization often with considerable loss. In view of this, coupled with the fact that the total volume of his mixture of bases was only 154 ml, his claim that no bases, except the ones isolated, were present was probably a rash one and the differences in results obtained by Bratton and by Hackman and Wibaut from California cracking-process gasoline may not indicate any essential difference between the two samples of bases.

The data obtained by Bratton and Bailey are shown in Table 54.

In several cases, two different salts were isolated for the same base. Filtrates and residues were reprocessed, but no new bases could be isolated.

Table 54. **Bases Isolated from Cracking-Process Gasoline by Bratton and Bailey**

Base	Pure Base Boiling Point °C (1)	Isolated from Fractions Boiling at °C	Isolated as	Melting Point of Salt, °C	Literature Bratton and Bailey
2-Methylpyridine	129	98–118	Picrate	165 [3]	165.5
				164 [1]	
2,6-Dimethylpyridine	144	128–135	Picrate	162 [3]	163
				161 [1]	
4-Methylpyridine	145	119–129	Picrate	167 [1,3]	167
2,5-Dimethylpyridine	157	142–150	$HgCl_2$ salt	197 [4]	168–169
				164 [1]	
2,4-Dimethylpyridine	158	137–140	Picrate	182 [3]	182–183
				179 [1]	
		142–156	$HgCl_2$ salt	131–133 [3]	131–132
2,4,6-Trimethylpyridine	170	155–167	Picrate	155–156 [1]	155–156
3,5-Dimethylpyridine	172	142–173	Picrate	244 [3]	245
				228–230 [1]	
Quinoline	238	199–220	Picrate	203 [1]	203
2-Methylquinoline	247	225–233	Picrate	191 [1]	193–194

In what appears to have been the second petroleum-base project supplied with an adequate volume of bases, Hackman and Wibaut [5] reported the isolation of thirty-one bases from a large supply of bases isolated from Dubbs cracking-process gasoline by the Shell Oil Company at their Shell Point refinery in California as shown in Flowsheet 3.

On the basis of the separations obtained in the pilot run they then worked up 100 kg of Dubbs "recovered oil" by analogous steps. They used a pilot-plant still for preliminary fractionation and finally an efficient laboratory column, repeating distillations until plateaus formed. If no plateau formed and no crystalline salt could be obtained, they employed Jantzen type [6] fractional liberation of bases from their hydrochlorides, using chloroform as solvent and liberating in ten stages. Most of the distillation and liberation fractions now yielded crystalline picrates or picrolonates. They found the picrolonates most convenient as far as laboratory operations were concerned, but could not employ them extensively in preliminary identifications because few picrolonates have been reported previously. Table 55 lists data reported by Hackman and Wibaut.

Table 55. **Bases Isolated from Cracking-Process Gasoline by Hackman and Wibaut**

Boiling Point of Fraction, °C	*at mm*	*Base Identified*	*Melting Point of Picrate °C*	*Melting Point of Picrolonate °C*	*Remarks*
129 (264°F)	760	2-Methylpyridine	168	214–215	
143–144	760	3-Methylpyridine	153	205–206	
(289–291°F)		2,6-Dimethylpyridine	164	206	
		4-Methylpyridine	167	262–263	
147 (297°F)	760	C_7H_9N	114–115	172	
155–156	760	2,4-Dimethylpyridine	184–185	207–209	
(311–313°F)		2,5-Dimethylpyridine	170	171–172	
157 (315°F)	760	2,3-Dimethylpyridine	186	223–225	
		$C_8H_{11}N$	121	183–184	
160 (320°F)	760	2-Methyl-6-ethylpyridine	130	165–166	
165 (329°F)	760	4-Ethylpyridine	168–169	241–242	
		2,4,6-Trimethylpyridine	156	229–230	
		3,5-Dimethylpyridine	244–245	239–240	
		$C_8H_{11}N$	143–145	Too soluble	
170–172	760	2-Methyl-5-ethylpyridine	168–169	153–154	(a) 1
(338–342°F)		2-Methyl-4-ethylpyridine	139–140	196	(a) 2
		2-Ethyl-4-methylpyridine	122–123	172–173	(a) 3
		3,4-Dimethylpyridine	163	227–228	(a) 5
		2,3,6-Trimethylpyridine	148–149	228–229	(a) 6
188 (370°F)	760				
95	30	$C_{10}H_{15}N$	——	176–178	
235 (455°F)	760				
126	30	Quinoline	201–203	219–220	
240 (464°F)	760				
126	30	Isoquinoline	224–225	250–252	
247 (476°F)	760				
136	30	2-Methylquinoline	194–195	235–236	
135–137	30	8-Methylquinoline	198–200	——	(b) 1
		8-Methylquinoline	——	——	(b) 2
		Pyridine bases	——	——	(b) 3
137–140	30	2,8-Dimethylquinoline	176–179	——	(c) 1 and 2
		7-Methylquinoline	238–239	——	(c) 3 and 4
		1-Methylisoquinoline	231–233	——	(c) 5 and 6
		3-Methylisoquinoline	197–198	——	(c) 7 and 8
		Mixed pyridines	——	——	(c) 9 and 10
142–145	30	2,8-Dimethylquinoline	——	——	(d) 3, 4, and 5 out of 36 cuts
		3-Methylquinoline	187–190	——	(d) 13–18
		4-Methylquinoline	215	——	(d) 23–25
		$C_{11}H_{15}N$	145–147	——	(d) 34–36

(a, b, c, d) were fractional liberation series and the numbers refer to cut numbers, 1 being the weakest of the series.

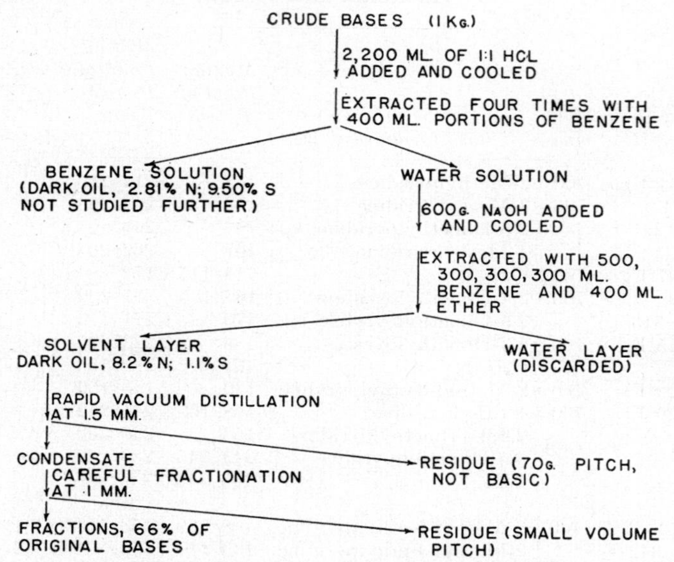

Flowsheet 3. PROCEDURE USED BY HACKMAN AND WIBAULT IN THE SEPARATION OF BASES FROM CRACKING-PROCESS GASOLINE

The very important work of Hackman and Wibaut widens the field of petroleum bases in a number of respects:

1. They isolated isoquinoline and 1- and 3-methylisoquinoline which had not been isolated before even though they are within the range of the kerosene bases.

2. They isolated 3-, 4-, and 7-methylquinolines. Previously there was a methyl group at position 2 in every quinoline isolated.

3. When they isolated 7-methylquinoline they showed the tentative rule that no substituents are found at position 5, 6, or 7 not to be valid for cracking-process bases, at least. Although cracking-process bases need not be the same compounds as those found in straight-run products, one would not expect to find extensive changes in the positions of alkyls or conversion of quinolines to isoquinolines to take place during cracking reactions. Since only a very small fraction of the compounds present in the boiling-point range of 230° to 260°C (446° to 500°F) has been isolated, future work may well show that most, if not all, of the compounds from this range isolated

by Hackman and Wibaut will also be found in straight-run kerosene bases.

Hackman and Wibaut, like the Texas group, encountered difficulties in identifying some of the pyridines isolated, because convenient and unequivocal methods of synthesis for these alkyl pyridines are not yet available and degradation methods may not be conclusive. They made use of one important modification in degradation technique when they employed electrolytic reduction of the pyridines instead of catalytic hydrogenation which does not always proceed smoothly in the presence of impurities like sulfur compounds.

A comparison of the list of compounds isolated from cracking-process gasoline with those from straight-run kerosene shows that, in the region in which they overlap, they are almost identical, and a comparison, in turn, of these bases with those isolated by Eguchi from shale oil again shows a striking similarity. Since the bases isolated by Bratton and Bailey were obtained in cracking gas oil boiling at 210° to 450°C (410° to 842°F), it seems unlikely that any of the low-boiling pyridines found could have been present, as such, in the gas oil stock and they, therefore, must have been formed during pyrolysis.

Bibliography

1. Beilstein, *Handbuch der organischen Chemie, 4th Ed.*, Volume 20, 1935.
2. Bratton, A. C., and J. R. Bailey, *J. Am. Chem. Soc.* **59,** 175 (1937).
3. Eguchi, T., *Bull. Chem. Soc. Japan* **2,** 176 (1927); **3,** 227 (1928); *Chem. Zentr.* **98,** 1223 (1927); **100,** 331 (1929).
4. Ganguli, S. K., and P. G. Guha, *J. Indian Chem. Soc.* **11,** 197 (1934).
5. Hackman, J. T., and J. P. Wibaut, *Recueil* **62,** 229 (1943).
6. Jantzen, E., *Das fraktionierte Destillieren und das fraktionierte Verteilen als Methoden zur Trennung von Stoffgemischen,* Dechema Mongraph No. 5, Verlag Chemie, Berlin, 1932.

CHAPTER TWENTY-SEVEN

THE ORIGIN OF PETROLEUM NITROGEN COMPOUNDS

Any theory on the origin of petroleum nitrogen compounds must be consistent with theories on the origin of petroleum itself. Workers generally hold that petroleum and its acids and nitrogen compounds are derived from animal and vegetable substances. The rather highly organized nature of some of the nitrogen compounds which have been isolated would present considerable difficulties for any purely inorganic theory, particularly when it is found that some of the compounds are optically active. Assuming then that the nitrogen compounds are derived from animal or vegetable substances, perhaps proteins, we must assume that the few bases which have been isolated and identified must have been formed from much more complex substances, since none of the bases identified appears to occur as such in nature When and where this conversion of complex high-molecular-weight substances to simple alkylated pyridines and quinolines took place cannot be decided, but we shall see that the basic compounds were formed underground.

Since the original complex substances were probably solids and insoluble in oil, any theory must assume that a considerable amount of degradation must have taken place between the time when the organic matter had been laid down and the time the oil was produced to convert the original material to liquid or at least oil-soluble substances. Whether this took place at only slightly elevated temperatures in the presence of microorganisms, or under pressure at higher temperatures of perhaps 200° to 300°C, or both degradations took place at different times can apparently not be decided.

There has been a considerable difference of opinion in regard to the extent to which this subterranean degradation proceeded. Bailey [8] was firmly convinced that both the simple petroleum bases and the acids now found were man-made. He held that pyridines and quinolines, which have been identified, were present in the form of much more complex substances in crude oil and were converted to the simple bases during refinery and laboratory operations. Bailey based this theory on the fact that only very small amounts of basic compounds could be extracted from California crude oils by means of dilute mineral acids.[48] In view of the high viscosity of the crude oil and the consequent difficulty in making good contact between the aqueous acid and the bases dissolved in the oil, it was felt that failure in extracting bases from crude oil might be explained on this basis. To prove or disprove this theory, quinoline was added to crude oil and readily extracted with dilute acids and Bailey concluded that bases were present only in very low concentration in crude oil.

This theory overlooked the fact that practically all workers prior to 1900 had obtained their supplies of petroleum bases by extraction of crude oil rather than of refinery fractions. Chlopin [14] must have had difficulty in extracting bases from crude oil because he shook his crude oil with dilute sulfuric acid for 12 and for 22 days to isolate the bases he studied.

The fact that certain gas wells produce small amounts of basic compounds which are tertiary in nature, boil above 200°C (393°F), and have the characteristic odor of petroleum bases and the fact that recent work in which the basic compounds in nonaqueous solution are titrated directly in crude oil as well as in its fractions [15,17,53,54] show that the 30 to 40% of basic compounds found by workers in this field were present as such, at least in large part, in crude oil. This still accounts for only some 30 to 40% of the nitrogen in petroleum and we know nothing about the nonbasic portion except that pyrroles, porphyrins, and a few other types of nonbasic compounds can be shown to be present.

Richter, Caesar, Meisel, and Offenhauer [53] recently studied the problem of the origin of the basic compounds using the new nonaqueous titration methods. Since Bailey's definition of basic nitrogen compounds as those which can be extracted by means of dilute mineral acids is difficult and uncertain in practice, they defined them as "those compounds which can be titrated by perchloric acid in a 50:50

solution of glacial acetic acid and benzene." Under this definition, the basic compounds include pyridines, quinolines, isoquinolines, acridine, anthranilic acid, and 2,4-dimethylpyrrole. The nonbasic compounds include such types as pyrroles, indole, carbazole, diphenylamine and 2,5-dimethylpyrrole.

They found that the ratio of basic to total nitrogen in crude oil or any fraction of it was not affected by simple heating to distillation temperatures. For example, the ratio in a sample of distillate was not affected by heating it to 315°C (600°F) for 24 hours. They found further that the ratio remained essentially constant when a Wilmington crude oil was distilled and when the residue was further fractionated through a molecular still. Therefore, they decided that the petroleum bases are present as such in crude oil.

Since we know very little about the 50 to 75% of nitrogen in petroleum which is nonbasic, these give us few important clues to the origin of nitrogen compounds in petroleum. Treibs [66, 67] showed that porphyrins are present in some fractions, and a considerable amount of work has been done since on these compounds.[20, 61] Their presence and the occasional discovery of fractions showing slight optical activity are, of course, excellent evidence in favor of the view that the nitrogen compounds are of animal or vegetable origin, but further than this they give us little information.

Mapstone [39] reviewed results which had been obtained in experimental work on pyrolysis of porphyrians and of proteins. He concluded that little is definitely known about the fate of porphyrins. Some seem to be very stable at high temperatures and have such a low vapor pressure that they should remain in the still pot in ordinary distillations. Pyrolysis would be expected to yield pyrroles and thus lead to part of the lower-boiling nonbasic nitrogen found.

The pyrolysis of proteins has been studied by a number of different workers. Ammonia seems to be one of the products always obtained in pyrolysis, while pyrrole is another commonly observed product. Sibata and Sioya [60] distilled sewage sludge and found that 7.3% by volume of the oil obtained consisted of basic compounds which, according to their statement, contained 34% quinoline, 18% collidine, and 5% pyridine. Collidine may represent a mixture of methyl and ethylpyridines and the exact concentration mentioned may not be significant, but the interesting part of their experiment is that in this sludge proteins should have been partially digested by bacteria under conditions somewhat similar to those under which proteins may have been changed underground in the formation of petroleum.

Parker, Gutzeit, and Bailey [44, 45] distilled 23 tons of cotton-seed meal in the presence of nitrogen-free lubricating oil to determine whether the protein of cotton-seed meal yields the same bases that had been isolated from petroleum by Bailey's group. They obtained 337 lb of crude bases of which about 50% were soluble in petroleum ether. Some of the compounds contained more than one atom of nitrogen and much of the material polymerized in storage or during distillation and other operations. They were able to identify most of the simple pyridines and some of the quinolines which had been isolated from petroleum, but were not able to isolate the very stable substituted pyridine, 2,4-dimethyl-6-(2,2,6-trimethylcyclohexyl) pyridine, which had been found in high concentration in California and which is easily isolated from petroleum by methods with which they were familiar. In this respect, as well as in the high loss due to gum formation and the presence of compounds containing more than one nitrogen atom per molecule, these nitrogen compounds more closely resembled material obtained in low-temperature cooking of coal or shale-oil bases than petroleum bases. In general, pyrolysis of proteins has not been very helpful in formulating theories on the origin of petroleum nitrogen compounds.

Table 56. **Bases Isolated from Petroleum, Shale Oil, and Coal Tar**

	Petroleum			
Pyridine	*Straight Run*	*Cracking Process*	*Shale Oil*	*Coal Tar*
2-Methylpyridine		12,28	9, 19, 25, 23, 31	65, 29
3-Methylpyridine		28	10, 19, 23	29, 42
4-Methylpyridine		12, 28	10, 23	29, 42, 65
2,3-Dimethylpyridine	40	28	10, 23, 25	
2,4-Dimethylpyridine		12, 28	10, 19, 23, 25	3, 29, 43
2,5-Dimethylpyridine		12, 28	19, 23, 25	3, 43
2,6-Dimethylpyridine		12, 28	10, 19, 23, 25	21, 29, 43, 65
3,4-Dimethylpyridine		28	10, 23	2
3,5-Dimethylpyridine	40	12	10, 23	3, 43
4-Ethylpyridine				16
2,3,5-Trimethylpyridine	40, 52		10, 23	43
2,3,6-Trimethylpyridine		28	10, 23	21, 43
2,4,5-Trimethylpyridine			23	2, 43
2,4,6-Trimethylpyridine		12, 28	9, 10, 13, 23, 25, 31, 63	21, 42
2-Ethyl-4-methylpyridine		28	10, 23	

Table 56 (Continued)

Pyridine	*Petroleum Straight Run*	*Petroleum Cracking Process*	*Shale Oil*	*Coal Tar*
2-Methyl-4-ethylpyridine		28		
2-Methyl-5-ethylpyridine	40	28	63	
2-Methyl-6-ethylpyridine		28	10, 23	
3-Ethyl-5-methylpyridine	47			
2,3,4,5-Tetramethylpyridine				1
2,3,4,6-Tetramethylpyridine			10, 23	
2,6-Dimethyl-4-ethylpyridine			23	
dl-2-*sec*-Butyl-4,4-Dimethylpyridine	36			
2,3-Dimethyl-6-isopropylpyridine	37			
2,4-Dimethyl-6-(2,2,6-trimethylcyclohexyl) pyridine	4, 38, 50, 59, 64			
Quinoline	68	12, 18, 28	9, 51	
Isoquinoline		18, 28	41	30
2-Methylquinoline		12, 28	9, 41, 51	32
3-Methylquinoline		28		
4-Methylquinoline		28		
7-Methylquinoline		28	41	
8-Methylquinoline		28	41	
1-Methylisoquinoline		28	41	
3-Methylisoquinoline		28	41	
2,3-Dimethylquinoline	11, 22, 46			
2,4-Dimethylquinoline	11, 22, 46			
2,8-Dimethylquinoline	46, 35	28	41	
5,8-Dimethylquinoline				24
2,4,8-Trimethylquinoline	22, 34			
2-Methyl-8-ethylquinoline	27			
2,3,8-Trimethylquinoline	49, 64	18	41	
2,3-Dimethyl-8-ethylquinoline	7, 22, 33			
2,4-Dimethyl-8-ethylquinoline	6			
2,4-Dimethyl-8-*sec*-butylquinoline	56			
2,4-Dimethyl-8-propylquinoline	7			
2,3-Dimethyl-8-propylquinoline	5, 7			
2,3,4,8-Tetramethylquinoline	5			
2,3,4-Trimethyl-8-propylquinoline	55			
2,3,4-Trimethyl-8-ethylquinoline	26, 56			
2,3,8-Trimethyl-4-ethylquinoline	27			
2,4,8-Trimethylquinoline	46			
2,3,4,8-Tetramethylquinoline	5			
2,3-Dimethyl-4,8-diethylquinoline	57			
2,3-Dimethyl-4-ethyl-8-propylquinoline	57			
2,3,4-Trimethyl-8-isopropylquinoline	57			
2,3-Dimethyl-7,8-benzoquinoline	58			
2,4-Dimethyl-7,8-benzoquinoline	58			
Pyrindane				23

Table 56 lists practically all of the basic compounds which have been isolated from straight-run refinery products, from thermal cracking-process fractions, and from shale oil. Some coal-tar bases which have been isolated from one of the other types of products are also included, but no attempt has been made to prepare a complete list of these bases. No recent list of coal-tar bases seems to have been compiled and the list of Spielman [62] is rather old.

General Conclusions

It is not yet possible to draw definite conclusions on the origin of petroleum bases and it may never be possible to do so since several different processes may be involved at different times or places, but we can see what must have taken place from the time the vegetable or animal substances were deposited and the time the oil was produced.

The organic matter in the course of ages of contact with moisture, part of the time in the presence of microorganisms and probably part of the time in contact with warm or hot inorganic solutions or solids was gradually degraded from solid to an oil-soluble liquid. This change yielded 30 to 50% of the nitrogen compounds that could be titrated in nonaqueous solutions while the remainder could be extracted with liquid sulfur dioxide or with concentrated sulfuric acid. Thermal cracking produces additional basic compounds probably partly from higher-boiling bases and partly from nonbasic compounds. In general, the bases produced during thermal cracking are similar to those isolated from shale oil. So little is known about the nitrogen compounds in both petroleum and shale oil that no conclusions should be drawn at this time. Possibly the nonbasic compounds in both petroleum and shale oil will give us more information on the origin of petroleum nitrogen compounds and permit formulation of more definite theories.

Bibliography

1. Ahrens, F. B., *Ber.* **28,** 795 (1895).
2. Ahrens, F. B., *Ibid.* **29,** 2996 (1896).
3. Ahrens, F. B., and R. Gorkow, *Ibid.* **37,** 2062 (1904).
4. Armendt, B. F., and J. R. Bailey, *J. Am. Chem. Soc.* **55,** 4145 (1933).
5. Axe, N., and J. R. Bailey, *Ibid.* **60,** 3028 (1938).
6. Axe, N., *Ibid.* **61,** 1017 (1939).

7. Axe, N., and J. R. Bailey, *Ibid.* **61,** 2609 (1939).
8. Bailey, J. R., Lectures and publications in the period 1933–1938.
9. Ball, J. S., G. U. Dinneen, J. R. Smith, C. W. Bailey, and R. van Meter, *Ind. Eng. Chem.* **41,** 581 (1949).
10. Benzie, R. J., J. N. Milne, and H. B. Nisbet, Paper read at Sceond Oil Shale and Cannel Coal Conf., Glasgow, 1950.
11. Biggs, B., and J. R. Bailey, *J. Am. Chem. Soc.* **55,** 4141 (1933).
12. Bratton, A. C., and J. R. Bailey, *Ibid.* **59,** 175 (1937).
13. Cane, R. F., *Papers and Proc. Royal Soc. Tasmania,* **55,** 1941, from Mapstone, Reference 39.
14. Chlopin, G. W., *Ber.* **33,** 2837 (1900).
15. Deal, V. Z., F. I. Weiss, and T. T. White, *Anal. Chem.* **25,** 426 (1953).
16. DeConinck, O., *Bull. soc. chim. France* **41,** 249 (1884).
17. Dinneen, G. U., and W. D. Bickel, *Ind. Eng. Chem.* **43,** 1604 (1951).
18. Dinwiddie, J. A., Ph.D. Thesis, The University of Texas, 1935.
19. Dodonov, I., and E. Soshestvenoskaia, *Ber.* **62,** 1348 (1929).
20. Dunning, H. N., J. W. Moore, and M. O. Denekas, *Ind. Eng. Chem.* **45,** 1759 (1953).
21. Eckart, A., and S. Loria, *Monatsh.* **38,** 225 (1917).
22. Edens, C. O., M.A. Thesis, The University of Texas, 1939.
23. Eguchi, T., *Bull. Chem. Soc. Japan* **2,** 176 (1927); **3,** 227 (1928).
24. Ganguli, S. K., and P. G. Guha, *J. Indian Chem. Soc.* **11,** 197 (1934).
25. Garrett, F. C., and J. A. Smythe, *J. Chem. Soc.* **81,** 449 (1902).
26. Glenn, R. A., and J. R. Bailey, *J. Am. Chem. Soc.* **61,** 2612 (1939).
27. Glenn, R. A., and J. R. Bailey, *Ibid.* **63,** 637 (1941).
28. Hackman, J. T., and J. P. Wibaut, *Recueil* **62,** 229 (1943).
29. Heap, J. G., W. J. Jones, and J. B. Speakman, *J. Am. Chem. Soc.* **43,** 1936 (1921).
30. Hoogewerff, S., and W. A. van Dorp, *Recueil* **4,** 125 (1885).
31. Horne, J. W., W. F. Finley, and C. B. Hopkins, *U. S. Bur. Mines, Bull.* **415,** Chapter 9, 1938.
32. Jacobsen, E., and C. L. Reiner, *Ber.* **16,** 1084 (1883).
33. Key, C. L., and J. R. Bailey, *J. Am. Chem. Soc.* **60,** 763 (1938).
34. King, W. A., and J. R. Bailey, *Ibid.* **52,** 1245 (1930).
35. Lake, G., and J. R. Bailey, *Ibid.* **55,** 4143 (1933).
36. Lochte, H. L., W. Crouch, and D. Thomas, *Ibid.* **64,** 2753 (1942).
37. Lochte, H. L., A. D. Barton, S. M. Roberts, and J. R. Bailey, *Ibid.* **72,** 3007 (1950).
38. Mahan, R. I., Ph.D. Thesis, The University of Texas, 1938.
39. Mapstone, G. E., *J. Proc. Roy. Soc. N. S. Wales* **82,** 91 (1948).
40. Meadows, J. L., Ph.D. Thesis, The University of Texas, 1937.

41. Meyer, H. W. H., Ph.D. Thesis, The University of Texas, 1953.
42. Mohler, J., *Ber.* **21,** 1006 (1888).
43. Oparina, M. P. *Ibid.* **64,** 562 (1931).
44. Parker, I., C. L. Gutzeit, A. C. Bratton, and J. R. Bailey, *J. Am. Chem. Soc.* **58,** 1097 (1936).
45. Parker, I., M.A. Thesis, The University of Texas, 1931.
46. Perrin, T. S., and J. R. Bailey, *J. Am. Chem. Soc.* **55,** 4137 (1933).
47. Pickard, P., and H. L. Lochte, *Ibid.* **69,** 16 (1947).
48. Poth, E. J., W. D. Armstrong, C. C. Cogburn, and J. R. Bailey, *Ind. Eng. Chem.* **20,** 83 (1928).
49. Poth, E. J., W. A. Schulze, W. A. King, W. C. Thompson, W. M. Slagle, W. W. Floyd, and J. R. Bailey, *Ibid.* **52,** 1239 (1930).
50. Prelog, V., and U. Geyer, *Helv. Chim. Acta.* **29,** 1587 (1946).
51. *Report of the Secretary of Interior on Synthetic Fuels Act* from January to December 1946, pages 58–59.
52. Rettig, W. A., M.A. Thesis, The University of Texas, 1939.
53. Richter, F. P., P. D. Caesar, D. L. Meisel, and R. D. Offenhauer, *Ind. Eng. Chem.* **44,** 2601 (1952).
54. Sauer, R. W., F. W. Melpolder, and R. A. Brown, *Ibid.* **44,** 2606 (1952).
55. Schenck, L., and J. R. Bailey, *J. Am. Chem. Soc.* **61,** 2613 (1939).
56. Schenck, L., and J. R. Bailey, *Ibid.* **62,** 1968 (1940).
57. Schenck, L., and J. R. Bailey, *Ibid.* **63,** 1364 (1941).
58. Schenck, L., and J. R. Bailey, *Ibid.* **63,** 2331 (1941).
59. Shive, W., S. M. Roberts, R. I. Mahan, and J. R. Bailey, *Ibid.* **64,** 909 (1942).
60. Sibata, S., and J. Sioya, *C. A.* **35,** 1913 (1941).
61. Skinner, D. A., *Ind. Eng. Chem.* **44,** 1159 (1952).
62. Spielman, P. E., *Constituents of Coal Tar,* Longmans Green, New York, 1924.
63. Stauffer, J. C., Ph.D. Thesis, Columbia University, 1926.
64. Thompson, W. C., and J. R. Bailey, *J. Am. Chem. Soc.* **53,** 1002 (1931).
65. Tiedcke, K., Ph.D. Thesis, Hamburg, 1928.
66. Treibs, A., *Ann.* **510,** 42 (1934); **517,** 172 (1935).
67. Id., *Angew. Chem.* **49,** 682 (1936).
68. Walker, C. A., M.A. Thesis, The University of Texas, 1940.

CHAPTER TWENTY-EIGHT

USES OF PETROLEUM BASES

Apparently of all of the oil fields of the United States, only those in California produce crudes containing sufficient concentrations of nitrogen compounds to merit serious consideration as sources of commercial supplies of bases. In addition to these, some of the Central and South American fields produce considerable amounts of nitrogen compounds and could perhaps be utilized if there were sufficient demand. Schenck [13] stated that the 1934 kerosene and gas-oil production of the Union Oil Company of California alone would have been sufficient to produce 6,500 barrels of crude bases. The total potential production of this type of material by California petroleum companies would have been perhaps ten times this amount. A California production of 50,000 barrels of crude bases of this type per year appears to be a conservative estimate. If we add to this the potential production of Central and South American bases, it is obvious that a very considerable supply of bases could be produced if the demand were sufficient.

Obviously, like coal-tar bases, the petroleum compounds could be marketed and used as crude mixtures, as narrow-boiling cuts, or in the form of pure pyridines and quinolines.

The use of crude bases as corrosion inhibitors in the pickling of steel, in acidizing oil wells, and other large-scale uses suggests itself and the patent literature indicates that experiments along this line have been made, but there seem to be no published reports on the results of such trials.

M. J. Komarova [8] claimed that crude-petroleum pyridine mixtures in 30 to 40% concentration would kill eggs of lice after $\frac{1}{2}$ hour contact and exposure in closed containers to vapors of pyridine at room temperature was said to kill lice in 2 hours. These estimates seem conservative as far as the alkylpyridines of petroleum are concerned. It has been observed that woolen clothing worn by laboratory workers in this field is not molested by moths and other insects, but the penetrating and very persistent odor of the compounds would be an almost insurmountable obstacle to any such use of the bases. Numerous attempts have been made to use solutions of crude petroleum bases in truck and orchard sprays, but even here, the modern synthetic compounds seem to be more popular.

In general, if the crude bases can be produced in volume, as they can in California, there seems to be no obvious reason why they should not become commercially important as coal-tar bases rise in cost. The west coast states should be the first region in which the bases should be able to compete because they could be produced there, whereas freight costs would be high on coal-tar bases. So far apparently none of the three large California companies that have been vitally interested in the production of bases has found it possible to produce and sell them in competition with coal-tar bases.

There is a considerable revival of interest in petroleum nitrogen compounds, not because of their importance as commercial products but because of their harmful effects as gum and color formers and as catalyst poisons. The nonbasic compounds may well be responsible for gum and color formation, but basic as well as nonbasic compounds may act as catalyst poisons.[7, 9, 11, 14] Possibly, nitrogen compounds removed as undesirable from petroleum products could be sold as crude mixtures and so reduce the over-all cost of their removal even if the isolation of the compounds to be sold is not by itself a paying proposition.

In the case of the analogous petroleum cresylic acids, there was little demand for them prior to World War II, but since then, the demand has increased, the quality has been improved, and coal-tar acids are running into competition from the petroleum compounds. They are in active demand in the open market in addition to extensive use within the petroleum industry.[1]

Laboratory experience in fractionating and refining the petroleum bases has not encountered more serious difficulties than those met in refining coal-tar bases.[3, 6] Consequently, there seems to be no technical obstacle now apparent to the production of any desired fractions of pyridines or pyridine-quinoline mixtures to compete with similar mixtures from coal tar. Whether such fractions could be produced commercially would then again depend on the crude oil being refined and the cost of isolation of the crude bases. In the case of the lower-boiling alkylpyridines it will probably be more economical to isolate the bases from cracking-process gasoline than from straight-run material, but these classes of products are usually mixed prior to further refining, which is an advantage rather than a drawback. In view of the relatively high concentration of nitrogen compounds in shale oil,[4, 15] this may well be the future source of pyridine bases as the shale oil industry develops. Whether synthetic alkylpyridines will be able to meet all demands for such bases remains to be seen. Since certain types can be produced readily, this may well be a serious complication because there may never be sufficient demand for special alkylpyridines to pay for isolating them when other types cannot be sold at a profit in competition with the synthetic compounds.

The production and sale of chemically pure pyridines or quinolines is probably not economically sound unless a very strong demand for some individual compound or series of compounds should develop in some field, such as the drug or dye industry, where the high cost of isolation and purification might be less important. In the first place, the production of bases that could meet pharmaceutical specifications would be very difficult and expensive and in the second place, the isolation and purification of any individual compound would ordinarily involve separation and purification steps on a number of similar bases only one of which might be salable and so would have to bear the whole cost of the operations. If such a special demand should develop, it would probably follow screening operations similar to those begun but not carried through by Axe, Henson, and Schuhardt.[2] They studied a series of available fractions for germicidal activity and found that certain high-boiling Edeleanu extracts obtained in refining transformer oil showed a high phenol coefficient when bases from these fractions were converted to their hydrochlorides and tested against *Ebertella typhi* and *Staphylococcus aureus* at 30° and 37°C. Since the bases had been only roughly fractionated, converted to hydrochlorides and extracted with chloroform, and again fractionally distilled as chloroform-and-water-soluble bases each of the final fractions probably consisted of a complex mixture of bases. The hydrochlorides from the water layer were found to have phenol coefficients of 2 to 7, while those from the chloroform layer showed values of 13 to 17. It was demonstrated that the activity was not due to the acidity and that the hydrochlorides of similar bases boiling higher or lower were essentially inactive. Without being aware of it, these workers evidently had found an early type of cationic disinfectant related to the modern quarternary ammonium compounds. Barkovsky [5] synthesized a series of pyridines with long side chains at position 2 and found that their hydrochlorides were moderately effective disinfectants. Whether the compounds of Axe and coworkers were simply members of this family and present in high concentration, or belonged to an entirely new type of disinfectant present perhaps in very low concentration, was not determined. If they had now proceeded to fractionate these fractions, using their disinfectant activity as a guide, the nature of the compounds involved could probably have been determined. Obviously this type of screening could be carried out in the hunt for other types of special compounds.

An entirely different line of attack is illustrated by a patent to Rutherford [12] in which a mixture of bases is converted to a new type with different and perhaps more valuable properties. In the Rutherford patent, a mixture of petroleum bases is refined to remove hydrocarbons and other impurities not soluble in 50% sulfuric acid. The refined mixture is then hydrogenated at 250° to 280°C (475° to 535°F) to yield a mixture of secondary amines which are said to be suitable for use as rubber vulcanization accelerators, antioxidants, insecticides, or additives for extreme-pressure lubricants. Since such conversion need not be an expensive step, the resulting stronger bases may provide a future outlet for the petroleum bases. Miller,[10] evidently interested in this outlet for bases, compared the sodium and the electrolytic reduction of petroleum bases to secondary amines and pointed out a number of uses for such secondary bases.

Whether, in the future, bases to supplement or compete with coal-tar and synthetic bases will be derived from petroleum or from shale oil remains to be seen. They could meet considerable demand, but so far petroleum bases have not been utilized except perhaps within the petroleum industry.

Bibliography

1. Aries, R. S., and S. A. Savitt, *Chem. Eng. News* **28,** 316 (1950).
2. Axe, N., D. D. Henson, and V. T. Schuhardt, *Ind. Eng. Chem.* **29,** 503 (1950).

3. Axe, N., and J. R. Bailey, *J. Am. Chem. Soc.* **60,** 3028 (1938).
4. Ball, J. S., G. U. Dinneen, J. R. Smith, C. W. Bailey, and R. van Meter, *Ind. Eng. Chem.* **41,** 581 (1949).
5. Barkovsky, C., *Ann. de chim.* **41,** 581 (1949).
6. Bratton, A. C., and J. R. Bailey, *J. Am. Chem. Soc.* **59,** 175 (1937).
7. Deal, V., F. T. Weiss, and T. T. White, *Anal. Chem.* **25,** 426 (1953).
8. Komarova, M. J., *C. A.* **41,** 6016 (1947).
9. Maxted, E. B., *J. Soc. Chem. Ind. (London)* **67,** 93 (1948).
10. Miller, R. J., *Ind. Eng. Chem.* **43,** 1410 (1951).
11. Mills, G. A., E. R. Boedecker, and A. G. Oblad, *J. Am. Chem. Soc.* **72,** 1554 (1950).
12. Rutherford, J. T., U. S. Patent 2, 320, 655.
13. Schenck, L. M., Ph.D. Thesis, The University of Texas, 1942.
14. Thompson, R. B., J. A. Chenicek, L. W. Druge, and T. Symon, *Ind. Eng. Chem.* **43,** 935 (1951).
15. Thorne, H. M., W. Murphy, K. E. Stanfield, J. S. Ball, and J. W. Horne, *Paper read at Second Oil Shale and Cannel Coal Conf.*, Glasgow, 1950.

INDEX